TOOLS FOR MAKING ACUTE RISK DECISIONS

with Chemical Process Safety Applications

Publications Available from the

CENTER FOR CHEMICAL PROCESS SAFETY

of the

AMERICAN INSTITUTE OF CHEMICAL ENGINEERS

Tools for Making Acute Risk Decisions with Chemical Process Safety Applications
Guidelines for Preventing Human Error in Process Safety
Guidelines for Evaluating the Characteristics of Vapor Cloud Explosions, Flash Fires, and BLEVEs
Guidelines for Implementing Process Safety Management Systems
Guidelines for Safe Automation of Chemical Processes
Guidelines for Engineering Design for Process Safety
Guidelines for Auditing Process Safety Management Systems
Guidelines for Investigating Chemical Process Incidents
Guidelines for Hazard Evaluation Procedures, Second Edition with Worked Examples
Plant Guidelines for Technical Management of Chemical Process Safety, Revised Edition
Guidelines for Technical Management of Chemical Process Safety
Guidelines for Chemical Process Quantitative Risk Analysis
Guidelines for Process Equipment Reliability Data, with Data Tables
Guidelines for Vapor Release Mitigation
Guidelines for Safe Storage and Handling of High Toxic Hazard Materials
Guidelines for Use of Vapor Cloud Dispersion Models
Safety, Health, and Loss Prevention in Chemical Processes: Problems for Undergraduate Engineering Curricula
Safety, Health, and Loss Prevention in Chemical Processes: Problems for Undergraduate Engineering Curricula—Instructor's Guide
Workbook of Test Cases for Vapor Cloud Source Dispersion Models
Proceedings of the International Symposium and Workshop on Safe Chemical Process Automation, 1994
Proceedings of the International Process Safety Management Conference and Workshop, 1993
Proceedings of the International Conference on Hazard Identification and Risk Analysis, Human Factors, and Human Reliability in Process Safety, 1992
Proceedings of the International Conference/Workshop on Modeling and Mitigating the Consequences of Accidental Releases of Hazardous Materials, 1991.
Proceedings of the International Symposium on Runaway Reactions, 1989

TOOLS FOR MAKING ACUTE RISK DECISIONS

with Chemical Process Safety Applications

CENTER FOR CHEMICAL PROCESS SAFETY

of the

AMERICAN INSTITUTE OF CHEMICAL ENGINEERS

345 East 47th Street, New York, New York 10017

Library of Congress Cataloging-in Publication Data
Tools for making acute risk decisions with chemical process safety applications.
p. cm.
Includes bibliographical references and index.
ISBN 978-0-8169-0557-7
1. Chemical plants—Risk assessment. I. American Institute of Chemical Engineers. Center for Chemical Process Safety.
TP155.5.T65 1994
660'.2804—dc20 94-2462
CIP

CONTENTS

PREFACE

The Center for Chemical Process Safety, first organized in 1985 as a Directorate of the American Institute of Chemical Engineers, has been responsible for the development of a significant number of Guidelines books in process safety. The first book in the Guidelines series, *Guideline for Hazard Evaluation Procedures*, was published in 1985. This book focuses on identifying hazards using tools that are familiar to the process industries. After achieving enormous success an updated second edition was published in 1992.

In 1989, another Guidelines book was published entitled *Guidelines for Chemical Process Quantitative Risk Analysis (CPQRA)*. The CPQRA book presents tools for the quantitative assessment of risk. At the time of publication, the use of quantitative tools in the process industries was much more limited than the use of subjective hazard evaluation procedures.

Our book serves to extend these two previous books. It addresses tools used for decision making where risks have been assessed. This book represents a departure from the guidelines series in that some of the tools and methodologies addressed have been applied to other industry problems, but only infrequently at best to process risk decisions.

Decisions in all facets of business, especially process risk decisions, have become more complex and more critical to the long term success of the industry. Better tools and methods are needed to help decision makers reach the best decisions.

Risk decision making has evolved in sophistication but is still constrained by the methods and tools currently in use. Historically, the process industries have based most of their risk decisions on standards, experience and good engineering practices without the need for formal decision processes. Process risk decisions were relatively straight forward. In many cases, process risk standards were easily met.

Some companies in the process industries have implemented quantitative criteria for assessment of risk tolerability, particularly with regard to the safety of their employees. In these instances a risk is considered either tolerable or not tolerable. The criteria can often be met simply and inexpensively. However, where risk to the public is concerned, establishing a single numerical criterion is difficult if not impossible. One could specify a threshold above which the risk is not tolerable under any circumstances. However, below this threshold it is extremely difficult to determine how safe is safe enough. Some governments are specifying thresholds for both individual and societal risk above which the risk of acute process incidents is considered not tolerable.

Comparison of one process risk with another process risk is an approach that has been tried in industry to a limited extent. There is a certain simplicity and philosophical appeal in attempting to keep all risks within a similar range. However, some risks will fall significantly out of the desired range even after available risk reduction measures are evaluated. For these risks a more thorough decision-making strategy or approach may be required.

In addition to chemical process risk, other factors may affect the decision. Possible factors include financial cost, corporate image, employment of workers, and many others. Addressing each of these factors and their associated uncertainties, however, adds complexity. In many instances the alternatives are costly and represent a broad spectrum of possible options. In addition, recent changes in the regulatory environment in which process industries must operate have made decision making even more complex. Public concerns, pressure from environmentally focused groups, and regulatory agencies may all have a bearing on the decision. Often, these concerns can conflict with one another.

A consistent and logically sound approach can help ensure that appropriate resources are made available and allocated effectively to risk reduction activities. Decision aids are tools to assist in these decisions. The process industry has limited experience in applying formal decision aids to the complex risk decisions it faces. This book provides a collection of decision aids that have been successfully applied to other problems such as strategic planning, R&D program design, and operations analysis. In using these approaches for decision making a major challenge has been to quantify subjective values for aspects such as fatality risk, esthetics, amenities, and public opinion. Many of the decision aids addressed here can technically overcome this challenge. The social and political acceptability of some of these subjective evaluations is not addressed in this book.

This book addresses both current industry practices and techniques to consider for future application to complex risk decisions. It also provides a stimulus for the process industry to continue to improve the quality of risk decision making.

Persons involved in all aspects of process safety decisions will find this book useful. Management science professionals and others who are interested in applying decision aids to risk decisions will also find value in this book.

ORGANIZATION OF THE BOOK

This volume provides an introduction to risk decision making and the decision aids available to support these decisions. The primary audience comprises those persons involved in analyzing risks and assisting risk decision makers in the process industry. Some familiarity with process risk analysis techniques is assumed. The book should also prove valuable to persons involved in similar types of decisions in other industries, including both professionals in the management science field and analysts and decision makers from a variety of organizations.

The material presented will, in general, not be detailed enough to permit an individual to become a practitioner of each of the decision aids. The book aims instead to familiarize the reader sufficiently with each decision aid so as to be able to assess its usefulness for specific problems in one's organization. Case studies are provided that apply each decision aid to a typical risk decision problem. The book also suggests readings, courses, and software that can help an organization develop the skill set necessary to begin to use each decision aid.

Chapter 1 provides an historical context, describing risk decision making methods as a natural extension of methodologies developed to identify hazards and assess risks. The difficulties in making these decisions and the benefits of a more formal decision-making process are described. A conceptual overview of the decision process and its key analysis steps is provided.

Chapter 2 discusses some key concepts that are common to the decision aids in this book, such as evaluation criteria, economic evaluation principles, value of life, and risk analysis.

Chapter 3 provides a framework for classifying the large number and variety of available decision aids. This framework is used to organize the discussion. Brief descriptions of 17 key decision aids are provided. The chapter concludes with a rationale for the choice of seven decision aids covered in succeeding chapters.

Chapter 4 provides guidance on how to select decision aids that are appropriate to a problem. Choosing an appropriate decision aid depends on an understanding both of the problem to be addressed, and of the strengths and weaknesses of the available decision aids.

Chapter 5 introduces three case studies from the chemical process industry. Two of these originally appeared in *Guidelines for Chemical Process Quantitative Risk Analysis* (AIChE/CCPS 1989). Each case study is used with two or three of the decision aids covered in succeeding chapters.

Chapters 6 through 12 each cover one decision aid (or class of decision aids) selected for more detailed treatment:

Chapter 6—Voting Methods
Chapter 7—Weighted Scoring Methods

Chapter 8—Cost–Benefit Analysis
Chapter 9—Mathematical Programming
Chapter 10—Payoff Matrix Analysis
Chapter 11—Decision Analysis
Chapter 12—Multiattribute Utility Analysis

Each of these chapters addresses the history of the decision aid, the decision situations for which the decision aid is most appropriate, the analytic approach and key steps, an application of the decision aid to a case study, related methods or extensions of the decision aid, data requirements and other implementation needs, and strengths and weaknesses of the technique. Suggestions for further learning and references are provided.

Chapter 13 highlights the case studies introduced in Chapter 5 and addressed again in Chapters 6–12. The discussion addresses important differences in how various decision aids handle a given problem.

Chapter 14 discusses issues surrounding the implementation of formal risk decision making in an organization.

Chapter 15 discusses research and development needs to improve risk decision making, including general research on decision making, improvements to existing decision aids, and new types of decision aids.

Following Chapter 15, appendices are provided that describe decision aid software and training programs. A topical bibliography provides references grouped by major topics.

HOW TO USE THIS BOOK

The chapters of this book generally build on previous ones. However, the reader can select a subset of the decision aid chapters (6–12) in order to focus on those of particular interest. For persons who help decide on which decision aids are most appropriate for their organizations, we suggest reading all of these chapters to build a good understanding of the range of available techniques.

For a management-level overview of the field, we suggest Chapters 1 through 4, and Chapter 14. These provide an orientation to the general risk decision process, the range of available decision aids, and the implications for an organization that seeks more formal risk decision making processes.

For a more practical orientation to the field, Chapter 5 introduces case studies, and the reader may wish to focus on the case studies in Chapters 6–12 using the "road map" outlined at the end of Chapter 5. Chapter 13 reviews the case studies.

For the reader interested in one or a few of the decision aids covered in Chapters 6–12, these chapters can be read by themselves. Each decision aid chapter suggests a course of additional reading and/or training for the reader to gain a deeper understanding of the technique.

ACKNOWLEDGMENTS

This volume was written by Applied Decision Analysis, Inc. (ADA) in support of a project funded by AIChE-CCPS. The CCPS Risk Assessment Subcommittee managed the project and provided direction and assistance to ADA.

The Applied Decision Analysis team was directed by Rick G. Schwartz, and included Gregory L. Hamm, Lawrence H. Gallant, and Katherine A. Weller. Miley (Lee) W. Merkhofer, John D. Kadvany, and others at ADA contributed valuable input to this project.

The CCPS Risk Assessment Subcommittee was chaired by Robert W. Ormsby (Air Products and Chemicals) and included (in alphabetical order): Robert E. De Hart II (Mobil Oil), Brian R. Dunbobbin (Air Products and Chemicals), Raymond A. Freeman (Monsanto), Raymond W. French (Exxon Chemical), S. Barry Gibson (formerly of du Pont), Richard M. Gustafson (Texaco), Dennis C. Hendershot (Rohm and Haas), Thomas Janicik (Solvay Polymers), Mark J. Katz (FMC), William K. Lutz (Union Carbide), Carmine A. Master (Fluor Daniel), Arthur G. Mundt (Dow Chemical), and Donald L. Winter (Mobil Oil). Felix Freiheiter provided staff support from CCPS. Bob Perry, Les Wittenberg, and Tom Carmody (formerly) of CCPS helped guide the project.

The substantial contributions, both in time and resources, of the employer organizations of the Subcommittee and of ADA are gratefully acknowledged.

A special acknowledgment is given to Joseph C. Sweeney (ARCO Chemical), who provided a great deal of insight into the concept for this book and assisted in the early stages of its development.

Our appreciation goes out to the following individuals who generously contributed their time in providing a peer review of this book: Ken Balkey (Westinghouse), Lee Brown (Los Alamos National Laboratory), Pat Clemens (Sverdrup Technology), Anirudh Dhebar (Harvard Business School), Robert Eddy (Hartford Steam), Thomas Gibson (Dow Chemical), John Hoffmeister (Martin-Marietta), Mike Hogh (BP Engineering), Craig Matthiessen (U.S. Environmental Protection Agency), Robert Perdue (Westinghouse), Irv Rosenthal (Wharton School), Randy Schwartz (Schoolcraft College), Wayne Simmons (Battelle Institute), and Detlof von Winterfeldt (University of Southern California).

Finally, a special acknowledgment is due to Dr. Elisabeth M. Drake (MIT), who volunteered her time to perform the technical editing of the book.

ACRONYMS

AHP Analytic Hierarchy Process
AIChE American Institute of Chemical Engineers
BLEVE Boiling Liquid Expanding Vapor Explosion
CBA Cost-Benefit Analysis
CCPS Center for Chemical Process Safety
CEA Cost-Effectiveness Analysis
CPQRA Chemical Process Quantitative Risk Analysis
CPSC Consumer Product Safety Commission
DA Decision Analysis
DOE Department of Energy
EPA Environmental Protection Agency
EPRI Electric Power Research Institute
FAA Federal Aviation Administration
FAR Fatal Accident Rate
HEP Hazard Evaluation Procedures
IIASA International Institute for Applied Systems Analysis
INFORMS Institute for Operations Research and the Management Sciences (formerly ORSA and TIMS)
KT Kepner–Tregoe
MUA Multiattribute Utility Analysis
NRC Nuclear Regulatory Commission
NSF National Science Foundation
OSHA Occupational Safety and Health Administration
OMB Office of Management & Budget
ORSA Operations Research Society of America, now merged into INFORMS
PMA Payoff Matrix Analysis
PRA Probabilistic Risk Assessment
SMART Simple Multiattribute Rating Technique
TIMS The Institute for Management Studies, now merged into INFORMS
UVCE Unconfined Vapor Cloud Explosion
WSM Weighted Scoring Methods

GLOSSARY

Acute risk: A risk associated with immediate effects of episodic events such as fire, explosion, and toxic material releases.

Alternative: One of several actions that can address a problem or opportunity. An alternative should involve a commitment of resources, and have consequences associated with changing the alternative in the future.

Attribute: A characteristic that can be measured to indicate the degree to which an objective is achieved or fulfilled.

Capturing techniques: A class of decision aiding techniques that infers the decision maker's underlying preferences based on personal ratings of the desirability of specific alternatives.

CATALYST: A process developed by the Electric Power Research Institute for conducting payoff matrix analyses with a group of participants.

Certain equivalent: The dollar amount for which a decision maker is indifferent between having the money for certain and undertaking a particular risky alternative. See "risk concepts" in Chapter 11 (Decision Analysis) for further explanation.

Compromise programming: A branch of mathematical programming that considers tradeoffs among multiple objectives explicitly.

Conditional probability: The probability of an uncertain event given the occurrence of a related event. For example, the probability of rain given that it is a cloudy day is a conditional probability.

Conjunctive ranking: A decision aid that reduces a set of alternatives by specifying minimum acceptable thresholds for each attribute used in the decision. The chosen alternative is determined by ranking the remaining alternatives on a single attribute chosen by the decision maker.

Constraints: In mathematical programming, the mathematical expressions that describe requirements or limitations affecting the range of alternatives available to the decision maker.

Contingent valuation: A method used to estimate an individuals' value for something by asking directly how much they are willing to pay for it. This approach is sometimes used to estimate a value of life.

Continuous variable: A variable that can take on any numerical quantity within a range.

Cost-benefit analysis (CBA): A decision aid that involves quantifying the costs and benefits of each alternative in dollar terms based on the preferences of individuals affected by the decision, and choosing the one that maximizes the net benefit (total benefit minus total cost).

Cost-effectiveness analysis (CEA): A variation of cost-benefit analysis that compares the dollar cost of each alternative against its benefits (e.g., reduced fatalities) without assigning a dollar value to these benefits.

Cumulative distribution function: A probability distribution of a variable in which each functional value along the y-axis represents the probability of occurrence at or at less than the corresponding value of the variable along the *x*-axis.

Decision: The choice of one alternative from among two or more available alternatives to address a problem or opportunity.

Decision aid: A tool or process used to select one alternative from a set of two or more available alternatives to address a problem or opportunity.

Decision analysis (DA): A decision aid based on utility theory and the laws of probability that provides a formal process for structuring, analyzing, and communicating decisions. The preferred alternative is the one with the highest expected utility.

Decision maker: The person or group that makes a decision, that is, commits resources to a particular alternative.

Decision policy: A set of alternatives that includes one alternative for each decision in the problem.

Decision rule: The rule used by a decision maker to select one from a set of available alternatives.

Decision tree: A graphical representation of the sequence of decisions, uncertain events, and outcomes a decision maker is facing.

Deterministic: Involving measures which are certain, or are treated as certain (e.g., deterministic model, deterministic phase, deterministic sensitivity analysis).

Discount rate: A factor used to compare costs and benefits at different years. If r is the discount rate, the value of x dollars received in year t is $x/(1+r)^t$ this year.

Discounting: The application of a discount rate to a stream of future costs and benefits to determine its value today.

Discrete variable: A variable that can take on only specific numerical quantities within a range.

ELECTRE: A decision aid that helps a decision maker better understand preferences and reduce the set alternatives. ELECTRE develops a partial ordering of the alternatives based on indices that capture the decision maker's expressed clarity of preferences among alternatives.

Expected utility: The mathematical expectation of the utilities associated with the outcomes of an uncertainty or series of uncertainties, calculated by summing the possible utilities times their probabilities, over all possible utilities.

Expected value: The mathematical expectation of the numerical values associated with the outcomes of an uncertainty or series of uncertainties, calculated by summing the possible numerical values times their probabilities, over all possible numerical values.

Externalities: Effects of an activity that are borne by persons other than those undertaking or benefiting from that activity, for example, the health risk that a chemical plant poses to a nearby population.

Game theory: A mathematical theory that addresses competitive situations by describing how each competitor responds to the actions of others, and using this to determine strategies and behaviors that result.

Goal programming: An extension of mathematical programming that addresses a decision maker's desire to attain pre-specified goals rather maximizing (or minimizing) a single particular quantity.

Hazard: A chemical or physical condition that has the potential for causing damage to people, property, or the environment.

Human capital: The value of a human life expressed as the net present value of an individual's future earnings.

Influence diagram: A graphical representation of the relationship between decisions, uncertainties, and values associated with a particular decision problem.

Integer programming: A branch of mathematical programming concerned with problems in which some or all of the alternatives are restricted to take on discrete (integer) values.

Joint probability: The probability of two or more uncertain events occurring.

Joint sensitivity analysis: Sensitivity analysis conducted to examine the result of changing two or more variables at once.

Linear programming: A branch of mathematical programming in which the objective function and constraints are linear functions of the decision variables.

Mathematical expectation: The sum of the possible numerical quantities that an uncertain variable could have, weighted by their probabilities of occurrence (also called the "mean").

Mathematical programming: A general class of quantitative decision aids that helps decision makers allocate resources in a manner that best achieves their desired objectives. Also referred to as constrained optimization.

Maximin: A decision criterion that associates each alternative with its worst possible outcome, then selects the alternative whose worst possible outcome is most desirable. When outcomes are expressed as costs, the corresponding criterion is called minimax.

Maximization: A decision criterion that selects the alternative that has the greatest value to the decision maker.

Mean: The sum of the possible numerical quantities that an uncertain variable could have, weighted by their probabilities of occurrence (also called "mathematical expectation").

Minimization: A decision criterion that selects the alternative that has the lowest value to the decision maker.

Minimum regret: A decision criterion that selects the alternative whose maximum possible regret over all future events is lowest, where regret is defined as the additional cost of having selected the chosen alternative instead of the alternative that, with perfect hindsight, would have been the best.

Minimum risk: A decision criterion that selects the alternative with the smallest probability for adverse outcomes.

Model: A selective representation of reality. A mathematical model utilizes mathematical symbols, relationships, or expressions.

Multiattribute utility analysis (MUA): A decision aid that extends decision analysis to explicitly incorporate the value judgments engendered in decisions with multiple objectives.

Multi-dimensional scaling: A technique used in mathematical psychology to model a decision maker's preferences by analyzing how the decision maker perceives different alternatives. Each alternative is a point in a multi-dimensional "space"; nearby points represent alternatives that the decision maker perceives as similar.

Multiattribute utility function: A function that assigns a numerical measure of the utility to a value measure for each of a set of attributes of concern.

Net present value (NPV): A measure of the value today of a stream of future costs and benefits, used to compare alternatives with differing streams of costs and benefits. If R_t are the benefits in year t, C_t are the costs in year t, n is the number of years, and r is the discount rate per year, the net present value is:

$$\text{NPV} = (R_0 - C_0) + (R_1 - C_1)/(1+r) + (R_2 - C_2)/(1+r)^2 + (R_3 - C_3)/(1+r)^3 + \cdots + (R_n - C_n)/(1+r)^n$$

Nominal group technique: A process designed to help generate ideas, set priorities, and reach decisions within a group context that strives to ensure equal contribution from all participants.

Nonlinear programming: A branch of mathematical programming concerned with solving planning problems in which either the objective function or some of the constraints are nonlinear functions of the decision variables.

Objective function: In mathematical programming, the mathematical expression that describes the goal (or objective) of the decision maker.

Objective: An expression of a decision maker's goal in terms of an object and a direction of preference (e.g., "minimize fatalities," where fatalities is the object and minimize is the direction of preference).

Objectives hierarchy: A graphical representation of the relationships among the objectives of a decision maker; each level of the hierarchy shows the objectives that contribute to the broader objectives shown at higher levels.

Opportunity cost: The implied cost of an investment due to the inability to use the capital invested in a particular, different way (e.g., the opportunity cost of a plant expansion might include the income from another, more productive use of the required land).

Outcome: The specific result of a decision and the resolution of an uncertainty or series of uncertainties.

Parameter: A numerical quantity that is assumed to be known.

Payoff matrix analysis: A decision aid that determines the expected value of distinct alternatives allowing for future uncertainty, using a simple tabular form. The problem must involve only one decision and one uncertainty, and the uncertainty must not depend upon the decision.

Point estimate: A single number used to summarize an uncertain quantity.

Portfolio analysis: A decision aid that helps a decision maker determine the combination of risky alternatives that offers the best combination of high return and low risk.

Preferential dependence: Preferential dependence exists between two attributes if the preference for one attribute depends upon the quantity of the second attribute.

Preferential independence: Two attributes have preferential independence if the preference for one attribute does not depends upon the quantity of the second attribute.

Probabilistic: Pertaining to the use of probability to represent uncertainty.

Probabilistic dependence: Probabilistic dependence exists between two uncertainties if knowing the resolution of one uncertainty gives information about the possible resolution of the other. For example, if the probability of rain is 20% when there is no information on cloud cover, and the probability of rain when the day is cloudy is 50%, rain is dependent on cloudy conditions and dependence exists between rain and cloudiness. Probabilistic dependence may also exist between a decision and an uncertainty. If a decision changes the probabilities of the states of an uncertainty, the uncertainty is dependent on the decision. For example, decisions to implement safety measures reduce the probability of accidents; therefore, accident uncertainty is dependent on decisions to implement safety measures.

Probabilistic independence: Probabilistic independence exists between two uncertainties if the resolution of one uncertainty gives no information about the possible resolution of the other. For example, if a pair of dice are rolled and then a coin is flipped, the probability of heads is assumed to be 50% regardless of the result of rolling the dice. The coin flip is probabilistically independent of the roll of the dice. An uncertainty can also be described as independent of a decision. For example, the chance of rain is independent of the decision to carry or not carry an umbrella.

Probabilistic Risk Assessment (PRA): A commonly used term in the nuclear industry to describe the quantitative evaluation of risk using probability theory. Also known as Probabilistic Safety Assessment (PSA).

Probability: A number that expresses the likelihood of occurrence of a possible state of an uncertainty. By definition, a probability must be a number between 0 and 1, and the sum of probabilities for all possible states of an uncertainty must be 1.

Probability density function: A mathematical description of the relative likelihoods of occurrence of the possible states of an uncertainty. For a continuous function, the integral of this function between two values is the probability that the true value lies in the interval between the values. For discrete variables, it is often called a probability mass function.

Probability distribution: A mathematical function that associates probabilities (e.g., weather is rainy, cloudy, or sunny) with uncertain events (e.g., amount of rainfall).

Rainbow diagram: A graphical output from a probabilistic sensitivity analysis used to examine the sensitivity of a decision to an input.

Rate of return: The discount rate at which the net present value of a stream of costs and benefits over time is zero.

Risk: A measure of economic loss, human injury, or environmental damage, in terms of both the incident likelihood and the magnitude of the loss, injury, or damage.

Risk analysis: The development of a quantitative estimate of risk based on engineering evaluation and mathematical techniques for combining estimates of incident consequences and frequency.

Risk assessment: A process by which the results of a risk analysis (i.e., risk estimates) are prepared for use in decisions, either through relative ranking of risk reduction strategies or through comparison with risk criteria.

Risk attitude: A decision maker's preferences towards facing variation in possible losses and gains. In decision analysis, risk attitude is expressed mathematically through a utility function.

Risk-averse: A description of a decision maker's risk attitude in which the value for a risky alternative is lower than the expected value of the alternative.

Risk-neutral: A description of a decision maker's risk attitude in which the value for a risky alternative is equal to the expected value of the risky alternative.

Risk-preferring: A description of a decision maker's risk attitude in which the value for a risky alternative is higher than the expected value of the risky alternative.

Risk profile: A cumulative probability distribution over the range of values associated with all possible outcomes resulting from selecting a given alternative.

Satisficing: A decision criterion that selects the alternative that satisfies as many of the decision maker's goals as possible.

Scenario: A set of states across several uncertainties. For example, two simple uncertainties are "how long will I wait for the bus," and "will it rain today." The "how long will I wait for the bus" states might be "10 minutes" and "20 minutes." The "will it rain today" states might be "rain" and "no rain." "Wait for the bus 20 minutes" and "rain" comprise one possible scenario. As a second example, consider a fire hazard. Two uncertainties associated with the hazard might be the chance of a flammable liquid leak and the presence of an igniting flame or spark. One scenario is that the flammable liquid leaks and there is no igniting flame or spark. An alternative scenario is that the flammable liquid leaks and is ignited by a flame or spark.

Sensitivity: The sensitivity of a measure to a parameter is defined as the change in the measure per unit change in the parameter. How much an output of a model changes with change in one or more inputs.

Sensitivity analysis: A technique in which one or more parameters are varied to examine their impact on a measure.

Shadow price: The decision maker's value for a unit of scarce resource. In economics, refers to the price at which something would sell if a competitive market existed for it. For example, the shadow price of public land might be estimated from the selling price for equivalent private property. In mathematical programming, the increase in the objective function per additional unit of a constrained resource.

Simple multiattribute rating technique (SMART): A weighted scoring technique that was created as a simplified alternative to rigorous multiattribute utility methods.

Stakeholders: Individuals or groups that are interested in and/or affected by a decision.

State: In a general decision context, the condition facing the decision maker. In payoff matrix analysis, decision analysis, and multiattribute utility analysis, one of the distinct possibilities associated with an uncertainty. In dynamic programming, an explicit description of this condition that contains enough information for the decision maker to make a decision.

Strategies: A set of alternatives across several decisions. For example, two simple decisions are, "how will I go to work today," and "should I take my umbrella." The alternatives for "how will I go to work today" might be "drive my car," "take the bus," and "ride with my neighbor." The alternatives for "should I take my umbrella" might be "take my umbrella" and "do not take my umbrella." "Take the bus" and "do not take my umbrella" represents one possible strategy. As a safety example, consider dealing with a fire hazard. Reducing ignition points, maintaining the availability of fire control equipment at current levels, and increasing the level of fire prevention training would comprise a strategy for controlling the risk. No action on ignition points, increasing the availability of fire control equipment, and increasing the level of fire prevention training would comprise a second strategy for controlling the risk.

Sunk cost principle: The concept that each new decision should be based on future costs and potential returns, not on prior expenditures or returns.

Taxonomy: A classification scheme.

Time value of money: The concept that possessing a dollar now is of more value than possessing a dollar at some time in the future.

Uncertainty: Something that is unknown and uncontrollable, or not perfectly known or controllable, such as a quantity or the outcome of a future event.

Utility: A quantity that expresses a relative strength of preference; the dimension of this quantity is sometimes called "utils."

Utility function: A function that assigns a utility to a value measure.

Value: What the decision maker views as important to the desirability of the outcomes of a decision.

Value function: A mathematical expression that assigns a value measure to each outcome of a series of uncertainties, to represent the desirability of each outcome.

Value measure: A single number and quantity (e.g., dollars) assigned to an outcome of a series of uncertainties that represents the desirability of the outcome.

Value of control: The value to the decision maker of being able to ensure that a particular uncertainty takes on a desired state.

Value of information: The value to the decision maker of obtaining more information about an uncertainty before making a decision.

Value of life: A monetary value placed on human life in order to trade off the objective to reduce risks to human life against other decision objectives.

Variance: A mathematical expression of the degree to which states other than the mean are likely to occur.

Voting methods: A class of decision aids in which a group selects alternatives by aggregating the preferences of the individual members of the group.

Weighted scoring methods: Decision aids in which alternatives are assigned numerical scores on various decision criteria that have been weighted according to their relative importance to the other decision criteria.

Willingness-to-pay: A measure that expresses the value for something in terms of how much money an individual would forego in order to have it.

1

INTRODUCTION

Because (risks) are not easy to balance we often do the job badly,
accepting unnecessary risk in some more familiar forms,
while grossly exaggerating it in others.
—H. W. Lewis, *Technological Risk*

1.1 THE CHALLENGE OF ACUTE RISK DECISION MAKING

This book is concerned with decision making for problems that involve acute risks in the process industry. Acute risks arise from episodic events such as fires, explosions, and toxic chemical releases that cause immediate harm to people or the environment. The analysis of long-term risks such as cancer and chronic disease arising from an episodic event is not the primary focus of this book.

Management has a responsibility to its stakeholders to effectively manage risks from the storage, handling, processing, and distribution of hazardous materials. A first step is to identify hazards and assess the risks. Doing this helps establish whether there is a threat to people, property, or the environment. If the risk is significant, then choices must be made on how to reduce or eliminate the risk.

Making these decisions is not easy. Risks can seldom be reduced to zero except by eliminating the activity, often at considerable cost to the organization. However, risks can always be reduced further. Many organizations try to prioritize the risks posed by their activities and then focus on reducing the highest risks first before moving on to reducing lower risks. The overall goal of this approach is to continually reduce risk whenever practical.

No organization—company, industry, government, or society—has infinite resources available for risk reduction. Yet each of these organizations must repeatedly make decisions on how much time, money, and other resources to invest towards reducing risks. Some typical risk decisions in the process industry are:

- Should any action be taken to reduce the risk of an existing process?
- Which one from a number of alternatives should be adopted to reduce the risk from an existing process?
- How should a company allocate resources among its plants and operations in order to reduce overall process risk?
- Should a company undertake a new venture that has significant process risk and financial uncertainty?
- When should a company get out of a business area based on process risks?

Some risk decisions are fairly easy to resolve, particularly when significant risk reduction can be achieved through simple and inexpensive measures. However, an organization may face some very difficult risk decision problems. Decisions in which the consequences have high stakes and low probability, and which involve large expenditures of resources, are particularly difficult. In making these decisions, an organization seeks logic and consistency, to ensure that its limited resources are used efficiently.

The chemical industry continues to evolve in its analysis and treatment of risk. Three key issues are to (1) identify the hazard, (2) assess the risks, and (3) decide on what action to take. Processes and tools to address the first two of these issues are becoming fairly well developed. The Center for Chemical Process Safety (CCPS) has published a book entitled *Guidelines for Hazard Evaluation Procedures*, 2nd edition (AIChE/CCPS 1992), which describes methodologies that can be used to identify hazards and qualitatively assess the associated risk. For many activities or studies this qualitative assessment process may suffice. At an early stage of projects, preliminary hazard reviews help identify risks and issues requiring further analysis.

For risks that require further attention, a quantitative risk analysis allows risks to be better defined and measured. The Center for Chemical Process Safety has published another book entitled *Guidelines for Chemical Process*

Quantitative Risk Analysis (CPQRA) (AIChE/CCPS 1989), commonly referred to as the *CPQRA Guidelines*, which describes techniques for quantifying risk and introduces a risk management framework.

Hazard evaluation procedures (HEP) and quantitative risk analysis help assess the risks. However, they only provide information. They do not identify a specific direction for action. Occasionally there are risk reduction measures that are immediately obvious, or measures that can be easily implemented. Management will often elect to carry out these measures without further study. However, in more complex situations tools are needed to help determine what action to take.

Methods that help an organization choose what action to take for complex risk decisions, are less proven in the process industry. The tools to help management resolve risk issues continue to be developed and, thus far, have not been applied extensively. Those tools or aids which have been applied in a limited fashion to risk decision making generally employ risk targets. A few targets have been proposed (e.g. fatal accident rate or FAR) and applied by industrial companies for risks to employees (Gibson 1976, Helmers and Schaller 1982). An employee risk level is set, above which the risk is unacceptable and below which the risk is considered acceptable. This approach has met reasonable success because the risk levels could be reached easily and inexpensively.

Methods to analyze and decide how to deal with risks to the public have not been so easily proposed and applied. Renshaw (1990) describes the use of societal risk targets by a U.S. chemical manufacturer. Industry cannot dictate what level of risk is tolerable to the public; establishing tolerability is inevitably a political process involving many different stakeholders. In the past, proposals were put forth which attempted to establish a tolerable lower limit of risk, and these have been useful in initiating further debate on the subject. The Dutch and British governments have grappled with the question of risk criteria for decisions that address risk to the public as the terms of individual and societal risk. They have approached the problem using the concept of two risk thresholds: one is a threshold above which the risks are considered to be intolerable under any circumstances, and the second is a threshold below which risks are considered trivial and therefore exempt from further action (Waterstact Groningen Provincial 1979, Van Kuijen 1988, U.K. Health & Safety Commission 1991, Netherlands Ministry of Housing 1989, Ale 1991). Between these two boundaries is a region in which the risks are to be reduced "to the extent practical." In many instances the minimal possible level of risk is one that can not practicably be attained. The real challenge facing management is then how far to reduce the risk below the intolerable level, such that the risk is "tolerable given the circumstances." It becomes evident that an approach using a risk target does not by itself provide the complete answer to the question of how much to invest toward reducing an identified risk.

Were cost and risk the only issues in risk decision making, the task would be difficult enough. But further complicating a risk decision is the multiplicity of other factors that may have a bearing. Such factors include:

- Alternatives available for reducing or eliminating the risks
- Availability of capital
- Codes, standards and regulations and good industry practices
- Company and/or personal liabilities
- Company image
- Costs of implementing available alternatives
- Economic impact of the activity in the local community
- Employment opportunities that the activity provides
- Frequency level(s) of the risk
- Inequities in how the risks and benefits are distributed among members of society
- Number of people at risk
- Perceived benefit of the activity and its impact on the public and/or stockholder image, such as whether the activity produces cigarettes or a miracle drug
- Profitability of the activity
- Societal component of the risk, such as the maximum number of people impacted by a single event
- Strategic importance of the activity to the company's growth and survival
- Type(s) of risk, such as human fatality, injury, and acute environmental damage.

It is increasingly important in today's society to make decisions on a more consistent, logical, and rigorous basis. This is especially true in acute risk decisions. Some reasons are the need for:

- *Quality*—A quality decision is one that results from properly accounting for management's values, uncertainties, creativity, and wisdom.
- *Cost Effectiveness*—Given the limited resources available to reduce risk, it is important to appropriate the right amount of resources toward risk reduction, and to allocate these resources effectively among competing risk reducing activities.
- *Completeness*—Relevant information needs to be used in the analysis, but the depth of analysis should not exceed what is appropriate given the stakes involved.
- *Recognition of Societal Concerns*—Societal concerns should be considered in risk decisions where appropriate.
- *Recognition of Other Stakeholders*—Concerns of workers, stockowners, and other affected parties should be considered.

- *Consistency*—Establishing agreement on corporate values can ensure that a consistent basis is used for the decisions of the organization.
- *Defensibility*—Decisions should withstand scrutiny and be easily communicated to others.

Decision aids that address these issues have been successfully applied by industry to a variety of problems including strategic planning, research and development program design, and operations analysis. However, these decision aids have been applied only to a very limited extent in risk decision problems in industry. Key obstacles have been the:

- Lack of familiarity with these decision aids among decision makers in the process industry
- Concerns that tools are either overly simplistic or overly complex
- Reluctance to use a more explicit and transparent basis for potentially sensitive decisions
- Limited track record for problems with complexities of the kind listed earlier.

This book attempts to overcome these obstacles. The chapters discuss in some detail decision aids that vary greatly in the time, resources, and data required to reach a decision and that vary in the degree to which group decision making and input from multiple stakeholders are accommodated. Case studies are provided that illustrate an application of each decision aid to a typical risk decision problem. For some of the decision aids addressed, the reader will gain sufficient understanding to begin using them immediately on decision problems. Other decision aids are more complex. We aim to make the reader familiar enough with each decision aid to be able to assess its usefulness for specific problems in his or her organization. The book also suggests readings, courses, and software that can help organizations to develop the skills necessary to implement each decision aid.

This book is aimed primarily at persons involved in risk decision making in the process industry, and presumes some familiarity with risk analysis techniques. However, the book should also prove valuable to persons involved in similar types of decisions in other industries, including professionals in the management science field as well as analysts and decision makers from a variety of organizations.

1.2 SOME KEY TERMS

The Glossary defines important terms as they are used in this volume. Some of the terms dealing with hazards and risks are taken directly from the discipline of quantitative risk analysis. Others are well known in the fields of management science and mathematics. A few are unique to the subject matter of this book.

Listed below are some essential terms for understanding the basic decision problem addressed in this book.

Alternative One of several distinct actions that can address a problem or opportunity, involving a commitment of resources and having consequences associated with changing the alternative in the future

Decision The choice of one alternative from among two or more available alternatives to address a problem or opportunity

Decision aid A tool or process used to select one alternative from a set of two or more available alternatives to address a problem or opportunity

Decision maker The person or group that makes a decision, that is, commits resources to a particular alternative

Outcome The specific result of a decision and the resolution of an uncertainty or series of uncertainties

Risk A measure of economic loss, human injury, or environmental damage, in terms of both the incident likelihood and the magnitude of the loss, injury, or damage

Stakeholders Individuals or groups that are interested in and/or affected by a decision

Value What the decision maker views as important to the desirability of the outcomes of a decision.

1.3 THE BASIC RISK DECISION PROCESS

Figure 1.1 illustrates the fundamental risk decision process addressed by this book. The figure identifies the key phases of most risk decisions. The decision process starts with a problem, and ends with action. The key phases are as follows:

- Define the problem
- Estimate and evaluate the baseline risk
- Identify and screen the alternatives
- Analyze the problem
- Make a decision.

1.3.1 Define the Problem

The first step involves defining the problem by establishing a problem statement and scope. Some problems will originate from risks identified and evaluated by HEP and by CPQRA procedures. Depending on how the prob-

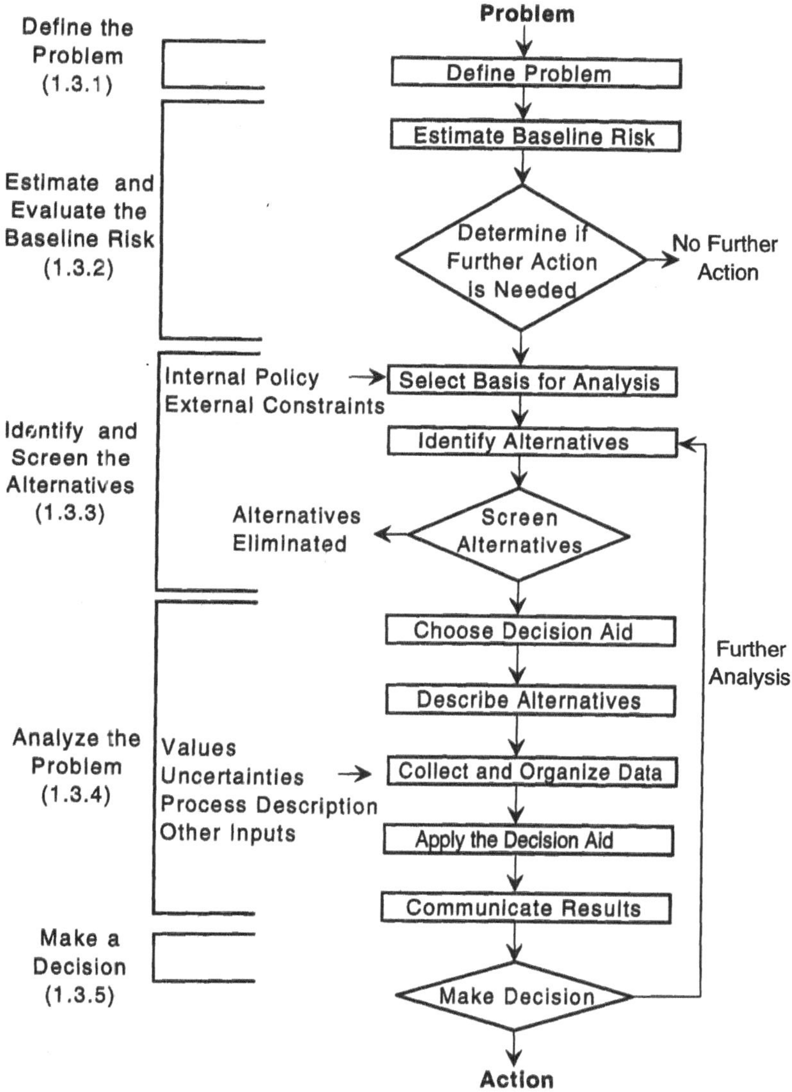

FIGURE 1.1. Risk decision process incorporating use of a decision aid, with section headings from this chapter.

lem is defined, the decision can be made in a matter of minutes or days, or could require weeks or months. Defining the problem will help determine the budget, staff skills, and other resource requirements. Some key issues to resolve at this stage are the:

- Goal and extent of the study
- Boundaries of the problem addressed, including the time frame and scope of effects to be considered
- Risks and uncertainties to be considered, such as risks to employees, risks to the public, and financial uncertainties
- Time and resources that will be directed toward analyzing the problem.

1.3.2 Estimate and Evaluate Baseline Risk

Next, a risk estimate is obtained of the system as designed or as being operated. Depending on the problem definition or previously studied information, a qualitative estimate of risk may be adequate, or a quantitative risk analysis may be required. Chapter 4 of the *CPQRA Guidelines* (AIChE/CCPS 1989) describes commonly used risk measures, formats used for presenting risk estimates, guidelines for selecting the risk measures and presentation formats to meet the objectives of a study, and the mechanics of estimating various risk measures. Once an estimate of the baseline risk has been obtained, the organization can begin to identify alternatives for reducing risk. Sometimes the risk estimate makes obvious an immediate decision to take no further action.

1.3.3 Identify and Screen the Alternatives

The process of developing a baseline risk estimate often identifies some alternatives for reducing the risk. For example, the risk analysis may highlight particular high-risk incidents that can be controlled through specific measures, or it may indicate some simple, inexpensive measures that can be taken immediately without formal analysis.

To fully develop a list of alternatives, one must first select the basis on which the alternatives will be analyzed. Organizational policies, whether expressed explicitly or experienced implicitly through the corporate culture, determine the evaluation basis. This can include factors such as those listed in Section 1.1. There may be internal or external requirements or constraints associated with these factors that govern the range of alternatives, impacts addressed, and processes employed when analyzing the alternatives.

It is generally useful to make the initial list of alternatives fairly broad. One of these alternatives is often to take no additional action. Group participation and structured brainstorming is often useful. Some of the decision aids addressed in this book have their own processes for generating alternatives. The use of *strategy tables* can also facilitate the process of generating alternatives (Kusnic and Owen 1992). Once the alternatives have been generated, a screening process may eliminate some of them from further consideration. This screening may be formal or informal depending upon the approach used. Conformance to codes and standards, both external and internal, and adherence to good engineering practices may be the basis for this initial screening. The culture and general decision process within the company may establish the way in which risk estimates and risk reduction alternatives are screened. For example, depending on the context of the decision, simple alternatives may be considered to be beyond the bounds of practicality or economic feasibility, eliminating these alternatives from further analysis. Likewise, there may be some simple alternatives that deserve immediate implementation. These alternatives can be executed without further analysis.

1.3.4 Analyze the Problem

1.3.4.1 Choose the Decision Aid

The first step in analyzing the problem, and a crucial one in terms of its impact on the decision process, is to select a decision aid. In Section 1.4 relevant issues are listed, and Chapter 4 of this book discusses these issues in more detail. The decision aid may affect:

- Resource requirements for the analysis, such as the kinds of participants and skills required (both technical and administrative)
- Issues to be addressed in the analysis, and therefore the information or data required
- Time and effort required to perform the analysis
- Form in which results will be communicated and used to determine the ultimate course of action.

Some problems may involve the use of more than one decision aid, or several passes through the analysis using one decision aid. The first pass through the analysis can be done at a high level of simplification, and the results can help guide subsequent effort. To illustrate, the decision might first be analyzed quickly by a group through use of a voting method (Chapter 6), weighted scoring method (Chapter 7), or a payoff matrix analysis (Chapter 10). The results may suggest new alternatives or reduce the list of alternatives initially considered. More detailed analysis using, for example, a decision analysis approach (Chapter 11) might then be employed to determine a preferred course of action. As another illustration, some of the decision aids that are generally used for more in-depth analysis (such as cost-benefit analysis, decision analysis, and multiattribute utility analysis, addressed in Chapters 8, 11, and 12, respectively) can also be used for "back of the envelope" analysis as a first pass. Such a first-stage analysis would involve simplifying the problem, performing a quick analysis using the decision aid, and gaining preliminary results and insights. Subsequent analysis would then refine the alternatives, collect additional required data, and again apply the decision aid.

1.3.4.2 Describe the Alternatives

Once the decision aid is selected, the alternatives to be analyzed need to be described or refined to be appropriate to that decision aid. For example, if the decision aid is a voting method (Chapter 6), then a large number of alternatives may need to be grouped into two main options, or distilled into a few representative alternatives, in order to make it feasible to take a vote. If the decision aid is a cost-benefit analysis (Chapter 8), then issues concerning a societal perspective used in the analysis may need to be considered, and the alternatives further described in terms of this societal scope.

1.3.4.3 Collect and Organize the Data

The next step is to collect and organize the data required for the analysis. Each decision aid helps define the type and amount of detail required to analyze the alternatives; the chapters of this book that describe each of various decision aids specify data requirements in some detail. Major categories of data include the:

- Description of the system or process
- Value judgments used in the analysis
- Risk and other uncertainties considered
- Other inputs to the analysis.

For some decision aids, these data can be qualitative, for example, the type of information exchanged in a group discussion prior to taking a vote. Other decision aids require more detailed and quantitative data.

1.3.4.4 Apply the Decision Aid

Analysis of the problem can vary in time and complexity depending on the decision aid applied. With a voting method, for example, applying the decision aid consists simply of tabulating the votes. Other methods involve numerical computations or specialized software. Quantitative methods usually accommodate sensitivity analysis and other types of analysis that provide insight into the results and support conclusions.

1.3.4.5 Communicate the Results

The results of the analysis are presented to those responsible for making and/or executing the decision. Depending on the audience and the stakes involved in the decision, the presentation will generally address one or more of the following:

- Description of the alternatives considered
- Logic and key assumptions used in the analysis
- Analytic approach and data sources
- Key results expressed in terms of the decision criteria of importance to management
- Sensitivity of the results to key inputs and assumptions

- Recommended alternative(s) and future action(s), possibly including further study.

1.3.5 Make a Decision

After analyzing the problem, a decision is made. A sound analysis of the problem can facilitate implementing a decision by providing a rationale that can be communicated to others. A decision might be to undertake one alternative or a combination of alternatives. However, there may be room to further refine the alternatives and input used in the analysis. A decision could be made to perform additional analysis, sometimes in addition to implementing one of the alternatives.

1.4 ISSUES IN SELECTING A DECISION AID

It is impractical to develop an explicit algorithm for selecting a decision aid. There is much flexibility in how the decision aids can be used, and most can be adapted to an organization's needs and to a wide range of problems. However, understanding key characteristics of the problem and of decision aids is basic to selecting an appropriate decision aid. Chapter 4 discusses a process for choosing which decision aid to use for a problem.

A key step in selecting a decision aid is to understand the aspects of the problem to be addressed. These aspects include:

- *Resource Availability:* The time and analytic resources to address the problem
- *Problem Complexity:* The number of alternatives, the complexity of the system, and the amount of uncertainty in the problem
- *Importance/Scrutiny:* How sensitive decision makers and stakeholders are to the decision, and how extensively the decision will be reviewed
- *Group Involvement:* The desire of the organization to involve multiple decision makers or incorporate input from multiple stakeholders in the decision
- *Need for Quantification:* The desire to be able to point to a quantitative basis for choosing one alternative over another.

In addition to these five problem aspects, there may be constraints that affect the selection of a decision aid. Constraints can include organizational guidelines on the type of analysis to be performed and the types of issues to be addressed.

Specific criteria can be used to evaluate the appropriateness of various decision aids to a given problem. Some key characteristics that differentiate decision aids include their:

- *Resource Requirements:* The time, budget, and effort required to use the decision aid
- *Depth of Analysis/Complexity:* The detail and explicitness with which important aspects of the problem are addressed, and the complexity of doing a complete and thorough analysis
- *Logical Rigor:* The mathematical soundness and logical rigor of the analysis
- *Group Focus:* The ability to incorporate group opinions, handle problems with more than one decision maker, and address competing objectives
- *Quantitativeness:* The ability to provide a quantitative basis for the decision, accommodate sensitivity analysis, and address inherently quantitative decisions such as resource allocation problems
- *Track Record:* How long the decision aid has been available and, preferably, demonstrated on actual risk decisions.

Each decision aid can be evaluated against these criteria to determine how well it addresses these concerns.

Depending on the distinguishing aspects of a problem, some of these characteristics may be more important than others. The final step in selecting a decision aid is to consider the relative importance of these characteristics and thereby identify decision aids that are best suited to a problem.

1.5 REFERENCES

AIChE/CCPS (1989). *Guidelines for Chemical Process Quantitative Risk Analysis.* New York: American Institute of Chemical Engineers.

AIChE/CCPS (1992). *Guidelines for Hazard Evaluation Procedures.* 2nd ed. with Worked Examples. New York: American Institute of Chemical Engineers.

Ale, B. J. M. (1991). Risk Analysis and Risk Policy in the Netherlands and the EEC. *Journal of Loss Prevention in the Process Industry* 4(January):58–64.

Gibson, S.B. (1976). Risk Criteria in Hazard Analysis. *Chemical Engineering Progress* 72(2):59.

Helmers, E. N., and L. C. Schaller (1982). Calculated Process Risks and Hazards Management. *Plant Operations Progress* 1(3): 190 .

Kusnic, M. W., and D. Owen (1992). The Unifying Process: Value Beyond Traditional Decision Analysis and Multiple Decision–Maker Environments. *Interfaces* 22(6):150–166.

Netherlands Ministry of Housing Physical Planning and Environment (1989). Premises for Risk Management: Risk Limits in the Context of Environmental Policy; Annex to the Dutch National Environmental Policy Plan. The Hague, Netherlands: Dept. for Information and International Relations.

Renshaw, F. M. (1990). A Major Accident Prevention Program. *Plant/Operations Progress* 9(3):194–197.

U.K. Health and Safety Commission (1991). "Major Hazards Aspects of the Transport of Dangerous Substances." London: UK Health and Safety Executive.
Van Kuijen, C. J. (1988). Prevention of Industrial Accidents in the Netherlands. *UNEP Industry and the Environment* (July/August/September):2–14.
Waterstact Groningen Provincial (1979). "Pollution Control and the Use of Norms in Groningen—Criteria for Risks Related to Dangerous Goods." Groningen, Netherlands: Provincial Waterstact Groningen (April).

2

KEY CONCEPTS

As soon as there is life there is danger (risk).
—R.W. Emerson, 1870

2.1 PURPOSE OF CHAPTER

This chapter introduces some key concepts of decision making. The concepts covered apply to decision making in general or are components of decision aids discussed in this book. The concepts discussed include economic evaluation principles, decision rules, externalities, value of life, uncertainty, and risk analysis.

2.2 ECONOMIC EVALUATION PRINCIPLES

This section presents two key economic evaluation principles that are important in making good decisions—sunk cost and the time value of money.

2.2.1 Sunk Cost

The *sunk cost* principle reflects a goal to maximize present or future benefits or wealth, while recognizing that past costs are irrelevant to this goal. To adhere to the sunk cost principle, the decision maker should always assess each new decision based only on future returns, not on past investments or goals. An example will help illustrate this principle.

Company A produces chemical X which is sold in either liquid or pellet form. The liquid sells for $2000 per batch, and the pellets sell for $1000 per batch. The cost of producing the liquid is $1400, and the cost of producing the pellets is $500. A decision is made to produce the liquid at a profit of $600 (versus a profit of $500 on the pellets). When the liquid process is partially completed at a cost of $300, a salesman calls. The buyer's needs have changed, and the liquid form is now worth only $1500 per batch. The salesman is sorry, but the new price has already been negotiated.

At this point, the firm has two possible courses of action. It can complete the processing at a cost of $1100 and sell the liquid for $1500, realizing a future profit of $400. Or, it can dispose of the partially processed material, start over with the pellet process and realize a future profit of $500. Under either alternative the customer will get the amount of the chemical ordered. We assume that there is no cost of disposing of the partially processed liquid.

Company A decides to start over and throw away the partially processed material, even though it represents a lost investment of $300. The firm wants to maximize its future profits and realizes that its past costs are irrelevant. The past costs are "sunk." There is no way to recover these costs.

2.2.2 Time Value of Money

A basic principle in the evaluation of investments is the *time value of money*. At its simplest, it means that because there are widespread opportunities to invest money safely and with assured interest payments, such as government bonds and federally insured bank accounts, a dollar paid to us today is worth more than a dollar paid to us tomorrow. For example, if the local savings bank is secure and offers 5% simple interest per year on savings accounts, we know that if we place $100 in the bank today we will have $105 at the end of a year. We would be unwise to lend money to others, or to undertake an investment, unless we could expect a similar or larger return over the year. A dollar now is worth one dollar and five cents a year from now.

A "discount rate" is a factor used to compare the value of money in different years. It represents the annual percent increase in a sum of money that would be necessary in order to make receiving a larger sum in the future worth the same as receiving the sum today. If r is the discount rate, the value today of x dollars received in year t is $x/(1+r)^t$.

The time value of money is often confused with inflation. However, they are separate issues. Almost invariably the interest offered by banks and government bonds exceeds the rate of inflation by a few percent. This reflects the fact that most of us would prefer to consume today rather than tomorrow, and we require some real growth in buying power if we are to defer present consumption and to save, invest, or loan our money.

Most investments involve expenditures and income at different periods in time. Because of this, and because the value of money evolves over time, we need methods to compare streams of payments and receipts that account for the time value of money. *Net present value* (NPV) analysis and *rate of return* analysis are the two most common methods. They are both based on the following formula:

$$\mathrm{NPV} = (R_0 - C_0) + \frac{R_1 - C_1}{(1+r)} + \frac{R_2 - C_2}{(1+r)^2} + \frac{R_3 - C_3}{(1+r)^3} + \ldots + \frac{R_n - C_n}{(1+r)^n}$$

where

NPV is the net present value
R_i are the revenues at time i
C_i are the costs at time i
r is the discount rate in each time period.

When we compare multiple series of revenues and costs, the series with the largest NPV represents the best investment. Only those series which have an NPV greater than 0 are worth considering; a negative NPV implies that we would do better investing in a bank account with the same interest rate.

The rate of return is defined by the same equation. In most investment problems, the costs exceed the revenues in early time periods and this pattern reverses in later time periods. This means that the first terms in the above expression are negative and later terms are positive. In this case there is a value of r such that NPV is zero. This value of r is the rate of return on the investment.

A major question in many companies is what discount rate, r, to use in comparing or evaluating investments. Some companies select a discount rate and publicize it within the organization, attempting to provide consistency across the organization. The discount rate typically reflects the rate of return that can be obtained with projects or investments that are similar to the one being evaluated.

Some persons understand the application of NPV analysis or *discounting*, as it is also called, to monetary investments, but doubt its applicability to investments to prevent future health and safety risks. An argument can be made that the same interest rate should be applied to both money and health and safety impacts. Suppose there is an opportunity to either save a life one hundred years from now by investing \$100 in improved toxic waste disposal, or to invest \$100 in government bonds that will accumulate 10% interest each year for 100 years and then pay out \$1.38 million to the charity of our choice. If we believe that the charity will save more than one life with the \$1.38 million, we would not be doing the best we could for future generations by investing in the toxic waste disposal alternative. The appropriate monetary investment would provide the future generation with greater opportunities to save lives or otherwise serve society. In this situation it is appropriate to discount both lives and dollar investments to determine what actions will have the most value to future generations. Of course, this argument is based on a very strong assumption that each subsequent generation will use resources in line with our perceptions of the value of reducing health and safety risks. See Redelmeier, Heller, and Weinstein (1994) for a discussion of this issue.

2.3 DECISION RULES

2.3.1 Evaluation Criteria and Decision Rules

A decision maker evaluates alternatives, or possible actions, based on one or more selected *decision criteria*. Next, valuations are integrated and the best alternative is chosen, using a *decision rule*. For example, in cost-benefit analysis the evaluation criteria include costs and benefits and the decision rule used is to select the alternative whose ratio of benefits to costs is highest.

The selected decision rule reflects the goal of the decision maker. What is chosen as the most appropriate decision rule can vary from one company to another and from one situation to the next. The goals of a conservative, well-established company may differ greatly from the goals of a young startup company. For instance, the first may want to minimize costs, while the second may seek to maximize market share. When presented with the same decisions, the same possible outcomes, and even the results of identical analyses, these companies will not necessarily choose the same course of action. Most likely the two companies' goals will differ, and these differing goals would be evident in the evaluation criteria and decision rule they select to evaluate alternatives.

In the remainder of Section 2.3, we present five common decision rules and explain how they relate to different goals of the decision maker. For ease of communication, we assume that the decision maker is making a decision whose outcome is measured in units for which more is better (profit, revenue, market share, etc.). Analogous situations exist in which the decision maker prefers a lower number (costs, etc.).

2.3.2 Maximization

The most common decision rule is *maximization*. The decision maker quantifies the net benefits with each alternative and selects the one for which the benefits are greatest. In order to apply this rule, each alternative must be associated with a single value that represents benefits. When uncertainty is ignored, applying the maximization rule is straightforward and expresses the decision makers simple desire for more of a good thing.

When considering uncertainty, the usual procedure for applying the maximization rule is to determine the likelihood of the possible outcomes of an alternative, and compute an expected value of each alternative. The expected value is the weighted sum of values for each possible outcome, with the weights determined by the likelihood (probability) of these outcomes. (Probability and expected value are discussed further in Sections 2.6.1 and 2.6.2, respectively.) The preferred alternative is the one with the greatest expected value. In this instance the maximization of expected value represents the decision maker's desire to receive the highest benefit on average. Due to uncertainty, a specific decision will not necessarily result in the most preferred outcome. But over the long run, maximizing expected value will return the highest benefit possible given this uncertainty. A further refinement is to maximize expected utility, where the utility is a number that reflects both the decision maker's preferences for outcomes and the decision makers desire to avoid risks. The concept of maximizing expected utility is described further in the chapter on Decision Analysis (Chapter 11).

2.3.3 Maximin and Maximax

Another common decision rule is the *maximin* rule ("maximize the minimum"). It can be applied in situations where multiple outcomes are possible, but whose likelihoods are not characterized. Using this rule the decision maker takes each decision and associates it with the value resulting from the worst possible outcome that could occur under that alternative. The alternative that provides for the highest value under its worst possible outcome is selected. The maximin rule is truly a pessimist's rule, for it ensures that the decision maker avoids choosing alternatives that could lead to the worst possible outcomes, even if these alternatives could yield attractive outcomes as well.

The corresponding optimist's rule is the *maximax* rule ("maximize the maximum"). Using this rule, the decision maker selects the alternative whose return under the most favorable circumstances is the greatest. With this rule the decision maker is "going for broke" by ignoring all possible downsides to each alternative, and focusing only on the possible benefits.

2.3.4 Minimum Regret

A decision maker may be criticized long after a decision has been made, by others who operate with perfect hindsight. Perfect hindsight may reveal that a different alternative than the one chosen would have led to a better outcome under the circumstances that occurred subsequent to the decision. For example, a plant manager chooses to delay a valve upgrade due to its high cost and a judgment that there is an extremely small chance of valve failure in the next five years. A valve failure occurs soon after this decision, which causes a half-day interruption of operations. Even though the decision was in fact a wise one given the decision maker's understanding of the costs and risks, the plant manager is criticized for not choosing the alternative that would have yielded the best outcome. The *regret* rule chooses to minimize the risk to the decision maker of being adversely confronted in hindsight. For each possible future scenario, the decision maker assesses the difference in value between each alternative and the best alternative given that the scenario were to occur. This difference is referred to as the "regret." Regret represents the "cost" of not having perfect foresight and selecting the best decision. This rule then applies a minimax rule, selecting the decision for whom the maximum regret is the smallest.

An example illustrating the minimum regret rule is shown in Table 2.1. A company faces the decision of how to fulfill its need for space to manufacture a new product. There are three alternatives, Use the Existing Plant, Build a New Plant, or Lease Additional Space, and three possible future events, *low sales*, *medium sales*, and *high sales*. The profits side of the table presents the profits under each alternative and event combination. The regrets side presents the maximum regret. For example, if we choose to Use the Existing Plant and we experience *low sales*, the maximum regret is 0, because Use the Existing Plant has the highest outcome when the event *low sales* occurs. However, if we choose to Build a New Plant and *low sales* occur, the maximum regret is the difference between the outcome under Build a New Plant (5) and the best outcome, 25, under Lease Additional Space, which equals 20. Scanning the columns on the profits side of the table, we see that the maximum regrets are 45 for Use the Existing Plant, 20 for Build a New Plant, and 25 for Lease Additional Space. Using the maximum regret decision rule, we choose Build a New Plant, because it has the lowest maximum regret.

TABLE 2.1
Minimum Regret Example

	PROFITS Alternatives			REGRETS Alternatives		
EVENTS	*Use the Existing Plant*	*Build a New Plant*	*Lease Additional Space*	*Use the Existing Plant*	*Build a New Plant*	*Lease Additional Space*
Low sales	25	5	15	0	20	10
Medium sales	45	55	60	15	5	0
High sales	45	90	65	45	0	25

2.3.5 Minimum Uncertainty

Most decision makers are risk-averse: they prefer to avoid choosing alternatives for which the outcomes are uncertain. When outcomes are presented to a decision maker in terms of the probability for each of the possible values that could result, another decision rule, called *minimum uncertainty,* can be employed. Minimum uncertainty may also be referred to as *minimum risk,* where risk refers not to physical or financial risks specifically but to aversion to uncertainty as described above. With outcomes quantified by a probability distribution, extremely risk-averse decision makers may wish to select an alternative that gives them the most certainty about future rewards by minimizing the variance on possible outcomes. Alternatively, some decision makers choose to account for only those outcomes associated with the most likely future event. A minimum uncertainty rule can lead the decision maker to reject attractive but uncertain alternatives, and instead choose a safer alternative with smaller potential rewards.

2.3.6 Satisficing

Not all decision rules yield a single best alternative. Many organizations are concerned with meeting many, sometimes contradictory, goals. Satisficing is a decision rule that is used to find one or more alternatives that can satisfy all (or most) of the organization's goals rather than determining the best alternative. When using a satisficing decision rule, a minimum level of achievement on each of several goals is set. For example, in a safety decision the goals may be to "reduce the risk of death below 10^{-7} per year" and "spend less than $100,000." All alternatives that meet these goals would be considered satisfactory.

2.4 EXTERNALITIES

As part of analyzing a decision, a decision maker will seek to describe the outcomes that will occur with each of the available alternatives. The description of outcomes can include a wide range of factors, including effects on the company (e.g., product sales, profits, and company size), effects on the companys workers (e.g., employment, salary, risk of fatality), effects on the community (e.g., jobs, tax revenues, and risk of fatality), and even direct effects on the decision maker (e.g., recognition, salary, job title).

An *externality* is a secondary or indirect effect of an activity that is borne by individuals or entities other than those that cause the effect. In some decision situations, it is important to characterize the externalities of a decision. In particular, public agencies often wish to formally recognize the spillover effects of their activities. In the private sector, corporate philosophy is more frequently leading companies to consider externalities. In addition, public policy sometimes attempts to *internalize* an externality by forcing an organization to bear the cost (or pay for the benefit) of an externality, for example through taxes, or licensing fees, or user charges.

Externalities can be of various types. Externalities such as pollution, chronic health effects, and accident-caused fatalities, are secondary effects of the production process whose costs are only partially (or not at all) borne by the producer. These types of externalities have been termed *externalities of production*. Externalities of production may also be desirable, such as the indirect contribution that a producer provides to a local economy through jobs, taxes, etc.

Externalities can also result from the consumption of the producer's products, and these effects have been termed *externalities of consumption*. Examples are the congestion caused by individuals choosing to use their own automobiles rather than public transit, or the depletion of landfill space from waste disposal. Externalities of consumption may be positive, such as the aesthetic improvement that one resident gains when a neighbor repaints the outside of the house next door.

By their very nature, chemical process safety decisions address a key externality: the risk that a chemical process imposes on workers or the public. The corporate philosophy will determine the extent to which this externality is addressed and given weight. The corporate philosophy will also determine whether and how other externalities will be addressed in the decision. Other externalities such as job creation and pollution are often considered relevant.

Externalities can be considered when using any of the decision aids in this book, most often just by setting a dollar value on recognized externalities. In cost-benefit analysis and multiattribute utility analysis the recognition of externalities is usually part of the analysis.

2.5 VALUE OF LIFE

A particularly difficult issue in evaluating chemical process safety decisions is how to appropriately determine the relative value of reducing risks to worker or public lives versus the value of other aspects of the decision such as financial ones. Applying the decision aids discussed in this book often requires assessing a *value of life*. Even when a value of life is not explicitly chosen, it is still implied by decisions involving the allocation of economic resources to activities that reduce risks to human life. Substantial research and discussion have focused on value of life issues, particularly in the last two decades. This section reviews some of the key issues and approaches for assessing the value of human life.

The idea of placing a monetary value on a human life strikes many as inhumane and morally unacceptable. Doniger (1980) states, "There is simply no morally and socially acceptable way to trade human life and such things as the survival of other species for conventional economic products. Any decision calculus which, even if it had precise numbers, attempted to put a price on human life would be surely rejected as morally repugnant and as practice unbecoming a democratic government."

However, as Barrager, Judd, and North (1976) explain, in social decisions, the dollar figure is not an attempt to place an absolute measure of worth on a life, but a means of representing the value of human life in decisions that unavoidably engender tradeoffs: "While some might see these procedures as callous, it is important to observe that we are not attempting to value a human life or injury in a moral sense, we are rather assigning values to life and other loss categories as a means of making social decisions. The values are not measurements but the results of a decision made by society concerning the balancing of social benefit and social risk...thus, the problem is to assign a value to human life (and to injury, and so forth) that reflects what we can afford to pay as a society to avoid a death from involuntary risk."

Despite the resistance to placing a monetary value on life, we must recognize it as necessary in order to make consistent decisions that most efficiently allocate resources to the reduction of risk. Weighing the monetary value of life into decisions enables us to spend money on those programs that provide a higher reduction of risk per dollar spent and avoid those that provide insufficient reduction in risk per dollar spent. For example, the *Budget of the United States Government, Fiscal Year 1992* (OMB 1992) presents the cost per premature death averted for selected regulations implemented by various government agencies. The cost of a given regulation per premature death averted ranges from $0 to $5.7 trillion. This very wide range of values provokes the question of whether there were policies or regulations not enacted that would have saved the same or more lives at an expenditure of less than $5.7 trillion per life saved. By using a single, consistent value of life and choosing to fund those policies and programs that have the lowest cost per death averted, the total number of deaths averted would have been significantly increased. The same can be done in funding measures to reduce chemical process risk.

Decision aids vary in the degree to which explicit valuation of life is required. For example, a vote among alternative safety programs need not consider an explicit value of life. However, the decision that results will indicate the organization's implicit willingness to spend money to reduce the risk of fatalities. Any effort to avoid explicitly assigning a value to human life must recognize public sensitivity to the values implicit in such decisions.

Several issues in risk perception complicate the task of assessing a value of life. One issue is whether or not the value of an individual life should be greater for risks to a large population than risks to a few individuals. The impact of a large-scale accident publicized by the media may lead society to place a higher value on society's loss than on the one-at-a-time loss of the same number of individuals. A second issue involves involuntary risks. Involuntary risks are often viewed differently than voluntary risks and can lead to relatively higher value-of-life estimates. Another complication in making value-of-life tradeoffs is that certain causes have more fear associated with them. For instance, studies suggest that people will pay more to avoid fatality from a nuclear power accident than from less dreaded causes such as automobile accidents. In addition, unfamiliar causes of risk are often perceived as more undesirable than familiar causes. Similarly, people will pay more to save an identified life, such as a child trapped under a collapsed building, than to save an anonymous life, such as a victim of a large-scale flood. These perceptual issues can complicate decision making.

Value-of-life measures are typically related to the number of lives saved (or fatalities avoided) or to the number of "life years" saved. The first view assigns a monetary value per life. For example, spending $1 million to reduce the number of fatalities from one to zero, would imply a value of life of $1 million dollars per life saved. Alternatively, value can be measured in dollars per expected year of life saved. In this case, the lives of those who have fewer expected years of life are valued lower than those with a higher expected lifetime. The latter approach is particularly relevant to programs aimed at prolonging life.

Several approaches have been used to estimate a value-of-life measure. The *human capital* approach values life by the present value of individuals' future earnings. This approach has been criticized for its assumption that an individual's life is worth the same as what society pays that individual for his work. This implies that a person's wages have something to do with how much the person, society, or the decision maker values life. It does not account for the individual's preference for life. In addition, this approach implies a zero or negative value of life for retired persons, those who are not employed, or those who consume resources in excess of what they produce.

A second approach uses decisions by government on safety-related issues as a basis for estimating a value of life. A criticism of this approach is that it is based on an unjustified assumption that past decisions resulted from appropriate reasoning and concern for human life. Empirically, values obtained under this approach vary widely depending upon which past decisions are considered. Figure 2.1 shows the value-of-life estimates used by several different government agencies for various decisions. For the interested reader, Graham and Vaupel (1981) review 57 government life-saving programs aimed at reducing fatalities, and estimate the cost and benefits of each program. The estimates range from approximately $55 thousand to $7 million.

A third approach is to use jury awards for compensation in cases involving wrongful death. This approach, like the previous one, assumes that past decisions had an appropriate basis. Further, awards have varied widely and do not provide a consistent value.

A fourth approach assumes that the amount of life insurance one purchases reflects the value placed on risks to one's life. However, the insurance is purchased for ones dependents, not for reducing or avoiding risk.

A fifth approach is to estimate what individuals are willing to pay for a statistical reduction in the risk of fatality. To do this, alternatives that increase or reduce the risk of fatality imposed on an individual are examined, and an

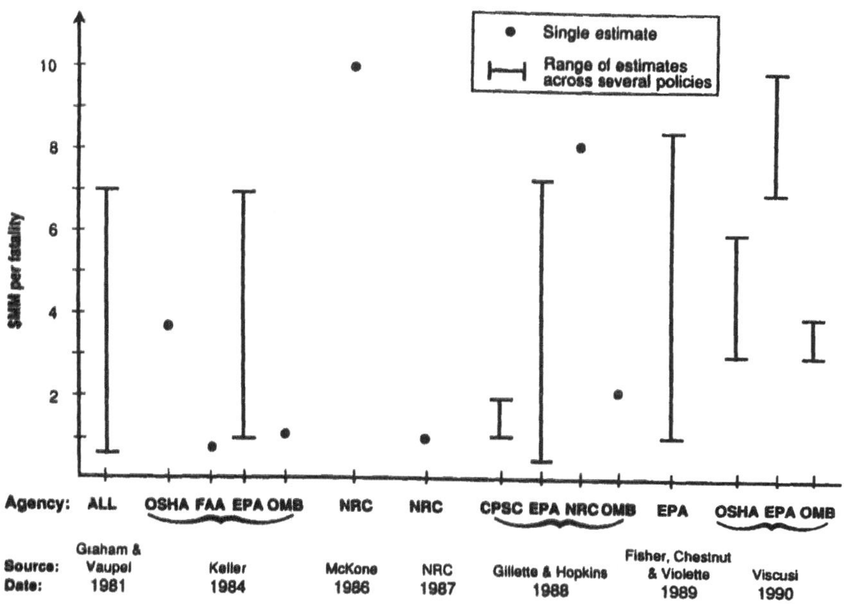

FIGURE 2.1. Value-of-life estimates used in government cost–benefit analysis.

individual's *willingness-to-pay* to obtain (or avoid) an alternative is taken as an expression of the value of life.

To estimate this willingness-to-pay, various techniques have been suggested. One of these is to use *revealed preferences*. This technique infers the value based on purchases that reduce risk or on wage differentials among risky occupations. For example, values for safety can be inferred from prices paid for automobile airbags and security services, and price differentials for housing in neighborhoods with different crime rates. In a job market context, wage differentials among risky occupations may reflect the extra pay that a worker expects as a reward for incurring an increased risk of injury or fatality. For example, if coal miners are paid more than factory workers for a job that requires similar skills and provides similar amenities, the pay differential may reflect the greater hazards of coal mining. This technique has drawbacks in that it assumes that workers are well-informed of the risks of their jobs and that the labor market allows workers to change occupations easily. Another drawback of the revealed preference approach is that the values are derived from a limited set of contexts such as the consumer market and the job market.

A second technique for measuring willingness-to-pay is to use a survey to infer how much persons are willing to pay for a statistical change in the risk of fatality. For example, a survey might pose hypothetical choices among alternatives that vary on cost and risk to personal life, and from these, the individual's values can be inferred. Figure 2.2 provides a question from a survey of this type, for a hypothetical decision by the government on which alternative hazardous waste sites should be cleaned up first. Responses to a series of questions like this in which costs and risks are varied can be used to infer the respondents value for risk reduction (this particular approach is called *conjoint* methodology and is described in most basic texts on market research methods; see also Maas and Stalpers 1992). These assessment approaches involve the challenges of avoiding biased or distorted answers. Moreover, conveying the meaning and implications of statistical risks to those who are affected is difficult.

Regardless of the approach used to estimate a value of life, it is advisable to conduct a sensitivity analysis to the estimate used in the analysis of a

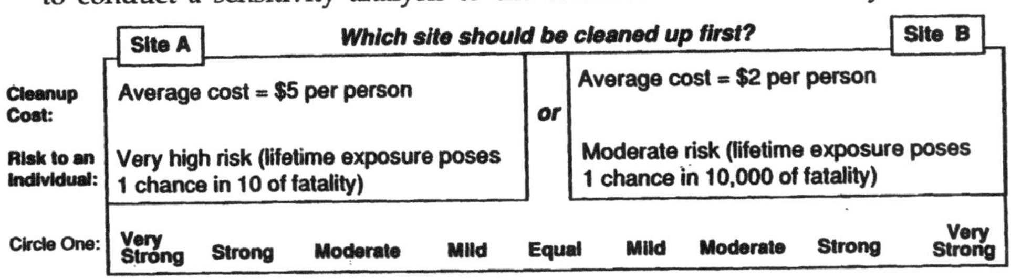

FIGURE 2.2. Illustration of survey assessment approach to estimate society's value for risk reduction.

decision. Instead of analyzing a decision using a single value-of-life estimate, the sensitivity analysis will examine the decision using a range of values. As the value of life is varied, one should note any points at which the preferred alternative changes. If the decision does not change within the range of values considered, then the decision is not sensitive to the value of life that is chosen within that range. This can eliminate the need to express a specific value of life used in the decision.

2.6 UNCERTAINTY

Our knowledge about the present and the future is never completely accurate. We do not know for certain all aspects of the world we live in, and we cannot perfectly predict the conditions or events that will occur in the future. Uncertainty complicates many of our decisions. If there were no uncertainty, our decisions would be reduced to simply choosing among alternatives whose future impacts are known. For instance, if we knew for certain when a pipeline was going to fail, the decision on when to replace it would be trivial. Similarly, if we knew for certain how various materials and thicknesses of pipelines would perform, the decision on which type of replacement pipe to install would be straightforward. Because of uncertainty, we are forced to make decisions with imperfect knowledge of the consequences.

We cannot avoid uncertainty. It is inherent in almost all decisions. For instance, we do not know perfectly the reliability of components that we install; we do not know how successful or unsuccessful investments will be; and we do not know how the economy will perform next month. Often, we attempt to reduce or eliminate uncertainty before making a decision, by gathering information or by taking measures to control the uncertainty. However, it is often costly or impossible to reduce or eliminate uncertainties. In reality, we must make decisions under uncertain conditions. Whether or not we consider uncertainty explicitly, both our decisions and uncertainties will determine future outcomes.

The decision aids described in this book vary in how they address uncertainty. Some ignore uncertainty, or assume that the decision maker will consider uncertainty informally when making a decision. Others are explicit in identifying and quantifying uncertainties, and determining the best alternative given these uncertainties.

2.6.1 Probability

In order to discuss uncertainty further, we introduce some basic concepts and terminology. An *uncertainty* is something that is unknown and uncontrollable, or not perfectly known or controllable. An uncertainty can have different possible *states* that can exist or may occur. For instance, in deciding whether or not to take an umbrella to work tomorrow, there is the uncertainty about the weather, which could be described as having two possible states: *Rain* or *No rain*. An uncertainty can have several or even infinitely many possible states. Tomorrows temperature is an example of an uncertainty with infinitely many possible states, or a continuous range of possible values. Examples of uncertainties affecting chemical process safety decisions are whether or not a failure in equipment will occur within a given time frame, the size of the area contaminated if a toxic release were to occur, and the cost of cleaning up an area if a chemical spill were to occur.

Each state of an uncertainty has an associated likelihood of occurrence. Characterizing the likelihood is useful in decision making and, with some decision aids, it is essential. There are several qualitative ways to represent the likelihood such as with words like "definitely," "probably," "unlikely," or "certainty." But these terms are vague, and open to misinterpretation. A precise and unambiguous way to represent uncertainty is to express the likelihood quantitatively using a numerical *probability*. We could make a qualitative judgment that rain tomorrow is "highly likely," but this will mean different things to different people. However, if we say that there is a 0.9 chance of rain tomorrow, our opinion about the likelihood of rain is unambiguous and precise.

A probability is a number that expresses the likelihood of occurrence of a possible state of an uncertainty. By definition, a probability must be a number from 0 to 1, and the sum of probabilities for all possible states of an uncertainty must be 1.

Probability can be characterized as either *subjective* or *objective*. The objective view characterizes uncertainty as a property of the physical world. A probability is obtained from historical frequency data. It is defined as the long-run frequency with which an event occurs in an infinite number of trials. Under this view, uncertainty is only relevant for repetitive events. For example, the probability of heads in a fair coin flip could only be determined by a series of coin flips. If in one hundred flips, 50 heads occurred the probability of a head would be 0.5.

The subjective view is that uncertainty is a degree of belief held by an individual. Probabilities are a way of communicating a person's (or a group's) knowledge or perceptions. Under this view of probability, uncertainty is relevant for both unique and repetitive events. This view of probability allows the uniqueness of almost all situations to be recognized. For example, the subjective view of probability would recognize that a pipe operating in a particularly harsh environment would have a higher probability of failure than indicated by failure frequencies in standard tests. A subjective probability of 0.5 indicates that a state is believed to have the same likelihood of occurrence as a head in a fair coin flip.

Decision analysis, payoff matrix analysis, and multiattribute utility analysis are decision aids that are founded on a subjective view of probability.

Debate over the nature of probability has existed for centuries. Philosophers and academics have supported both objective and subjective views. In practice, both views are usually relevant to assessing the probability of an unknown event or quantity. A common approach for assessing probabilities is to obtain all of the available "objective" data and adjust it subjectively to fit the situation and represent the degree of belief in the data. If the quantity assessed is one for which substantial relevant data and experience are available, little or no subjective adjustment may be required. However, in order to quantify the uncertainty around an unprecedented future event, a primarily subjective assessment of probabilities is necessary.

Two types of functions, a *probability density function* and a *cumulative distribution function*, are used to express probabilities for the states of an uncertainty. A probability density function describes the relative likelihood of the occurrence of the possible states. Figure 2.3 shows a *discrete probability density function* $f(x_i)$. A discrete density function is also referred to as a *probability mass function*. There are ten possible occurrences, x_i, each with an associated probability. As noted before, the probabilities of the possible states must sum to 1. Figure 2.4 shows a probability density function, $f(t)$, for a continuous variable t. The distribution represents a continuum of possible

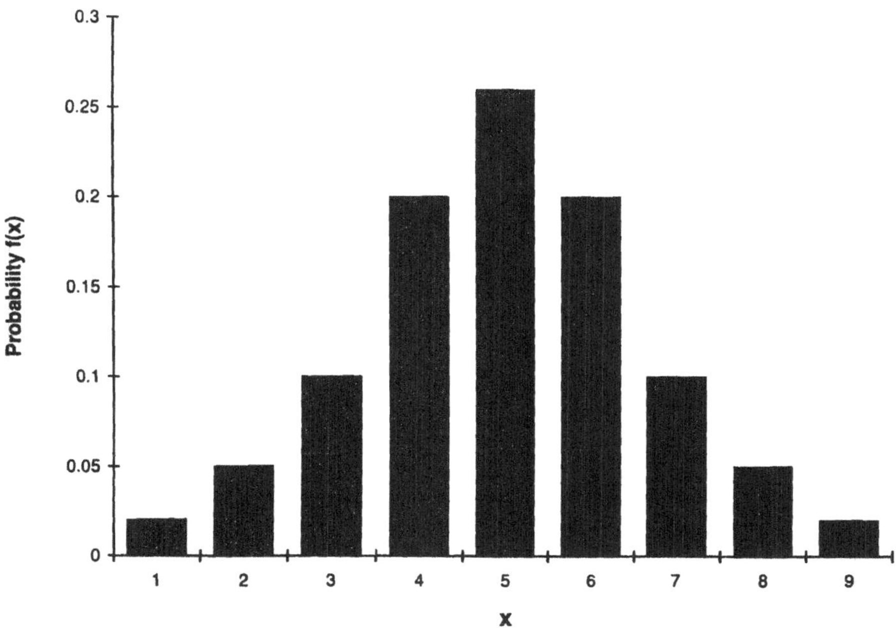

FIGURE 2.3. A discrete probability density function.

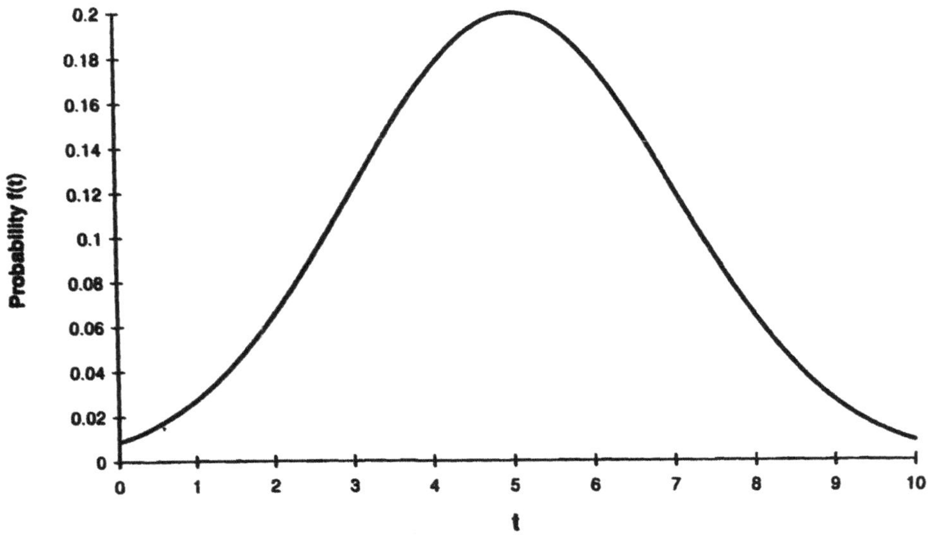

FIGURE 2.4. A continuous probability density function.

states. Any area under the distribution between two values of the variable corresponds to the probability of occurrence within the range defined by the two end values. Therefore, the total area under the probability density function is 1.

In a cumulative distribution function, each functional value represents the probability of occurrence at a value equal to or less than the corresponding value of the variable. Cumulative distribution functions can be constructed for uncertainties that have states that can be ordered numerically. Each value on the cumulative distribution, therefore, is a cumulative representation of the probabilities of the previous states. A discrete cumulative distribution function associated with the density function in Figure 2.3 is shown in Figure 2.5. The final value of the cumulative distribution must have a probability of one, since all possible values of the uncertain variable are less than or equal to the final value. Similarly, Figure 2.6 shows the cumulative distribution function associated with the continuous probability density function shown in Figure 2.4. Again, the cumulative probability distribution function increases from zero to one as the uncertain variable increases from its smallest to its largest value. Each functional value of the uncertain variable represents the probability that the occurrence of the variable will be at or at less than the corresponding value of the variable.

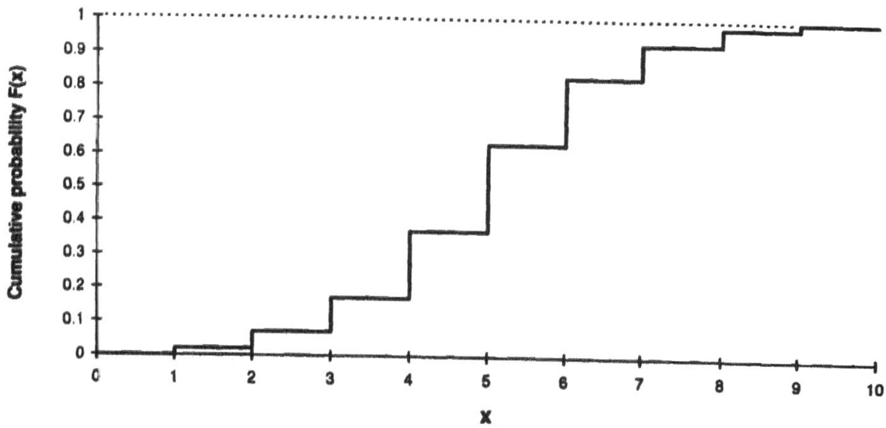

FIGURE 2.5. A discrete cumulative probability distribution function.

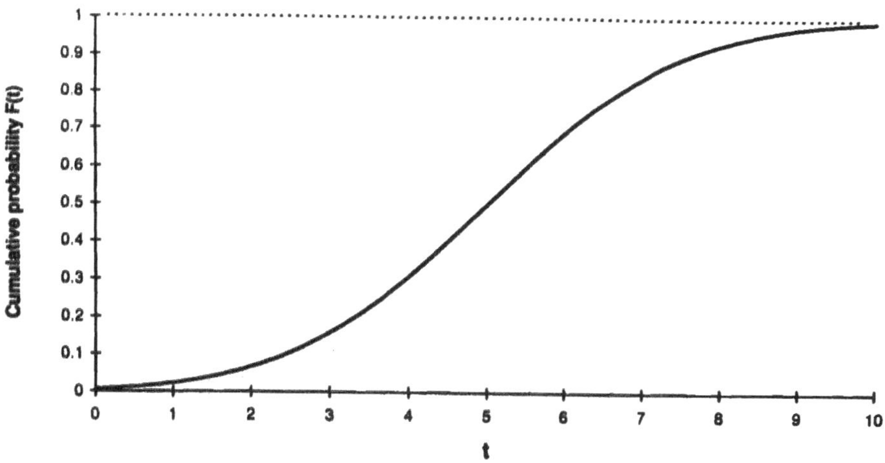

FIGURE 2.6. A continuous cumulative probability distribution function.

2.6.2 Mean and Variance

Summary measures are often used to describe probability distributions. Two commonly used measures are the *mean* or *mathematical expectation* (or *expected value*) and the *variance*. The mean or mathematical expectation, μ, can be thought of as a weighted average of possible states, with the weights determined by the probability of these states. For a discrete function, $f(x_i)$, the mean is determined by the following rule:

$$\mu = \sum_i x_i f(x_i)$$

For a continuous function, $f(t)$, the mean is determined by the following rule:

$$\mu = \int_{-\infty}^{\infty} t f(t)\, dt$$

The variance, v, expresses the variation of possible states about the mean. The variance is the mean of the squares of the variations from the mean of the distribution. For a discrete function, $f(x_i)$, the variance v is determined by the following rule:

$$v = \sum_i (x_i - \mu)^2 f(x_i)$$

For a continuous function, $f(t)$, the variance v is determined by the following rule:

$$v = \int_{-\infty}^{\infty} (t - \mu)^2 f(t)\, dt$$

The relationship between two uncertainties of a decision problem can show either probabilistic dependence or probabilistic independence. Probabilistic dependence exists between two uncertainties if knowing the resolution of one uncertainty gives information about the possible resolution of the other. For example, if the probability of rain is 20% when there is no information on cloud cover, and the probability of rain when the day is cloudy is 50%, rain is dependent on cloudy conditions and dependence exists between rain and cloudiness. Probabilistic dependence may also exist between a decision and an uncertainty. If a decision changes the probabilities of the states of an uncertainty, the uncertainty is dependent on the decision. For example, decisions to implement safety measures reduce the probability of accidents; therefore, accident uncertainty is dependent on decisions to implement safety measures.

Probabilistic independence exists between two uncertainties if the resolution of one uncertainty gives no information about the possible resolution of the other. For example, if a pair of dice are rolled and then a coin is flipped, the probability of heads is assumed to be 50% regardless of the result of rolling the dice. The coin flip is probabilistically independent of the roll of the dice. An uncertainty can also be described as independent of a decision. For example, the chance of rain is independent of the decision to carry or not carry an umbrella.

Probabilities are used in payoff matrix analysis (Chapter 10), decision analysis (Chapter 11), and multiattribute utility analysis (Chapter 12). Since probability is not intuitive to most individuals, accurately expressing a probability distribution that reflects one's beliefs about an uncertainty is not a trivial task. Decision analysis employs formal interviewing techniques that help an analyst motivate individuals to estimate probabilities, adjust for biases that lead to inaccurate assessments, and provide an accurate representation of the individual's knowledge through a probability distribution (Merkhofer 1987). This process of assessing probabilities is referred to as *probability encoding*. For best results, the encoding process should always be guided by an independent, unbiased analyst.

2.7 RISK ANALYSIS

2.7.1 Risk

Risk may be defined as a measure of economic loss, human injury, and/or environmental damage in terms of both the incident likelihood and the magnitude of the loss, injury, or damage (AIChE/CCPS, 1989). There are many ways to measure risk, depending on the type of loss, injury, or damage being considered. Some description of risk, whether qualitative or quantitative, is necessary in evaluating risk reduction strategies for risk management decisions.

Chemical Process Quantitative Risk Analysis (CPQRA) is a series of methods for the quantitative estimation of risks in a chemical storage or processing facility. CPQRA techniques are described in detail in the *CPQRA Guidelines* and are summarized by Arendt, Lorenzo, and Lusby (1989) and Arendt (1992). Several common CPQRA risk measures and their calculation methodologies are discussed in detail in Chapter 4 of the *CPQRA Guidelines*. Depending on the scope and objectives of the CPQRA, one or more of the following risk measures may be developed:

- *Individual Risk*—A measure of the risk to a person in the vicinity of a hazard, including the nature of the injury or other undesired outcome and the likelihood of occurrence. Individual risk is generally expressed in terms of a likelihood or probability of a specified undesired outcome per unit of time. For example, the individual risk of fatality at a particular location near a hazardous installation might be expressed as 1 in 350,000 per year, or 2.9×10^{-6} per year. Individual risk is sometimes presented graphically in the form of individual *risk contour maps*. These maps are generated by calculating individual risk at various points surrounding a hazardous facility, assuming that people are present at each location 100% of the time. The risk contours are then drawn to connect points of equal individual risk ("iso-risk" contours). These risk contour maps are analogous to topological maps on which the contours connect points of equal elevation. Figure 2.7 is an example of an individual risk contour map.

- *Societal Risk*—A measure of the risk to a group of people, including the risk of incidents potentially affecting more than one person. Individual risk is generally not significantly affected by the number of people involved in an incident. The risk to a person at a particular location depends on the probability of occurrence of the hazardous event, and on the probability of adverse impact at that location should the event occur. The number of people at the location does not usually change either of these probabilities. However, the impact on a defined group of people (a plant work force, a town, a nation, etc.) is clearly larger if more people are involved in a hazardous event. Societal risk measures this risk to a group of people (a "society"). One measure of societal risk is the Average Rate of Death (ROD), an estimate of the average number of fatalities resulting from all accidents, large and small, that might occur in a facility. Another measure, which provides more information about the distribution of large and small potential accidents, is the societal risk F-N (Frequency-Number) Curve, as illustrated in Figure 2.8. The F-N Curve is a cumulative plot, usually on a logarithmic scale,

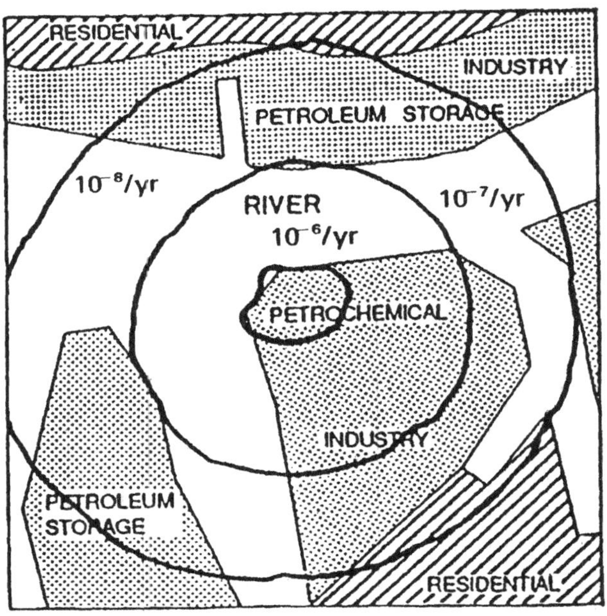

FIGURE 2.7. Example of an individual risk contour plot. Contours indicate individual risk of fatality per year.

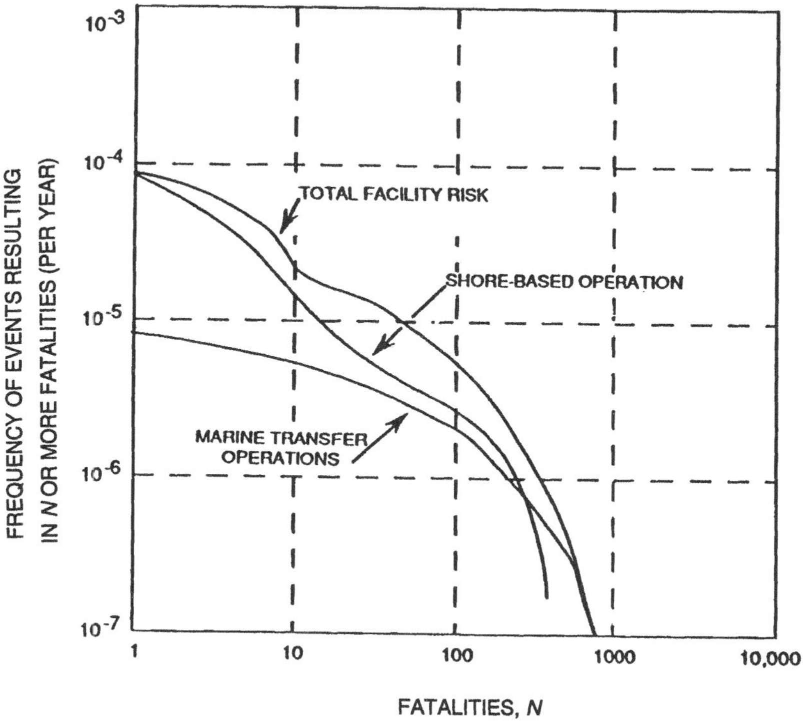

FIGURE 2.8. Example of a societal F-N curve.

of the frequency of incidents resulting in N or more fatalities as a function of the number of fatalities, N.

- *Risk Indices*—A risk index is a single-number measure of the risk associated with a facility. Some risk indices are qualitative or semi-quantitative, ranking risks in various general categories (for example, the Dow Chemical Company Fire and Explosion Index [1994]). Risk indices may also be quantitative averages or benchmarks based on other risk measures. For example, the Average Rate of Death (ROD) may be considered as a risk index that represents societal risk as a single number. Similarly, an Average Individual Risk in a population can be calculated. Risk indices can simplify risk comparisons and communication, but do not convey as much information about the distribution of risks as the more complex risk measures. Table 2.2 gives several examples of risk indices, in addition to examples of presentation formats for individual and societal risk.

TABLE 2.2
Presentation of Measures of Risk

*Risk measure**	*Presentation format*
Indices	
Equivalent social cost index	A single-number index value representation
Fatal accident rate	A point estimate of fatalities/10^8 exposure hours
Individual hazard index	An estimate of peak individual risk or FAR
Average rate of death	A number representing the estimated average number of fatalities per unit time
Mortality index	A single-value representation of consequence
Individual risk	
Individual risk contour	Closed curves of equal individual risk values superimposed over a local map
Individual risk profile	A graph of individual risk as a function of distance from the plant in a specified direction
Maximum individual risk	A single numerical value of individual risk corresponding to the person at highest risk
Average individual risk (exposed population)	A single numerical value estimating the average risk to a person in the exposed population
Average individual risk (total population)	A single numerical value estimating the average risk to a person in a predetermined population, whether or not all members of that population are exposed to the hazard
Societal risk	
Societal risk curve (F-N curve)	A graph of the cumulative probability or frequency of events causing *N* or more fatalities, injuries or exposures versus *N*, the number of fatalities, injuries, or exposures
Average societal risk	Another term for average rate of death

* See Chapter 4 of the *CPQRA Guidelines* for a more detailed discussion of these measures of risk.

2.7.2 Risk Analysis

Risk analysis is the process of estimating the magnitude of the risk arising from a potentially hazardous activity. Risk analysis can be qualitative or quantitative, depending on how the results are to be used. CPQRA refers to the quantitative estimation of the risks arising from the immediate impact of fires, explosions, or toxic material releases from a chemical processing facility.

The depth and complexity of a CPQRA study can vary broadly depending on the requirements of the situation. In some cases, a limited analysis considering the likelihood and consequences of only a few worst-case incidents and providing a simple risk model will produce sufficient information for risk management and decision making. In other cases, a major study using detailed models and requiring many person-years of effort may be needed. The factors which should be considered in defining the scope of a risk analysis study are discussed in Chapter 1 of the CPQRA Guidelines. Some of the decision aids discussed in this book may be useful in defining the depth of study required for a risk analysis which provides information for use in the decision-making process.

In general, CPQRA consists of the following steps:

1. *Formulation*: Define goals, establish scope, describe system and assemble background information.
2. *Hazard Identification*: Identify hazards and potential incidents and select those which will be considered in the study.
3. *Consequence Estimation*: Estimate consequences of incidents considered in the study, including both the physical effects and the impact of those physical effects on people, the environment and property.
4. *Likelihood Estimation*: Estimate the likelihood of the incidents considered in the study.
5. *Risk Estimation*: Combine the estimates of consequences and likelihood into appropriate measures of risk.

CPQRA Guidelines provides a more detailed description of the overall process. Below we summarize the procedures involved in Steps 2 through 5.

2.7.2.1 Hazard Identification

Hazard identification involves enumerating and selecting hazards and potential incidents to include in the risk analysis. Enumeration attempts to ensure that no significant incidents are overlooked; selection tries to reduce the incident outcome cases to a manageable number. Hazard evaluation and analysis techniques can provide the basis for hazardous identification. A number of these techniques are used in the chemical industry, and many of the common methods are described in the *Guidelines for Hazard Evaluation Procedures* (AIChE/CCPS 1992). These methods can also be adapted to provide a qualitative evaluation of risk, which sometimes is sufficient for a decision to be made. However, in other cases a quantitative estimate of risk is desired.

2.7.2.2 Consequence Estimation

Consequence estimation characterizes the physical effects resulting from a specific incident and the impact of these physical effects on people, the environment, and property. The consequence models used in CPQRA to estimate the potential for damage or injury fall into several categories:

- *Source models* describe the release rate of material from the process equipment into the external environment, and the rate of release of spilled vapors and volatile liquids into the atmosphere.

- *Dispersion models* describe the behavior of the released material in the atmosphere following its release.
- *Fire and explosion models* describe the magnitude and physical effects (heat radiation, explosion overpressure) resulting from a fire or explosion.
- *Effect models* describe the impact of the physical effects of a fire, explosion, or toxic gas release on exposed people, the environment or property, based on the results of the source, dispersion, and fire and explosion models.

Other models may be used to consider the effects of escape or evacuation, sheltering, protective equipment, or other factors (e.g., water contamination) that may be considered in a risk study. Extended discussions of consequence models can be found in Chapter 2 of the *CPQRA Guidelines*.

2.7.2.3 Likelihood Estimation

Likelihood estimation, sometimes called *frequency estimation*, characterizes the probability of occurrence for each potential incident considered in the analysis. The major tools used for likelihood estimation are listed below:

- *Historical data* are used for facility types where there is extensive experience available from similar or identical installations.
- *Failure sequence modeling techniques* such as fault tree analysis or event tree analysis are used to estimate the likelihood of incidents in facilities where historical data is unavailable, or is inadequate to accurately estimate the likelihood of the hazardous incidents of concern. Other modeling techniques may be required to consider the impact of external events (earthquakes, floods, etc.), common cause failures, and human factors and human reliability.
- *Expert judgment* quantifies an expert's state of knowledge or perceptions of the likelihood of an incident. This knowledge may be based on historical data, insights gained from models, experience, or a combination of these factors.

Likelihood estimation techniques are discussed in Chapter 3 of the *CPQRA Guidelines*.

2.7.2.4 Risk Estimation

Risk estimation combines the consequence and likelihood estimates of all incident outcomes for all selected incidents to provide a measure of risk. The risks of all selected incidents are individually estimated and summed to give an overall measure of risk. The risk estimate will require specific information about the plant site such as:

- The number of people in the vicinity of the plant, their specific geographical location, and the times they are present at that location
- Details of the location of local environmental features potentially subject to damage
- The location and value of property in the vicinity of the plant, possibly including loss of business resulting from property damage or loss
- Time distribution of weather conditions at the site, including wind speed and direction and atmospheric stability class.

Various methods, ranging from simple graphical techniques giving a rough estimate of risk to complex computer models that consider all of the factors impacting risk in great detail, are described in Chapter 4 of the *CPQRA Guidelines*. In general, the results of these risk estimation techniques will be in the form of individual and societal risk measures and risk indices such as those described in Figures 2.7 and 2.8 and in Table 2.2.

2.7.3 Risk Decision Making

Once the risk analysis has been completed, it is necessary to determine what action, if any, must be taken based on the results. The risk analysis will be extremely useful in identifying the major sources of risk associated with the facility. With this information, one can identify alternatives for risk reduction. Evaluating these alternatives and determining which ones, if any, should be pursued or implemented, is the subject of this book.

2.8 REFERENCES

AIChE/CCPS (1989). *Guidelines for Chemical Process Quantitative Risk Analysis*. New York: American Institute of Chemical Engineers.

AIChE/CCPS (1992). *Guidelines for Hazard Evaluation Procedures. 2nd ed. with Worked Examples*. New York: American Institute of Chemical Engineers.

Arendt, J. S. (1992). Managing Risk. *Chemtech* (August):490–497.

Arendt, J. S., D. K. Lorenzo, and A. F. Lusby (1989). *Evaluating Process Safety in the Chemical Industry: A Manager's Guide to Quantitative Risk Assessment*. Washington, D.C.: Chemical Manufacturers Association.

Barrager, S. M., B. R. Judd, and D. W. North (1976). "The Economic and Social Costs of Coal and Nuclear Electric Generation," Decision Analysis Department, SRI International. Menlo Park, CA: SRI International.

Doniger, D. (1980). Statement of David D. Doniger, Natural Resources Defense Council. In *Comparative Risk Assessment, Hearings before the Subcommittee on Science and Technology*, No. 129, May 14–15. Washington, D.C.:Government Printing Office.

Dow Chemical Company (1994). *Fire and Explosion Index—Hazard Classification Guide*. 7th ed. CEP Technical Manual. New York: American Institute of Chemical Engineers.

Fisher, A., L. G. Chestnut, and D. M. Violette (1989). The Value of Reducing Risks of Death: A Note on New Evidence. *Journal of Policy Analysis and Management* 8(1):88–100.

Gillette, C. P., and T. D. Hopkins (1988). Federal Agency Valuations of Human Life: A Report to the Administrative Conference of the United States, July.

Graham, J. D., and J. W. Vaupel (1981). Value of a Life: What Difference Does It Make? *Risk Analysis* 1(1):89–95.

Keller, B. What is the Audited Value of Life? New York *Times*, October 26, 1984.

Maas, A., and L. Stalpers (1992). Assessing Utilities by Means of Conjoint Measurement: An Application in Medical Decision Analysis. *Medical Decision Making* 12:288–297.

McKone, J. E. (1986). The Implicit Valuation of Environmental Cancer by the United States Regulatory Agencies. *Toxics Law Reporter* 1(16): (September 24).

Merkhofer, M. W. (1987). Quantifying Judgmental Uncertainty: Methodology, Experiences, and Insights. *IEEE Transactions on Systems, Man, and Cybernetics* 17(5):741–752.

NRC (1987). "Reactor Risk Reference Document," NUREG–1150, Vol. 1. U.S. Nuclear Regulatory Commission, Office of Nuclear Regulatory Research, Washington, D.C. (February).

OMB (1992). "Budget of the United States Government, Fiscal Year 1992." Section IX.C. Part Two–370. Washington, D.C.: U.S. Office of Management and Budget.

Redelmeier, D. A., D. N. Heller, and M. C. Weinstein (1994). Time Preference in Medical Economics: Science or Religion? *Medical Decision Making* 14(3):301–303.

Viscusi, W. K. (1990). Conversation with Applied Decision Analysis, Inc., April.

3

CLASSIFICATION AND DESCRIPTION OF RECOGNIZED DECISION AIDS

Deliberate as often as you please, but when you decide it is once for all.
—***Publilius Syrus***, *Sententiae*, ***No. 132***

3.1 PURPOSE OF CHAPTER

This book describes a wide range of decision aids and provides guidance on selecting decision aids that are appropriate for a particular problem. This chapter and the next provide a framework for thinking about decision problems and the decision aids that can be applied to them. Chapter 3 introduces a series of well-established decision aids. Chapter 4 describes the distinguishing characteristics of the decision aids and important aspects of decision situations that can help one choose appropriate decision aids.

This chapter presents an overview of many decision aids, each one described in a series of paragraphs. The descriptions cover a brief history, the key concepts, and strengths and weaknesses of each decision aid, with references for further study. The brief descriptions are sufficient for understanding how the aids work, without providing detailed instruction on how to use them. Seven of the decision aids described in this chapter have been selected for more in-depth treatment in later chapters of the book.

We present the decision aids in the context of a classification scheme that highlights some key similarities and differences among the aids. The classification scheme is admittedly incomplete. It helps place the decision aids into logical groups, but does not provide enough detail on the aids to be sufficient for selecting a decision aid. Chapter 4 suggests how to select decision aids by matching more detailed decision aid characteristics with aspects of the problem to be addressed.

3.2 DESCRIPTIONS OF RECOGNIZED DECISION AIDS

In this section we describe many recognized decision aids. We refer to them as "recognized" decision aids to emphasize that they are well-established methods for making decisions. In selecting this list of aids we balance the conflicting goals of completeness and economy. A survey of the literature of this field reveals hundreds of decision aids. However, only a small fraction of decision aids has a significant body of research and applications. We have chosen to focus on these more proven and popular aids. We have filtered out most of the decision-making techniques that:

- Have few or no published applications
- Are highly specialized to particular decision situations
- Are minor variants of other, more recognized, techniques
- Are inappropriate for process safety decisions.

Although the selection of decision aids for inclusion in this chapter is necessarily judgmental, we feel that most of the important decision aids are represented.

3.2.1 A Classification Scheme for Decision Aids

Even when the list of decision aids is reduced to the more recognized ones, it still encompasses a wide variety of approaches to, and philosophies of, decision making. This section introduces a simple classification scheme. The scheme groups similar decision aids and highlights important differences among them.

The classification scheme is not meant to be a complete taxonomy of all possible decision aids; rather, it is a simple way to introduce the decision aids by grouping them by their more significant characteristics. Its goal is practical, not theoretical. As new or refined decision aids are developed, the classification scheme allows one to categorize them relative to other, more familiar, decision aids.

Several authors have tackled the difficult problem of completely classifying decision aids (Brown and Ulvila 1976, Buede 1992, Fischhoff et al. 1981, Hwang and Lin 1987, Hwang and Masud 1979, Hwang and Yoon 1981, Merkhofer 1987). Taxonomies that aim for completeness necessarily rely on technical aspects of the decision aids, either with regard to their theoretical underpinnings, analytical approach, or the manner in which they are used.

Figure 3.1 presents our classification of decision aids according to fundamental differences in capabilities and philosophies. The classification scheme consists of five branches that separate the aids into smaller groups sharing similar characteristics. The five branches characterize the decision aids as follows:

Branch 1: The decision aid considers competitive response
Branch 2: The decision aid addresses many or infinite alternatives
Branch 3: The decision aid requires objective inputs
Branch 4: The decision aid does not treat uncertainty and value separately
Branch 5: The decision aid treats uncertainty and value separately.

The following sections introduce each branch of the classification and describe the decision aids characterized by that branch.

3.2.2 Branch 1: Considers Competitive Response

- *Game Theory*

Most decision aids are designed primarily to support decisions by a single decision maker or by a group of decision makers with consistent goals. For the sake of completeness, we also include decision aids that address the need for competitive response. The first branch separates those decision aids that deal explicitly with decision makers with competing goals. These decision aids address decision situations in which opponents with conflicting or incompatible goals interact. Situations in which these aids are typically applied include:

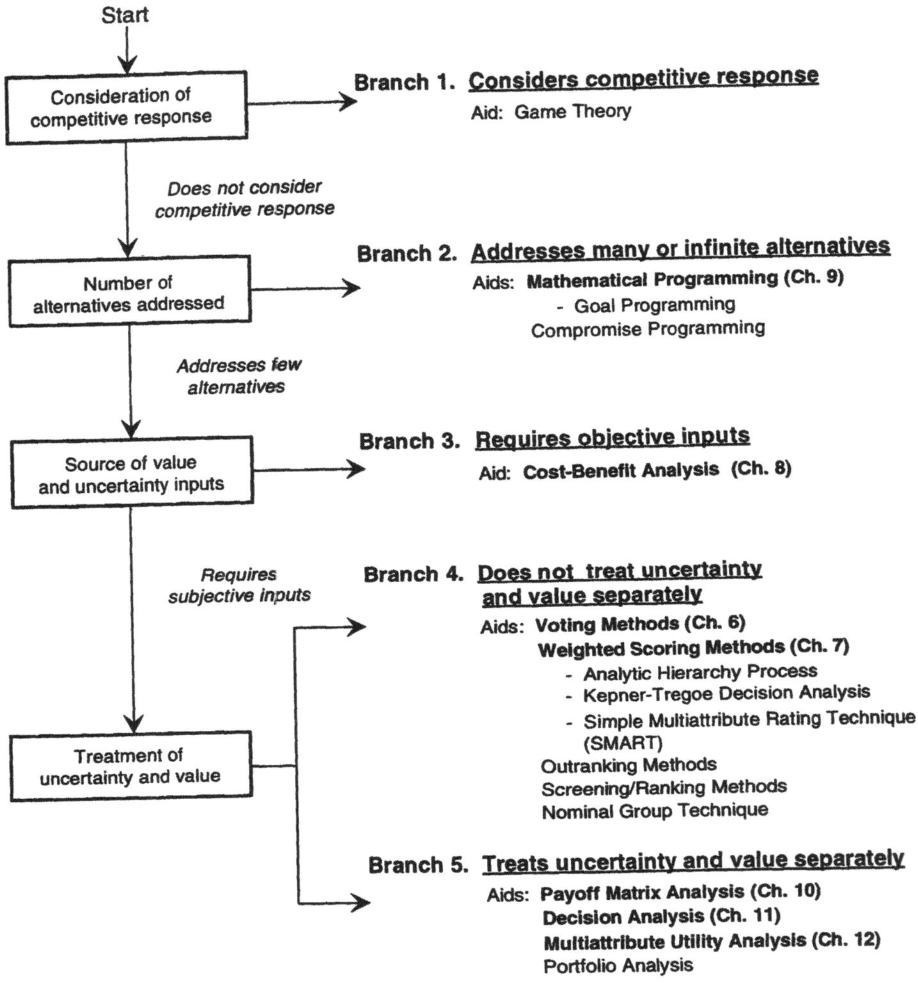

FIGURE 3.1. Decision aids classified according to methodological traits. Techniques in **bold** are described in separate chapters.

pricing and marketing decisions, election campaign decisions, employer-employee negotiations, and war games.

Acute risk decisions are typically not made to gain competitive advantage. Therefore, this text does not deal with game theory or other approaches that are directed specifically towards competitive situations. Even though they are not focused explicitly on competitive situations, the decision aids described in this text can usually be applied to competitive situations. In such applications, assumptions are required about how competitors will respond to the actions taken by the decision maker.

3.2.2.1. Game Theory

Among the decision aids described in this book, game theory alone is geared specifically toward situations that involve competitive response. Game theory is an approach to analyzing competitive situations that has been developed by economists and mathematicians interested in studying the theory of strategy. Game theory does not refer to any specific technique, but rather to an approach to decision-making situations that tries to understand the nature of competitive response.

The "game" in game theory refers to a competitive situation. The participants in the game, whether people, corporations, countries or other entities are called "persons." Persons are usually assumed to be motivated by self-interest. Games are classified both by the number of competitors (e.g., two-person game for two competitors or *n*-person game for multiple participants) and by the nature of the competition. An important class of games characterize situations in which one competitors gain is necessarily another competitors loss. These are called "zero-sum" games. Games are called "cooperative" if participants are allowed to engage in bargaining or other forms of communication before the game commences. One objective of game theory is to discover strategies that are best for each person in the game. Typically, the best strategy is defined to be the one which maximizes the expected gain for that person.

An advantage of game theory is that it is designed specifically to consider competitive situations. Unfortunately, there is a wide gulf between the state of the theory and the needs of todays decision makers. Even the most sophisticated game theory approaches can handle only the simplest of competitive situations, and even in those cases can provide only limited prescriptions for what actions might be best. No general approaches have emerged from game theory that have been successfully applied to the complex competitive situations that arise in practice. However, game theory has been useful as a tool for understanding the nature of conflict. A good reference is Davis's *Game Theory: An Introduction* (1983).

3.2.3 Branch 2: Addresses Many or Infinite Alternatives

- *Mathematical Programming (Chapter 9)*
- *Goal Programming (Chapter 9)*
- *Compromise Programming*

The second branch considers whether the decision aid is designed to address a large number of alternatives. Decision aids that address many (hundreds or thousands) or an infinite number of alternatives require strong, quantifiable assumptions about the constraints on the alternatives, the relationships among alternatives, and the relationships among alternatives and outcomes. These assumptions allow a systematic search that can find the best alternative without examining every possibility. Linear optimization or linear programming (a form of mathematical programming) is probably the best known technique for examining an infinite number of alternatives. The topic of mathematical programming is discussed in more detail in Chapter 9 of this book.

Most of the decision aids in this book are best equipped to examine relatively few alternatives, where "few" can be from two to a few dozen alternatives. Some decision aids can be modified to address larger numbers of alternatives, for example, by using mathematical programming to reduce an essentially infinite number of alternatives by finding those that are the most efficient. Decision aids that can address large numbers of alternatives are designed to exhaustively compare all alternatives or to use rules of thumb to limit the search.

3.2.3.1 Mathematical Programming

Mathematical programming is a class of decision aids that addresses the problem of how to allocate resources. Mathematical programming can consider very complex planning problems and determine an allocation of resources that best helps the decision maker reach a stated goal, called the *objective*. Objectives correspond to quantities that are to be maximized or minimized; typical objectives are *Maximize revenues*, *Minimize cost*, or *Minimize travel time*. Because it can solve large resource allocation problems, mathematical programming is used extensively in industry on problems as diverse as chemical blending, job assignment, truck routing, and aircraft scheduling.

The word "programming" in mathematical programming refers to "planning." A mathematical program is a mathematical statement of the decision-makers planning problem. Developing a mathematical program is a process in which an analyst strives to understand the key parts of a planning problem and represent them using mathematical expressions. The resulting mathematical program is often called a "model" because the analyst tries to develop a simplified representation, or model, of the problem addressed by the decision, just like an architect might create a model of a proposed building. The model simplifies the problem by capturing only the most important parts of the complex planning situation. Expressing the model in mathematical terms enables the application of sophisticated solution techniques—called algorithms—that solve the problem very efficiently.

All mathematical programming models divide the decision problem into three main components. These are the:

- Objective—what the decision maker wants to achieve (e.g., least cost, highest profit)
- Decisions to be made (e.g., how much resource to be allocated to each activity)
- Constraints—those external or internal conditions that define which allocations are possible.

In a mathematical programming model, each of these elements is expressed in mathematical terms. The practicality of mathematical programming techniques stems from the existence and implementation of efficient algorithms that solve these resource allocation problems. Algorithms are available in computer software for many types of mathematical programs, especially linear programs, the most common and widely used class of mathematical programming models.

Due to its reliance on mathematical modeling, mathematical programming is best suited to problems in which the three components of the planning problem—the objective of the decision maker, the decision to be made, and the constraints on the decision—are unambiguous. The key benefit of mathematical programming is that it can search efficiently through thousands, millions, or even an infinite number of possible allocations to find the one that helps the decision maker best achieve the objective. Mathematical programming can also reveal the marginal benefit of scarce resources and the sensitivity of results to decision parameters. However, the effective use of mathematical programming techniques requires special training and computer resources. Most introductory management science or operations research textbooks give an adequate introduction to mathematical programming. One such textbook is *Introduction to Mathematical Programming* by Hillier and Lieberman (1990).

3.2.3.2 Goal Programming

Goal programming is an important extension of mathematical programming that addresses planning situations in which the decision maker has multiple goals. Conventional mathematical programming considers a single objective (profit, cost, etc.) that the decision maker seeks to maximize or minimize. Goal programs, in contrast, allow the decision maker to balance the attainment of several objectives simultaneously. A corporation, for example, may wish to obtain a target market share while maintaining certain levels of both profit and revenue. While ordinary mathematical programming is concerned with optimizing, goal programming is concerned with satisfying as many goals as possible ("satisficing"). In addition to considering the desired attainment level for each separate objective, goal programming can also take into account the relative importance the decision maker places on each goal.

Procedures have been developed to solve goal programming problems, but there are fewer commercial software programs than for other mathematical programming approaches. Consequently, training and expertise is even more important for solving goal programs. Many would argue that the key benefit of goal programming is that it accommodates more realistic planning models, based on satisficing rather than optimizing. However, the solution techniques are generally more difficult than those for ordinary mathematical programming. A good introduction to goal programming is provided in Hillier and Lieberman (1990).

3.2.3.3 Compromise Programming

Compromise programming techniques extend mathematical programming to the situation where a decision maker wants to maximize (or minimize) several objectives simultaneously. Whereas goal programming considers predefined target goals and relative goal priorities, compromise programming tends to be more interactive. Rarely will one alternative be optimal for all of the decision-maker's objectives. Compromise programming defines the set of alternatives which do well on many of the objectives. It determines which alternative is best based on the decision-makers willingness to trade off achievement on one objective with achievement on another. Compromise programming techniques vary in how they obtain the tradeoff information from the decision maker. One representative technique has the decision maker define an "ideal" solution and then determines which solution is closest to the ideal. Other techniques require interacting with the decision maker and asking tradeoff questions.

Compromise programming attempts to do more than ordinary mathematical programming and hence requires more analytic resources. Finding the best compromise solution involves determining the decision-makers relative preferences for different achievement levels of the various objectives. Eliciting this information can be difficult and time-consuming. Also, no commercial software packages currently exist to assist in this process. This technique definitely requires an expert. For more information, see Tabucanon's *Multiple Criteria Decision Making in Industry* (1988).

3.2.4 Branch 3: Requires Objective Inputs

- ***Cost–Benefit Analysis (Chapter 8)***

The third branch involves a distinction in the source of inputs to the decision. Some decision aids emphasize objectivity in the analysis while others assume that subjective values and information are necessary and/or useful in analyses of problems. The objective approach strives to reach decisions that are independent of the preferences or judgments of the decision maker. The objective approach favors the use of data from recorded experience or experimentation, avoiding the use of expert opinion and a decision-maker's personal values. Typical data resources for this approach include historical failure rates of equipment, statistical analyses of past practices, prices in actual markets, laboratory tests of components, and other objective observations. Cost–benefit analysis (Chapter 8) is an example of the objective approach to decision making.

Alternative approaches prefer subjective input to the decision. These approaches assert that the decision-makers beliefs should govern the information and preferences used in an analysis. Subjective approaches recognize that hard data will often not be available, and that experts can provide valuable data not available from statistics. To ensure decisions are consistent with a decision-maker's beliefs, a combination of expert judgment, experience, and other data may be appropriate.

3.2.4.1 Cost–Benefit Analysis

Cost–benefit analysis (CBA) is often used to support decision making in the public sector, where an agency seeks a decision that is as beneficial as possible to the constituency that it serves. Using cost–benefit analysis, the agency decides among alternatives by quantifying their costs and benefits in dollar terms. It chooses the alternative that fulfills a decision criterion to maximize net benefit (benefit minus costs). A key principle driving the analysis is that the inputs should be objective ones that reflect the preferences of the individuals in society affected by the decision. Variations of CBA differ in how they quantify these inputs and on the decision criterion employed. Despite its roots as a tool for making public decisions, CBA is also used widely in private organizations, particularly on decisions that have impacts outside of the organization. CBA has a theoretical underpinning in economic theories of consumer surplus developed in the 1800s. Its application was stimulated by federal legislation from the early 1900s that has stipulated its use on public works projects. Key steps of a CBA are to:

- Decide on the scope of the analysis
- Assess the costs of each alternative
- Assess the effects (both benefits and adverse impacts) of each alternative
- Apply the decision rule for choosing among the alternatives.

The analysis will usually include indirect effects such as short-term jobs created while undertaking the projects. To estimate the costs and effects, society's "willingness to pay" is the driving principle. Specifically, the analysis is based on how much the individuals affected by a decision are willing to pay to avoid the costs and negative effects of the decision. Similarly, the willingness to pay to receive the positive effects of a decision is included in the analysis. Possible inputs to this willingness-to-pay estimate are the prices actually charged on the market for the product or service involved, the "shadow price" or price charged for a similar or analogous product or service, and willingness-to-pay information obtained through surveys. Different approaches have been suggested for the decision criterion used to compare costs and effects among alternatives, with the most common being the time-discounted benefits minus the time-discounted costs. Extensions to cost–benefit analysis attempt to overcome some key limitations of the method such as its failure to:

- Account for uncertainty in the costs and effects
- Address distributional issues arising from one group bearing the costs while another receives the benefits
- Address programs in which the benefits are difficult to quantify in dollar terms.

Cost-effectiveness analysis (CEA) addresses the latter issue, by comparing alternatives based on the cost required to obtain a given level of (non-dollar) benefit such as reduced health risks or increased life expectancy.

The main inputs to a CBA or CEA are the estimates of costs and effects. In large-scale problems it can be difficult to determine what to include in the analysis so as to ensure completeness without double-counting the costs or benefits. It can also be difficult to estimate these inputs. It is often helpful to confer with an economist experienced in applying CBA to similar problems. CBA has a long track record of application. It is perceived to provide an objective basis for decision making by using society's values rather than those of the decision maker(s). Some persons view this objectivity as a drawback. Other expressed concerns are with the simplified handling or nonrecognition of uncertainties, risk aversion, and distributional effects, with possible biases in how costs and benefits are quantified, and with the relatively greater sophistication, time, and resources required for its application. A good source on CBA is Sassone and Schaffer's *Cost–Benefit Analysis: A Handbook* (1978).

3.2.5 Branch 4: Does Not Treat Uncertainty and Value Separately

- *Voting Methods (Chapter 6)*
- *Weighted Scoring Methods (Chapter 7)*
- *Analytic Hierarchy Process (Chapter 7)*
- *Kepner-Tregoe Decision Analysis (Chapter 7)*
- *Simple Multiattribute Rating Technique (SMART)*
- *Outranking Methods*
- *Screening/Ranking Methods*
- *Nominal Group Technique*

Branch 4 and Branch 5 include decision aids that require subjective inputs to the analysis. Key inputs are the uncertainties in the problem and the values or preferences of the decision maker.

The fourth branch separates methods that do not distinguish uncertainty and value from methods that do. To illustrate this, consider an investment problem with two alternatives. One alternative offers a certain, but moderate, gain on the investment The other alternative is risky; it has the possibility of both gains and losses. Methods that do not distinguish uncertainty and value treat the risk and return aspects of such an investment problem similarly, and evaluate them with the same methods. Methods that differentiate risk aspects of a problem from other characteristics of the problem use special techniques to quantify the impact of risk and are therefore viewed as more logically rigorous. Proponents of methods identified by Branch 4 view the risk (or uncertainty) versus value distinction as an unnecessary complication. Examples of approaches fitting in this category are voting methods (examined in detail in Chapter 6) and weighted scoring methods (Chapter 7).

3.2.5.1 Voting Methods

Voting is the most common approach used by groups to decide among alternatives. Voting is attractive because it is quick and easily understood. Little time or data is required to take a vote. While it is tempting to think of voting as an inherently "fair" method for aggregating opinion, each voting method has its own potential for bias or manipulation.

Several voting methods exist. They differ in how the votes are conducted, what preference information is required of the voter, or how the votes are aggregated to determine a winner. The most common method of voting is the plurality method. In this method, each voter selects the alternative they most desire among a list of possible alternatives. The alternative with the largest number of votes wins. The Borda count method has each voter rank the alternatives in order of preference. Alternatives are given points based on the rankings of each voter, with a larger number of points indicating a lower ranking. The alternative with the fewest points summed across all voters wins. Because the Borda count method takes into account the ranking of the entire set of alternatives, it considers more preference information than does the plurality method. An additional method, the Copeland method, assigns each alternative a score based on how it would perform in head-to-head competition with other options. The score given is the number of alternatives it outranks (again, summed over all voters) minus the number of options that outrank it.

The key advantages of voting are its simplicity, its minimal resource requirements, and its acceptability. However, unless well documented, the voting process can obscure the true basis for the decision because each participant can vote according to unexpressed thoughts or preferences. There is no way to conduct sensitivity analysis using voting, and it is difficult or impossible to know how the decision might change if the participants knowledge or values were different. Another unfortunate aspect of voting is that it has been shown that all voting systems are, at least in theory, subject to manipulation by astute voters or coalitions. For a good introduction to voting systems, see Niemi and Riker's article in *Scientific American* entitled "The Choice of Voting Systems" (1976).

3.2.5.2 Weighted Scoring Methods

Weighted scoring methods are a group of both informal and formal decision aids that have been developed over the years, for the most part independently of one another. These decision aids share the same general process: decision alternatives are scored on a set of decision criteria that are weighted by relative importance. These methods are particularly well-suited to decision situations in which formal or explicit value judgments are desired and situations in which time, money, or analytical expertise are constrained.

At the core of weighted scoring methods are the decision criteria, the factors upon which the alternatives will be judged. Once the analyst determines the decision criteria, the criteria are individually weighted according to their relative importance to the overall decision. Assigning weights quantifies the value judgments that underlie the tradeoffs among the decision criteria. First, each alternative is assigned a series of scores that characterize how well the alternative performs on the decision criteria. The scores are then weighted by multiplying by the weight of the corresponding criteria and summing across the criteria. The decision maker chooses the alternative with the highest total score. Many variations of this general process exist. Two of these methods, Kepner-Tregoe decision analysis and the Analytic Hierarchy Process, have been broadly applied and are discussed here separately in Sections 3.2.5.3 and 3.2.5.4.

Although the soundness of their theoretical foundations has been questioned (e.g., see Section 7.7.1), weighted scoring methods have a role in some decision situations. They make value judgments explicit through quantification, can be executed quickly and easily, and require relatively few analytic resources. One of their main drawbacks is that decision makers must rely on intuition in dealing with uncertainty and the tradeoffs between potential positive and negative consequences. Both the likelihood of consequences and their desirability are not made explicit, but must be dealt with implicitly when assigning scores. In addition, the scores may not have a physical meaning based on how they are constructed. The scores are not tied to concrete measures or expressed in terms that are natural, so it may be hard for the decision maker to maintain consistency and to represent feelings accurately. The analyst must also be aware that the results are easy to manipulate through the selection of the criteria. Despite their drawbacks, weighted scoring methods can have a role in decisions that require quick analysis, or as a first-cut analysis before using a more sophisticated analytic technique.

3.2.5.3 Analytic Hierarchy Process

The Analytic Hierarchy Process (AHP) is classified in this book as a weighted scoring method. AHP starts by breaking decisions into their components. It then builds a structure that captures important relationships among components. This structure facilitates the process of quantifying subjective judgments and synthesizing them into overall measures of the attractiveness of the various alternatives. AHP was developed in the 1970s by Thomas Saaty at the University of Pennsylvania. Since its introduction, it has been the subject of some debate regarding its theoretical basis, but it has gained wide acceptance in government, industry, and academia. Its structure and formality make it well-suited for decisions in which explicitness is desired. Since AHP is less resource-intensive, it may be used when analytic resources are constrained.

A "decision hierarchy" provides the structure for the AHP process. Each level of the hierarchy consists of goals, decision criteria, or alternatives from which a choice will be made. By eliciting pairwise comparisons between the criteria on one level of the hierarchy with respect to achieving the goals or criteria on the next higher level, numerical indications of the priority of criteria on the lower level are obtained. A synthesis of the priorities on all levels leads to a priority ranking of the alternatives with respect to the overall goal or criteria. Additional analysis can produce indicators of how consistent judgments are within the hierarchy.

Proponents of AHP like its structured framework, its ability to easily elicit judgments among criteria in the hierarchy, and its computational speed. Critics claim that it can produce arbitrary results that do not necessarily reflect the preferences of the decision maker. In addition, AHP does not perform well in situations where formally accounting for risk is important, since it does not explicitly address risk and uncertainty. Those who choose to use AHP find that the technique is not difficult to use after an initial learning period. However, the analyst must thoroughly understand the mathematics and the implicit assumptions before interpreting the results. For a good introduction written by AHP's creator, consult Saaty's *Decision Making for Leaders* (1990) or *The Analytic Hierarchy Process* (1980).

3.2.5.4 Kepner–Tregoe Decision Analysis

Kepner–Tregoe decision analysis is one of the components of the Kepner–Tregoe Rational Management Process for organizational management. It was developed by Charles Kepner and Benjamin Tregoe in the 1950s. The Kepner–Tregoe decision analysis process is categorized here as a weighted scoring method. A Kepner–Tregoe decision analysis centers around decision objectives that are assigned scores for each decision alternative. These scores are synthesized to obtain an overall quantitative measure for each alternative. Kepner–Tregoe decision analysis is a formal structured process that is well-suited to situations in which resources are constrained and judgments need to be made explicit. However, it is less desirable when a rigorous, quantitative, systematic process for making tradeoffs between risks and benefits is needed.

In performing a Kepner–Tregoe decision analysis the analyst initially creates a decision statement, a list of decision objectives, and a list of decision alternatives. Objectives are divded into "MUSTS," or mandatory objectives, and "WANTS." Alternatives that satisfy all MUSTS are evaluated by scoring each alternative on its performance regarding each WANT objective. Each WANT is weighted according to its relative importance. The scores for each WANT are multiplied by the respective weight of the WANT, and all of these weighted scores for an alternative are summed to get a total score. The total weighted score and a list of the potential negative consequences of each alternative are used in making the final decision for each alternative.

Although Kepner–Tregoe decision analysis is easy to use, it does require training. Often this training is part of an organization-wide effort to implement the Rational Management Process. Kepner-Tregoe decision analysis derives one of its benefits from being part of an entire management process that encourages teamwork and consistency in decision making. Other strengths of

Kepner–Tregoe decision analysis are that it avoids complicated mathematical computations, facilitates the communication and justification of decisions, and provides a step for examining potential negative consequences of the decision. However, although the negative consequences are examined, Kepner–Tregoe decision analysis does not incorporate a way to explicitly trade off the total weighted score against these consequences. Also, in many situations, the total weighted scores may not be a meaningful measure of the desirability of the outcomes of a decision. A final weakness is that the numbers can be easily manipulated to favor one particular alternative. Nevertheless, if used appropriately, the process can provide insight into the important elements of a decision and can give the decision maker a feel for which alternatives are the best or the worst. A good introduction to the technique is provided by Kepner and Tregoe in *The New Rational Manager* (1981).

3.2.5.5 Simple Multiattribute Rating Technique (SMART)

The Simple Multiattribute Rating Technique (SMART) was developed as an alternative to multiattribute utility analysis (Section 3.2.6.3) for problems involving multiple objectives. SMART is designed to be more transparent and easier to implement than the standard multiattribute utility analysis. SMART is a weighted scoring method (Section 3.2.5.2).

The basic SMART technique consists of a series of steps designed to identify alternatives under consideration, identify and rank the important attributes of the alternatives, and rate each alternative on the attributes. Different versions of SMART vary in how weights for attributes are developed and how ratings are incorporated. In all versions, however, the overall rating is a weighted sum of the attribute scores for each alternative. The chosen alternative is the one with the highest score. For more information see von Winterfeldt and Edwards, *Decision Analysis and Behavioral Research* (1986).

3.2.5.6 Outranking Methods

While many decision aids are concerned with a complete and distinct ordering of alternatives by preference, outranking methods have a more modest, and perhaps more realistic, aspiration. These methods recognize that some preferences are clear and distinct, while others are not. For example, a decision maker may feel that one alternative is definitely better than a second, but may not be able to clearly distinguish a preference between the first alternative and a third. Outranking techniques seek to delineate the portion of a decision maker's preferences that is clear. Outranking does not rank all possible alternatives, but instead aims to help the decision maker understand which alternative (or group of alternatives) is clearly preferred to others. Outranking methods accomplish this aim by developing an "outranking" relationship over the set of alternatives considered. One alternative is said to outrank another if it is clearly superior.

One of the more widely used outranking methods is the ELECTRE approach, developed by B. Roy at the University of Paris. To develop the outranking relationship, ELECTRE utilizes two indices: concordance and discordance. Concordance is a measure of the degree to which the decision maker is comfortable preferring one alternative to another. Discordance is the a measure of the discomfort the decision maker feels with that same preference. In the ELECTRE technique, concordance and discordance measures are calculated for each pair of alternatives. By developing minimum thresholds of concordance and maximum thresholds of discordance, a decision maker specifies the test for when one alternative will be considered preferred to another. The preferences defined by these thresholds are assembled to yield an ordering of the alternatives. ELECTRE may produce a partial ordering, that is, a subset of alternatives may exist that can be ranked as a group relative to the other alternatives, but whose elements cannot be ranked relative to each other.

The ELECTRE technique has been used in Europe but not significantly in the United States. It has been applied to a number of multicriteria problems in Europe including selection of Paris Metro stations for renovation (Roy, Present, and Silhol 1986). The popularity of outranking methods (such as ELECTRE) in Europe seems to indicate that they have significant practical appeal. However, they do have some drawbacks. First, they may not select a unique alternative for the decision maker. And second, the methods are based more in experience than rigorous theory. The reader is referred to Vincke's article "Analysis of Multicriteria Decision Aid in Europe" (1986) for more discussion and further references.

3.2.5.7 Screening/Ranking Methods

Screening/ranking methods are extremely simple techniques for reducing a set of alternatives to a single alternative. As their name implies, they reduce the set of alternatives available by screening alternatives that do not meet minimal criteria and ranking the remaining alternatives. Conjunctive ranking is a typical screening/ranking method. To use conjunctive ranking the decision maker sets minimum thresholds for all attributes. Any alternative that does not meet the minimum threshold on all of its attributes is eliminated. The remaining alternatives are then ranked on a single attribute identified by the decision maker. The alternative that is ranked first with respect to the attribute is chosen.

The benefit of screening/ranking methods such as conjunctive ranking is simplicity. The key disadvantage is that they do not allow for tradeoffs between attributes, e.g., one alternative may do poorly (but well enough to pass) on almost all attributes and still be chosen. Screening/ranking methods are more appropriate to decisions where one can clearly delineate between alternatives that are acceptable and those that are unacceptable and where the decision maker is comfortable making the final decision based on scores for possibly a few attributes. For more information see Hwang and Yoon, *Multiattribute Decision Making Methods and Applications* (1981).

3.2.5.8 Nominal Group Technique

Nominal Group Technique is a process for generating ideas, setting priorities, and making group decisions. The focus of Nominal Group Technique is on maintaining a decision-making environment in which all participants can contribute equally, without the usual pressures and influences of social interaction that often accompany debates.

Nominal Group Technique accomplishes its goal by limiting interaction among members of the group. Participants silently and anonymously generate ideas about the question or problem in writing without any group discussion. These ideas are recorded and discussed among the group for the purpose of clarification only. The group then participates in a preliminary secret ballot voting procedure that ranks the ideas presented. The group reviews the preliminary vote and discusses ideas that received many or few votes. Then, without attempting to persuade any changes of vote, the group conducts a more sophisticated, final voting procedure that determines the selection of an idea or strategy.

Nominal Group Technique provides a highly controlled process for integrating the ideas of many individuals. It generally works best with small groups, from five to nine members. Nominal Group Technique can be very useful in heated situations for it is designed to eliminate conflict, remove domination by individual members, and provide an atmosphere in which judgments can be expressed without hesitation by all participants. The technique requires a leader who is knowledgeable about the process and is skilled in leading group interaction. Also, the final decision making in Nominal Group Technique comes from voting, and so, as a decision aid, it shares the benefits and disadvantages of voting (described in Section 3.2.5.1). Nominal Group Technique works best for situations requiring creative decision making, in which a significant portion of the task is idea generation, setting priorities, and evaluating possible solutions. A good introduction can be found in Hwang and Lin, *Group Decision Making Under Multiple Criteria* (1987).

3.2.6 Branch 5: Treats Uncertainty and Value Separately

- ***Payoff Matrix Analysis (Chapter 10)***
- ***Decision Analysis (Chapter 11)***
- ***Multiattribute Utility Analysis (Chapter 12)***
- ***Portfolio Analysis***

Branch 4 methods do not distinguish between uncertainty and value, whereas Branch 5 methods do. Proponents of Branch 5 methods feel that this distinction clarifies the analysis. Further, this distinction allows decisions to be based on a firm theoretical foundation. This foundation is provided by expected utility theory as developed by von Neumann and Morgenstern and others and on the mathematics of probability theory, described further in Chapter 11. Examples of the Branch 5 approach examined in this text are payoff matrix analysis (Chapter 10), decision analysis (Chapter 11), and multiattribute utility analysis (Chapter 12).

3.2.6.1 Payoff Matrix Analysis

Payoff matrix analysis is probably the simplest method that explicitly recognizes the differences between probabilities and values. It has been used for over 20 years in introductory texts on quantitative decision making. Its simplicity makes it easy to learn and apply, but also hampers its applicability to more complex problems. It is most appropriate for problems in which the uncertainties are not affected by the action a decision maker takes. For example, it is easily applied to decisions on how best to protect against the damage from accidents; it is more difficult to apply to decisions on how best to prevent accidents.

In a payoff matrix analysis a two-dimensional matrix describes the key elements of the problem (for an example, see Table 10.2). The columns of the matrix refer to actions, and the rows of the matrix refer to different outcomes of uncertain events. The probability of each outcome appears at the start of each row. Each box of the matrix contains the value of the columns action, assuming occurrence of the rows outcome. To determine the recommended action, the decision maker weights the value within each box by the probability of the row, and totals each column. The column with the highest total is the favored action. This simple approach has been extended by the Electric Power Research Institute to analyze strategies and scenarios in a group setting. This group-oriented version of payoff matrix analysis is called CATALYST.

Only a rudimentary understanding of mathematics and probability is required to apply payoff matrix analysis in its simplest form. To apply the CATALYST process for group decision making, skill and training in group facilitation is required. The decision aid is well-suited for the analysis of very simple problems or for initiating a more detailed analysis in a group setting. It is not appropriate for the detailed analysis of complex problems in which the action taken can substantially modify the likelihood of different outcomes. Useful introductions to the payoff matrix method include an article by M. Martin, "Decision Making: The Payoff Matrix" in the *Journal of Systems Management* (1979), and the "CATALYST Facilitator's Guidebook" (EPRI 1992).

3.2.6.2 Decision Analysis

Decision analysis (DA) is a systematic, logically rigorous technique for analyzing decisions involving complex uncertainties. The key feature that differentiates DA from other decision aids is that it explicitly treats uncertainty by accounting for the potential consequences of a decision and the likelihood of occurrence of the consequences. The theory upon which this is based can be traced back to the early 1700s, when Daniel Bernoulli introduced the idea that decisions should incorporate the likelihood of the possible consequences and the relative desirability of the consequences to the decision maker. In the past 30 years, DA has grown into a formal discipline, with practical tools that are used in private industry, academia, and the government. These tools lend themselves especially well to situations that involve uncertainty, numerous interrelated factors, complex value tradeoffs, and the need for formal justification.

Decision analysis provides a formal process for structuring, analyzing, and communicating decisions. This process is called the "decision analysis cycle." In the first of five phases in the cycle, the formulation phase, the analyst defines the basic elements of the decision:

- *Alternatives*—possible actions
- *Uncertainties*—aspects that influence the decision, but which are not perfectly known
- *Outcomes*—possible consequences of the alternatives
- *Values*—the decision-maker's preferences for the outcomes.

In the deterministic phase, a deterministic decision model is constructed to represent the relationships between the alternatives, uncertainties, outcomes, and values. In the probabilistic phase, uncertainty is incorporated into the model by assessing the probabilities of the various outcomes. The probabilistic decision model is then analyzed in the evaluation phase to determine the optimal alternative. The optimal alternative is the alternative with the highest expected utility, or sum of the decision-maker's values for possible outcomes times their probabilities. The analysis can also determine the value of gathering more information, the value of controlling an uncertain variable, and the recommended course of future action. In the final phase of the decision analysis cycle, the results and insights are communicated to those involved in making the final decision.

The structured process of DA and its axiomatic basis ensure a logically sound analysis that explicitly incorporates uncertainty and risk. The resulting decisions are documented and defensible. However, DA requires a knowledge of the methodology and, possibly, a sizable commitment of resources unless a decision model has been developed previously and is being used for repetitive decisions. In addition, it requires that less tangible issues such as community concern and employee morale be quantified, and that a single set of value judgments be used. Assessing a single set of value judgments may be difficult in situations that involve multiple decision makers. A good introductory book is Clemen's *Making Hard Decisions: An Introduction to Decision Analysis* (1991).

3.2.6.3 Multiattribute Utility Analysis

Multiattribute utility analysis builds upon the principles and methods of decision analysis (DA) to explicitly address value judgments associated with multiple, competing objectives. The fundamental problem addressed by MUA is how to determine the overall value of an alternative, and compare this alternative with others, based on how well the alternatives perform on multiple objectives. The process used for an MUA ensures that the multiple objectives are defined appropriately, that the assessment of the value of each alternative on each objective is accurate, that the uncertainties are factored in correctly, and that the computation of the overall value of each alternative based on its value for each objective is done properly. MUA retains the key strength of DA: a logically rigorous and explicit basis for the results of the analysis.

MUA has the same theoretical underpinning as DA. Pioneering work by von Neumann and Morgenstern in the 1940s justified the concept of a utility function and the use of expected utility to determine the preferred alternative of several risky alternatives. Howard Raiffa is credited with many of the key ideas that, in the 1960s, extended DA techniques to problems with multiple objectives. With Raiffa, Ralph Keeney developed practical approaches to assessing utilities and applying MUA techniques. MUA is suited to problems that involve complex uncertainties, difficult value tradeoffs, and the need to justify the resulting decision. It has been successfully applied to numerous public policy decisions addressing environmental, political, social, and economic concerns.

The building blocks of an MUA are the same as those described for a DA: alternatives, uncertainties, and values. The primary task of an MUA is to model and assess the decision-maker's values. To accomplish this, the key steps are:

- Develop an objectives hierarchy that identifies and organizes the goals of the decision
- Specify an attribute and measurement scale that quantifies how well each alternative fulfills each objective
- Define a utility function that characterizes preferences for each attribute
- Define a multiattribute utility function that captures the tradeoffs for different objectives.

Once the multiattribute utility function is specified, the utility for each alternative can be determined. If necessary, the analysis can account for uncertainties in the same way as in a DA. The preferred alternative is the one with the highest expected utility, where the expected utility is computed by taking values times probabilities, summed over possible outcomes. The utility function in an MUA is sometimes similar to that used in weighted scoring methods, but can also involve more complex mathematical forms. Extensions of MUA enable it to be used in group situations or facilitate the gathering of input from outside stakeholders.

Key inputs to an MUA are the:

- Attributes and measures for each objective
- Assessments of possible outcomes and their probabilities with each alternative on each attribute

- Utility functions and weighting factors for each attribute
- Mathematical form of the multiattribute utility function.

Specialized training in both decision analysis and in MUA is required to apply the technique, and on larger problems an organization will often use outside help from experienced practitioners. Although MUA provides a rigorous, defensible basis for complex decisions, it often requires more time and resources than other methods. Suggested texts on MUA are von Winterfeldt and Edwards's *Decision Analysis and Behavioral Research* (1986) and Keeney and Raiffa's *Decisions with Multiple Objectives: Preferences and Value Tradeoffs* (1975).

3.2.6.4 Portfolio Analysis

Portfolio analysis is a technique originating in investment-type decisions. When selecting a portfolio of securities, investors commonly balance two goals: to maximize expected return from their investment while simultaneously minimizing the risk. The goal of portfolio analysis is to help the decision maker find the bundle of securities that best satisfies these often-contradictory goals.

A key concept in portfolio analysis is the idea of a "nondominated" portfolio. A portfolio is said to be nondominated if no other portfolio exists that has an equal rate of return with a smaller risk. A portfolio is also nondominated if no other portfolio with an equal level of risk has a higher return. The set of nondominated portfolios is the set for which there is not an obviously better allocation of investment funds (often called the "efficient frontier"); the best portfolio is known to reside in this set. Portfolio analysis finds this set of nondominated solutions and tries to determine which solution among the nondominated ones is most preferred by the decision maker. This decision depends heavily on the decision-maker's relative desire for high return and low risk. Finding the best solution requires eliciting the decision-maker's preferences for return and risk. Portfolio analysis can also be applied to situations that involve goals in addition to expected return and risk. In these cases, a multiattribute utility function must be assessed. For a simple introduction to portfolio analysis, see *Normative Decision Making* (1970) by Kassouf. A classic text on portfolio analysis is Markowitz's *Portfolio Selection* (1959).

3.3 RATIONALE FOR CHOOSING DECISION AIDS TO BE TREATED IN DETAIL

Chapters 6–12 provide a more detailed discussion of some of the decision aids presented in this chapter. These chapters serve two purposes. First, they serve as an introduction to the selected aids. Each chapter covers the history, key concepts, methodology, and in-depth evaluation of the decision aid. Second, these chapters illustrate through case studies and other examples how each decision aid is used. This gives the reader an understanding of how the technique can be applied to a problem of chemical process safety.

Due to constraints of space and resources, the in-depth treatment has been restricted to only a subset of the aids introduced. In selecting which aids would be addressed more extensively, the authors were guided by four criteria. These criteria are listed in Table 3.1 along with the associated decision aids that fit the criteria. Clearly, such judgments reflect the opinions of the authors. The absence of any aid from the brief descriptions in this chapter, or from the more detailed discussions in Chapters 6–12, does not imply anything about its quality or potential applicability to problems of process safety.

3.4 A WORD ABOUT DECISION AIDS BASED ON THE THEORY OF FUZZY SETS

In closing this chapter we would like to say a brief word about decision aids based on the theory of fuzzy sets. Fuzzy set theory is an alternate approach to characterizing uncertainty that was first developed by Lofti A. Zadeh. Fuzzy sets are designed to address uncertainty in meaning, rather than uncertainty in the state of the world.

This distinction is best illustrated with an example. Consider the uncertainty contained in the statement "Bob is young." Standard probability theory would attribute uncertainty in this statement to our uncertainty about whether Bob is young, assuming that there is clarity in the meaning of what it means to be "young." We must define what it means to be young before trying to assess a likelihood that we think Bob is young. Fuzzy set theory makes no such

TABLE 3.1
Criteria for Choice of Decision Aids Treated in Detail

Criteria	*Decision Aids*
Is the decision aid **well-established in the areas related to process safety?**	Cost–Benefit Analysis Voting Methods
Is the decision aid a **promising new addition to the collection of decision aids available for process safety?**	Decision Analysis Multiattribute Utility Analysis Weighted Scoring Methods
Does the decision aid have a **particularly wide area of applicability?**	Mathematical Programming Voting Methods Weighted Scoring Methods
Is the decision aid **representative of an important class of decision aids?**	Mathematical Programming Payoff Matrix Analysis Voting Methods Weighted Scoring Methods

concession. It argues that there is additional uncertainty—perhaps more appropriately called vagueness—as to what is meant by "young." This type of uncertainty is addressed by a fuzzy set—the set of "young" people—to which individuals belong in varying degrees. A 20-year-old is not necessarily "young" or "not young" but rather belongs in the fuzzy set of "young people" to some degree, a degree less than that of a 10-year-old, but more than that of a 30-year-old.

Fuzzy set theory has produced an impressive volume of literature in the rather short amount of time—25 years—since it was established. It is not yet clear whether fuzzy set theory will make a significant contribution to practical corporate decision making. Decision aids based on fuzzy sets are not addressed in this volume for several reasons:

- Most "recognized" decision aids utilizing fuzzy sets are variants of more traditional decision-making approaches.
- Introducing decision aids based on fuzzy sets would necessitate including introductory material on the calculus of fuzzy sets, something beyond the intended scope of this text.
- There is still considerable debate on the usefulness and appropriateness of fuzzy set approaches for improving human decision making. The more celebrated applications of fuzzy set theory have been in the field of engineering control, where the aim is to make machines mimic human decision-making behavior rather than to guide human decision making.

3.5 SUMMARY

This chapter introduced the range of available decision aids. To assist in understanding similarities and differences among the aids, they were described in the context of a classification scheme whose various branches highlight salient characteristics of groups of decision aids. We then presented the criteria used to select the decision aids that are given detailed treatment in subsequent chapters of the book.

3.6 REFERENCES

Brown, R. V., and J. W. Ulvila (1976). "Setting Analytic Approaches for Decision Situations; A Matching of Taxonomies." McLean, VA: Decisions and Designs, Inc.

Buede, D. M (1992). Software Review: Overview of the MCDA Software Market. *Journal of Multi-Criteria Decision Analysis* 1(1):59–61.

Clemen, R. T. (1991). *Making Hard Decisions: An Introduction to Decision Analysis*. Boston: PWS-KENT.

Davis, M. (1983). *Game Theory: An Introduction*. New York: Basic Books.

EPRI (1992). "CATALYST—A Group Process for Strategic Decision Making: Facilitator's Guidebook." Prepared by Applied Decision Analysis and The Brattle Group. TR-100583, Research Project 2631, Final Report. Palo Alto, CA: Electric Power Research Institute (September).

Fischhoff, B., S. Lichtenstein, P. Slovic, S. L. Derby, and R. L. Keeney (1981). *Acceptable Risk.* New York: Cambridge University Press.

Hillier, F. S., and G. J. Lieberman (1990). *Introduction to Mathematical Programming.* New York: McGraw-Hill, Inc.

Hwang, C., and M. Lin (1987). *Group Decision Making Under Multiple Criteria; Methods and Applications.* New York: Springer-Verlag.

Hwang, C. L., and A. S. M. Masud (1979). *Multiobjective Decision Making Methods and Application: A State-of-the-Art Survey.* New York: Springer-Verlag.

Hwang, C., and K. Yoon (1981). *Multiattribute Decision Making Methods and Applications.* New York: Springer-Verlag.

Kassouf, S. (1970). *Normative Decision Making.* Prentice-Hall Foundations of Administration Series, Ed. H. A. Simon, Englewood Cliffs, N.J.: Prentice-Hall.

Keeney, R., and H. Raiffa (1975). *Decisions with Multiple Objectives: Preferences and Value Tradeoffs.* New York: John Wiley and Sons.

Kepner, C. H., and B. B. Tregoe (1981). *The New Rational Manager.* London: John Martin Publishing Ltd.

Markowitz, H. M (1959). *Portfolio Selection.* New York: John Wiley & Sons.

Martin, M. P. (1979). Decision Making: The Payoff Matrix. *Journal of Systems Management* 30(1):14–18.

Merkhofer, M. W. (1987). *Decision Science and Social Risk Management: A Comparative Evaluation of Cost–Benefit Analysis, Decision Analysis, and Other Formal Decision-Aiding Approaches.* Dordrecht, Holland: D. Reidel (Boston: Kluwer Academic Publishers Group).

Niemi, R. G., and W. H. Riker (1976). The Choice of Voting Systems. *Scientific American* 234(6), 21–27.

Roy, B., M. Present, and D. Silhol (1986). A Programming Method for Determining Which Paris Metro Stations Should Be Renovated. *European Journal of Operations Research* 24:318–334.

Saaty, T. L. (1980). *The Analytic Hierarchy Process.* New York: McGraw-Hill.

Saaty, T. L. (1990). *Decision Making for Leaders.* Belmont, CA: Lifetime Learning Publications.

Sassone, P. G., and W. A. Schaffer (1978). *Cost–Benefit Analysis: A Handbook.* New York: Academic Press.

Tabucanon, M. T (1988). *Multiple Criteria Decision Making in Industry.* New York: Elsevier.

Vincke, P. (1986). Analysis of Multicriteria Decision Aid in Europe. *European Journal of Operational Research* 25(2):160–168.

von Winterfeldt, D., and W. Edwards (1986). *Decision Analysis and Behavioral Research.* Cambridge: Cambridge University Press.

4

EVALUATING AND SELECTING DECISION AIDS

Where there is much desire to learn, there of necessity will be much arguing, much writing, many opinions; for opinion in good men is but knowledge in the making.
—***Milton***, *Areopagitica*

4.1 PURPOSE OF CHAPTER

The previous chapters laid the groundwork for a closer study of decision aids. Chapter 1 introduced the basic decision problem, key analysis steps, and issues of particular concern for risk decision making. Chapter 2 focused on the key concepts of risk decision making. Chapter 3 provided a classification scheme and descriptions of the decision aids discussed in this book. This chapter deals with the selection of the appropriate decision aid(s) to resolve a particular problem.

There is no simple way to answer the question, "Which decision aid is the best one for my problem?" A decision aid is simply a tool used to solve a problem. Selecting a decision aid requires finding one that is consistent with the resources available for the decision problem. This suggests that a prudent way to proceed is to determine the salient characteristics of both problems and decision aids. Decision aids should be considered when their strengths match the requirements of the problem.

In some cases, the choice of decision aid is clear. For example, some decision aids were created specifically to handle certain types of problem complexities. Corporate guidelines may govern the depth and scope of analysis of a decision or impose time and/or resource constraints on the analysis, limiting which decision aids can be considered. More typically, the problem will not point to one decision aid, and several good decision aids could be used. Related to the choice of decision aid for a specific problem, a more general issue is what decision aid or set of decision aids an organization should use for its risk decisions. This question is partly addressed through the approach outlined in this chapter, and partly by understanding the organization's goals, philosophy, and capabilities in risk decision making.

For a given problem, more than one decision aid may be appropriate, and selecting an aid will depend on the objectives and experience of the organization. But more importantly, it may be useful to employ more than one decision aid in one problem. For example, one aspect of a problem might be best addressed by a vote (e.g., a vote on whether to undertake any analysis, on which consultant to use to help with the analysis, or on which alternatives to subject to further analysis). Another aspect might best be addressed by mathematical programming (e.g., how to reduce infinitely many combinations of possible actions to a manageable set of alternatives). A preliminary analysis using a decision analysis approach might be undertaken, followed by a more in-depth multiattribute utility analysis. For a given problem, the decision-making process is best viewed as iterative. Different decision aids may be useful at various points in the analysis, or the analysis may be further refined in later stages using a more sophisticated decision aid.

In summary, there is no precise rule for selecting a decision aid that applies to the wide range of risk decision problems. However, to guide the reader on this issue, we define some key aspects on which problems may differ. We also identify characteristics on which decision aids may differ. We discuss how problem aspects may be matched to decision aid characteristics. The chapter concludes with some principles for screening and selecting appropriate decision aids for a problem.

4.2 SELECTING DECISION AIDS

Each of us makes decisions every day, using a variety of methods. Without always realizing it, we tailor our approach to making decisions to the type of problem at hand. For example, you would likely apply different approaches to decisions on which car to buy, and on which person to marry.

When making a decision, it makes sense to use a decision aid that is equipped to address the unique characteristics of the problem. For example, some decision aids are well-suited to problems that have few alternatives that can be easily characterized. Other decision aids are geared toward more complex problems. Using a simple tool for a complex problem can lead to an incomplete analysis that does not consider all relevant factors. Using a complex tool for a simple problem can waste resources and add confusion to the decision-making process. This section provides guidance on how to characterize a decision problem to help identify appropriate decision aids.

The suggested approach for selecting decision aids is to match the needs of the decision problem to the strengths of particular aids. The process for doing this is diagrammed in Figure 4.1 and consists of the following five steps:

1. *Describe the problem.* Characterize the decision problem using the broad aspects introduced in Section 4.3.
2. *Identify the distinguishing aspects of the problem.* Which aspects make the problem unique? Examples are: unusual complexity, need for group input, and limited time and staff available to work on the problem (Section 4.4).
3. *Identify the problem class.* Match the problem to one of the sample problem classes described in Section 4.7. Problem classes were devel-

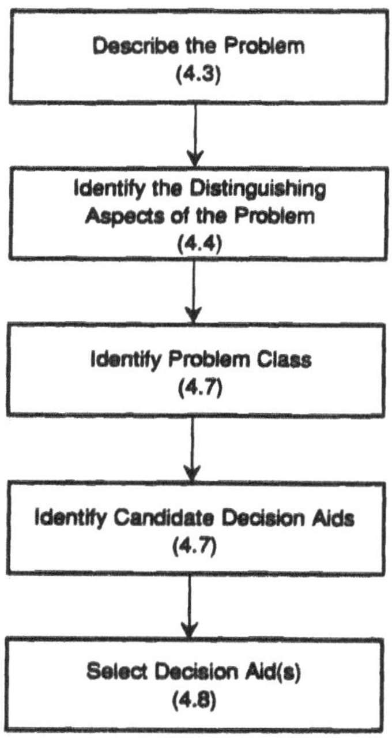

FIGURE 4.1. Process for selecting a decision aid, with section headings from this chapter.

oped by the authors based on an understanding of problem aspects (Section 4.3) and decision aid characteristics (Sections 4.5 and 4.6), and the relationship between the two (Section 4.7).

4. *Identify candidate decision aids.* Section 4.7 suggests decision aids for each problem class (summarized in Table 4.4).
5. *Select the decision aid(s).* Select the particular decision aid or aids that best match the problem (Section 4.8). If the problem class points to more than one decision aid, compare the decision aids on the other aspects of the problem, as described in Section 4.5.

4.3 DESCRIBE THE PROBLEM

Process safety decisions are multifaceted. Decisions must take into account the goals and needs of the organization, the characteristics of the problem at hand, and other factors. These can be grouped into the following six broad aspects:

- Resource Availability
- Problem Complexity
- Importance/Scrutiny
- Group Involvement
- Need for Quantification
- Other Constraints on Decision-Making Basis.

We discuss each of these aspects in turn.

Resource Availability concerns the amount of resources that the organization can bring to bear on the solution of the decision problem. Corporate resources include budget, human resources, expertise in decision-making techniques, and calendar time. Information about the problem is also a resource.

Problem Complexity refers to how complicated the decision problem is. Many factors can complicate a decision situation, including a high degree of uncertainty about key aspects of the problem (e.g., has the pipeline deteriorated?), the presence of many alternatives, a complex underlying physical or chemical process, and the need to consider competing objectives.

Importance/Scrutiny concerns the sensitivity of the decision maker and other stakeholders to the decision. This category considers how many people or entities are affected by the decision and how they are affected.

Group Involvement refers to the degree to which it is necessary or desirable to involve multiple decision makers or stakeholders in the decision process.

Need for Quantification arises when results need to be expressed numerically. This need usually comes from one of two places. Either the decision problem concerns a process that lends itself to quantification, or there is a desire for an "objective" basis for the decision.

Other Constraints may also exist, in addition to the above five problem aspects. Decisions can be constrained in various ways. For example, organizational or legal constraints can require that certain types of decision-making approaches be considered or can influence the types of issues addressed in an analysis.

Table 4.1 summarizes these key considerations. Note that these aspects describe both the nature of the problem itself and of the decision-making process or environment. For example, importance and scrutiny are included as one characteristic, since the combination of the two determines the rele-

TABLE 4.1
Key Problem Parameters Affecting Process Risk Decisions

Problem Aspect	*Considerations*
Resource Availability	Calendar time available Budget available Staffing available Availability of expertise on the use of particular decision aids Availability of information on risks, uncertainties, and preferences Availability of information/model of the underlying process
Problem Complexity	Degree of uncertainty Number of alternatives Complexity of the underlying process Need to address multiple competing objectives
Importance/Scrutiny	Population at risk Severity of risk Financial cost of adverse events Financial costs and impacts of alternative decisions Review by stakeholders
Group Involvement	Impact of decision on different parts of the organization Involvement of multiple decision makers Involvement of multiple stakeholders
Need for Quantification	Need for quantitative solution to problem, e.g., resource allocation Organization's desire to use quantitative basis for decision
Other Constraints on Decision-making Basis	Organization's attitude toward use of subjective information Organization's constraints on evaluation basis (e.g., Value of Life, Return on Investment) Organization's desire to ensure confidentiality Legal constraints

vance of decision aids that help address problems having this characteristic. For each aspect, the factors included with it are not necessarily independent or related. For example, there may be plenty of calendar time available to reach a decision, but little staff or budget available to do any analysis.

Our approach to characterizing a risk decision problem parallels a scheme by Ozernoy (1992). Ozernoy describes decisions by characteristics of the decision problem, the decision maker, and the resource constraints. Brown and Ulvila (1976) provide a more complete listing of problem factors, organized in a "situation taxonomy."

4.4 IDENTIFY THE DISTINGUISHING ASPECTS OF THE PROBLEM

To select the most appropriate decision aid, it is useful to identify the aspects that best distinguish the problem at hand. This second step of the process will help determine the problem class into which our problem falls. Is resource availability particularly constrained? Is the problem especially complex? Especially simple? Is group involvement particularly crucial in this decision? The answers to these questions are helpful in determining the type of problem to be addressed.

4.5 DECISION AID CHARACTERISTICS

As with a decision problem, there are numerous aspects on which decision aids can differ. We focus on six major characteristics, related closely to the problem aspects described earlier. The previous chapter classified decision aids mainly on methodological or theoretical traits. Here, greater importance is placed on other characteristics that help to clarify the situations in which certain aids are most appropriate.

The following six major characteristics help us evaluate the decision aids:

- Resource Requirements
- Depth of Analysis/Complexity
- Logical Rigor
- Group Focus
- Quantitativeness
- Track Record.

A description of each characteristic follows, along with levels used to characterize decision aids on that aspect.

Resource Requirements include the amount of budget, time, and effort required to perform an analysis with the decision aid.

- *Limited resources*—The decision aid requires few resources and is suitable for problems where the decision maker has little time, budget, or other corporate resources to pursue an analysis.
  Example: Aside from a brief amount of time for participants to congregate, voting requires virtually no resources.
- *Moderate resources*—The decision aid requires a nontrivial amount of time or budget for analysis. A typical analysis can be performed in an average amount of time.
  Example: A typical application of weighted scoring methods requires a nontrivial amount of time to rank and rate alternatives.
- *Extensive resources*—The decision aid requires a significant amount of resources to perform a thorough analysis. Complex problems will tend to involve relatively more calendar time and resources for analysis than other methods require.
  Example: Multiattribute utility analysis typically requires time and effort to structure the problem, assess utility functions, and obtain expert knowledge about the problem.

Depth of Analysis/Complexity refers to the detail and explicitness with which important aspects of the problem are addressed, and the simplicity with which the aid can perform a complete and thorough analysis of the decision. We characterize the aids using three levels:

- *Limited depth*—The decision aid is most appropriate for decision problems that are relatively small, involve few alternatives, or are in other ways less complex, or problems that do not require a thorough analysis of the underlying values, uncertainties, and relationships among these aspects. Little training is required to use the decision aid. The aid tends to examine the decision problem as a whole, rather than decomposing the problem into its component factors. Explicit description or quantification of the aspects influencing the decision is not required.
  Example: Voting assumes that the participants mentally consider and integrate relevant factors prior to the vote, but does not explicitly analyze the underlying factors.
- *Moderate depth*—The decision aid involves some analysis of the aspects influencing the decision, but does not require that the relationships between the factors and their comparative weights be quantified. Some training or experience is helpful in applying the aid, particularly when it is used on more complex problems.
  Example: Weighted scoring methods address the key aspects that determine the desirability of various alternatives, but do not explicitly account for the uncertainties associated with each alternative.

- *Extensive depth*—The decision aid both encourages a detailed decomposition of the problem, and requires that all relationships among factors be expressed and that the factors themselves be characterized in detail. The aid employs explicit rules for determining the optimal decision taking explicit account of the information about each alternative. Training and experience are usually needed to apply this technique.
  Example: Decision analysis addresses the decision-makers preferences and uncertainties to whatever level of detail is considered appropriate to the problem.

Logical Rigor is concerned with the logical and mathematical soundness of the analysis and the ability to avoid incorrect conclusions. We rate the decision aids as follows:

- *Limited rigor*—The decision aid has little theoretical underpinning and the results may be subject to bias or manipulation.
  Example: Voting can be subject to manipulation or can result in a choice that does not necessarily reflect the preferences of the group.
- *Moderate rigor*—The decision aid is neither inconsistent nor entirely sound; the methodology does not violate any principles, but it may not reflect or consider all of the relevant factors in a decision, or may oversimplify the problem and result in less than optimal solutions.
  Example: Cost–benefit analysis, as we define it, does not explicitly consider choices under uncertainty.
- *Extensive rigor*—The decision aid is grounded in theory and is based on and fully reflects relevant mathematical or logical principles. The methodology properly addresses all factors considered relevant to a decision, and can indicate the sensitivity to factors that have been ignored or approximated.
  Example: Decision analysis and multiattribute utility analysis are founded on choice theory and designed to fully reflect mathematical and logical principles.

Group Focus refers to the ability of the decision aid to incorporate group judgments, handle group decision making, and address widely divergent concerns or objectives.

- *Limited group focus*—The decision aid is not designed to be used in group decision making; very few tools for using the aid within groups have been developed.

Example: Cost–benefit analysis tends to avoid subjective group inputs and strives to identify market values; therefore, cost–benefit analysis has not developed a specific set of techniques for its application within groups.

- *Moderate group focus*—The decision aid, while developed for use by single decision makers, has a history of use in group decision making and has developed specific techniques for aiding group decision making.
  Example: Kepner–Tregoe decision analysis is often used in group decision-making situations.

- *Extensive group focus*—The decision aid is explicitly designed to be used in group settings.
  Example: Voting, by definition, is a group process.

Quantitativeness is the ability of the decision aid to provide a quantitative basis for the decision, to address sensitivity analysis, and to address decisions such as resource allocation problems, that are inherently quantitative.

- *Limited quantification*—The decision aid does not attempt to quantify the desirability of each alternative, is limited to discrete choices, or is limited in its ability to address the sensitivity of the decision to differing characterizations of the alternatives.
  Example: Voting.

- *Moderate quantification*—The decision aid quantifies the desirability of each alternative and allows some sensitivity analysis. However, the quantification may be arbitrary, incomplete, or inaccurate.
  Example: Weighted scoring methods score the desirability of various alternatives, but the scale used to evaluate alternatives is not tied to a physical characteristic.

- *Extensive quantification*—The decision aid quantifies all relevant aspects of the decision, and the quantification is not arbitrary.
  Example: Decision analysis involves quantification of the uncertainties and values of the decision maker, and the quantities represent physical properties of the alternatives or the decision-makers beliefs about these aspects.

Track Record has to do with how long the decision aid has been available, how many times it has been applied to process safety decisions and similar types of decisions, how widely the aid is accepted, and how much documentation is available on the quality and practicality of the method.

- *Limited track record*—The technique is new and has not been used in documented practical applications.
  Example: Game theory has no discernable track record in process decision making.
- *Moderate track record*—The technique has only a few documented applications to process safety decisions, but a significant record of application in other fields.
  Example: Decision analysis has some applications in the areas of nuclear safety and chronic health risks and has been used extensively for business decision making.
- *Extensive track record*—Applications of the decision aid to chemical process safety decisions are frequently reported.
  Example: Cost–benefit analysis has a long history of applications to chemical process safety decisions.

Table 4.2 summarizes the major characteristics and example decision aids just described.

TABLE 4.2
Summary of Major Characteristics of Decision Aids

Major Characteristic	*Level*	*Example Decision Aid(s) with This Characteristic*
Resource Requirements	Limited Moderate Extensive	Voting Weighted Scoring Methods Multiattribute Utility Analysis
Depth of Analysis/Complexity	Limited Moderate Extensive	Voting Weighted Scoring Methods Decision Analysis
Logical Rigor	Limited Moderate Extensive	Voting Cost–Benefit Analysis Decision Analysis Multiattribute Utility Analysis
Group Focus	Limited Moderate Extensive	Cost–Benefit Analysis Kepner–Tregoe Decision Analysis Voting
Quantitativeness	Limited quantification Moderate quantification Extensive	Voting Weighted Scoring Methods Decision Analysis
Track Record	Limited Moderate Extensive	Game Theory Decision Analysis Cost–Benefit Analysis

In most decision contexts, we would like to use a decision aid that scores well on all six of these characteristics. If no resources were required, we would probably subject every decision to a thorough, quantitative analysis, and use a method that is as logically sound as possible and has a solid track record. If no resources were required, we would also like to allow for group input whenever the decision is made by, or affects, more than one person. This applies to most corporate or governmental decisions, and particularly to chemical process safety decisions.

Unfortunately, no decision aid excels on all six dimensions. In practice, the decision maker needs to make tradeoffs in choosing among decision aids that are less than ideal on one or more characteristics. For example, aids that provide depth of analysis are generally more resource-intensive. A desire for a deep analysis without any resource investment is unrealistic.

Generally there is an organizational cost to ensuring that decisions fulfill as many of the ideals as possible. This cost can include the time and resources required to analyze the decision, the discomfort or legal implications resulting from explicit, documented analysis or from accommodating group participation. For many problems, it may be sensible to use a simple decision aid. For example, an organization may accept use of a voting approach for a decision that involves a simple problem for which there is limited time for analysis. As mentioned earlier, it may also be useful to apply a voting approach as part of a larger decision analysis, to address some aspects of the problem.

In summary, the decision maker may be willing to compromise on one or more of the qualities of the decision aid given the nature and context of the decision. Section 4.7 addresses the problem situations in which certain types of decision aids may be more or less appropriate.

4.6 DECISION AID CHARACTERIZATIONS

Table 4.3 characterizes the decision aids presented in Chapter 3 on each decision aid characteristic described in Section 4.5, considering typical applications of each decision aid. The results are admittedly imprecise and somewhat judgmental. Moreover, the way in which a decision aid is used can affect its characterization. For example, many of the more quantitative decision aids can be applied in a quick, less-detailed analysis. In such a context, the decision aid would be characterized as being less resource-intensive and providing less depth of analysis than holds for the typical applications addressed in the table.

4.7 IDENTIFY THE PROBLEM CLASS AND CANDIDATE DECISION AIDS

Having characterized both decision problems and decision aids, we can now better put the two together. Realistically, there are too many problem aspects

TABLE 4.3
Overview of Decision Aids and Their Characteristics[a]

	Resource Requirements	*Depth of Analysis/ Complexity*	*Logical Rigor*	*Group Focus*	*Quantitativeness*	*Track Record*
Analytic Hierarchy Process	Moderate	Moderate	Moderate	Moderate	Moderate	Limited
Compromise Programming	Extensive	Extensive	Moderate	Limited	Extensive	Limited
Cost–Benefit Analysis	Moderate	Moderate	Moderate	Limited	Extensive	Extensive
Decsion Analysis	Extensive	Extensive	Extensive	Moderate	Extensive	Moderate
Game Theory	Moderate	Limited	Moderate	Limited	Moderate	Limited
Goal Programming	Extensive	Extensive	Moderate	Limited	Extensive	Limited
Kepner–Tregoe Decision Analysis	Moderate	Moderate	Moderate	Moderate	Moderate	Moderate
Mathematical Programming	Extensive	Extensive	Moderate	Limited	Extensive	Limited
Multiattribute Utility Analysis	Extensive	Extensive	Extensive	Moderate	Extensive	Limited
Nominal Group Technique	Moderate	Limited	Limited	Extensive	Limited	Limited
Outranking Methods	Moderate	Moderate	Limited	Limited	Moderate	Limited
Payoff Matrix Analysis	Moderate	Moderate	Moderate	Moderate	Moderate	Limited
Portfolio Analysis	Extensive	Moderate	Moderate	Limited	Extensive	Limited
Screening/ Ranking Methods	Limited	Limited	Moderate	Moderate	Moderate	Limited
SMART	Moderate	Moderate	Moderate	Limited	Moderate	Limited
Voting	Limited	Limited	Limited	Extensive	Limited	Limited

to explicitly address all possible combinations. Instead, this section focuses on major problem classes. Problem classes are categories of typical risk decisions characterized by the problem aspects presented earlier. What follows is a description of each problem class, how the aspects of the problem influence the choice of decision aids, and a short list of decision aids that are candidates for problems in each class.

To help make explicit the relationships between the problem aspects and the decision aid characteristics, Figure 4.2 shows the key influences between the two. Arrows show important links between a problem aspect and decision aid characteristic. For example, if a decision situation is dominated by a need for Group Involvement, then the obvious decision aid characteristic to consider would be whether it provides Group Focus.

Most problem aspects match one-for-one with a decision aid characteristic. Exceptions are Resource Availability and Importance/Scrutiny. The Resource Availability associated with the decision situation affects both the amount of resources you can invest in using a decision (Resource Requirements) and the Depth of Analysis, because decision aids that provide more analysis depth usually require more resources.

Problems that have less importance or sensitivity to the company and stakeholders, or for which the decision will not be scrutinized by stakeholders other than the decision maker, may be amenable to decision aids that are less than ideal in their Depth of Analysis/Complexity, Logical Rigor, and Track Record. Problems with greater importance or sensitivity, or for which the decision is likely to be subjected to outside scrutiny, may require the use of a decision aid that is closer to ideal in these characteristics, which lend credibility and defensibility to the decision.

In the paragraphs that follow, we describe six problem classes that are useful for identifying the appropriate decision aids for a problem and then summarize this information in Table 4.4.

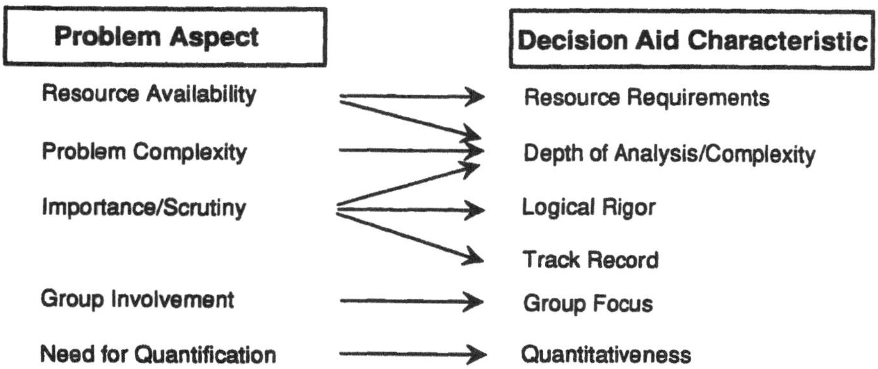

FIGURE 4.2. Relating problem aspects to decision aid characteristics.

Problem Class 1: A Quick, Simple Decision

Problems in Class 1 are most distinguished by simplicity across all problem aspects. Specifically, these problems involve low complexity, low resource availability, low importance/scrutiny, low group involvement, and low need for quantification. Problems in Class 1 include many of the quick, single-decision-maker decisions that are encountered on a daily or weekly basis. An example is selecting from among alternative types of safety valves for a pipeline. Another example is choosing from alternative hazard evaluation techniques to apply to a problem.

Screening/Ranking procedures are well-suited for addressing problems of this type.

Problem Class 2: A Quick Group Decision

Problems in Class 2 are most distinguished by low resource availability, high need for group involvement, and low need for quantification. Problems in this class include many of the decisions typically encountered by a group on a recurring basis, and which require a quick group decision. An example is a group decision on the organization's priorities for hazard analysis when the degree of hazard is unclear but relatively low.

Voting methods are well-suited to address problems of this type.

Problem Class 3: A More Thorough Group Decision

Problems in Class 3 are most distinguished by high resource availability, a high need for group involvement, and a low need for quantification. Problems in this class include many of the decisions that a group wishes to make using a qualitative basis, but which it would like to spend more time or resources to carefully analyze the decision. An example is a group decision on which of a number of proposed risk-reduction alternatives to pursue in more depth, when the alternatives are relatively unknown or poorly understood.

Nominal Group Technique is suggested for addressing problems of this type.

Problem Class 4: A Quick, Quantitative Non-Group Decision

Problems in Class 4 are distinguished by low resource availability and a high need for quantification, but a low need for group involvement. Examples include problems in which there is a need for quantitative analysis, but a desire to avoid the time and resources required to apply a more sophisticated decision aid. An example is a decision on whether chlorine or sodium hypochlorite should be used for water treatment in a plant's water cooling system.

A number of decision aids have been developed for decisions of this type. Analytic Hierarchy Process, Kepner–Tregoe Decision Analysis, Payoff Matrix Analysis, Outranking, and SMART are suggested for consideration in these situations.

Problem Class 5: A Quick, Quantitative Group Decision
Problems in Class 5 are similar to those in Class 6 in their low resource availability and high need for quantification, but are also distinguished by a high need for group involvement. Examples are problems faced by a group, but for which the group wishes to use a quantitative basis for the decision without necessitating the investment of significant time and resources. An example is a decision on how to reduce risk from an existing chlorine loading facility located next to a neighboring residential area. This decision is the subject of a case study used later in the book.

Payoff Matrix Analysis, Kepner–Tregoe Decision Analysis, and Analytic Hierarchy Process are decision aids well-suited to this class of problems.

Problem Class 6: A Complex, Quantitative Decision with High Importance and Resource Availability
Problems in Class 6 include many of the most challenging risk decisions: the ones that are highly complex and for which there is a need to provide a quantitative basis for the decision. These problems involve high importance/scrutiny. The organization, in consideration of the complexity and importance of the decision, has allocated significant resources and time to the analysis of the decision. An example is a decision of how to reduce risk from a distillation column that separates hexane and heptane near a residential area. This decision is the subject of a case study used later in the book.

Decision Analysis, Multiattribute Utility Analysis, Cost–Benefit Analysis, Mathematical Programming, Goal Programming, and Compromise Programming are suggested for consideration in addressing these problems.

Table 4.4 summarizes the key distinguishing characteristics of each problem class and the decision aids that are suggested for consideration with these problems.

4.8 SELECT THE DECISION AID(S)

As can be seen, there are usually several decision aids available to address a given class of problems. An organization may already possess experience with one of these decision aids, or wish to develop experience with one such aid, and therefore choose to use that one. A broader understanding of the characteristics of decision aids, as summarized in Table 4.2, can also guide the selection of a decision aid from the set of aids that is appropriate to the problem at hand.

4.9 SUMMARY

The number of available decision aids seems large until one realizes the diversity of problems and environments in which decisions are made. Even

TABLE 4.4
Problem Classes and Appropriate Decision Aids

Problem Class	*Distinguishing Problem Aspects*					*Well-Suited Decision Aids*
	Resource Availability	*Problem Complexity*	*Importance/ Scrutiny*	*Group Involvement*	*Need for Quantification*	
1 Quick, simple decision	Low	Low	Low	Low	Low	Screening/Ranking
2 Quick group decision	Low	—	—	High	Low	Voting
3 More thorough group decision	High	—	—	High	Low	Nominal Group Technique
4 Quick, highly quantitative nongroup decision	Low	—	—	Low	High	Analytic Hierarchy Process Kepner–Tregoe Decision Analysis Payoff Matrix Analysis Outranking SMART
5 Quick, highly quantitative group decision	Low	—	—	High	High	Payoff Matrix Analysis Kepner–Tregoe Decision Analysis Analytic Hierarchy Process
6 Complex, quantitative decision with high importance and resource availability	High	High	High	—	High	Cost–Benefit Analysis Compromise Programming Decision Analysis Goal Programming Multiattribute Utility Analysis Mathematical Programming

when we restrict ourselves to those decision aids for which there is a large body of practitioners and practical experience, there is still much leeway in the choice about which decision aid to use. There is no definitive answer to the question, "Which decision aid should I use?" Each decision aid has its own strengths and weaknesses, and the appropriateness of a decision aid will depend on the specific problem and on the goals of the decision maker. Moreover, refinements are possible with most any decision aid to enable it to better fulfill the goals of the decision maker.

This chapter provides guidance both to help an organization determine a subset of decision aids that it considers useful for its problems, and to help an organization choose an aid that is appropriate to a specific problem or aspect of a problem. To do this, we have developed a taxonomy that categorizes decision aids based on the types of problems they are equipped to address. We identified some key dimensions on which problems can be characterized (Table 4.1) and some key dimensions on which decision aids can be characterized (Table 4.2). We provided an overview of the characteristics of the major decision aids presented in Chapter 3 (Table 4.3). We indicated how problem characteristics can relate to decision aid characteristics (Figure 4.2). Finally, we suggested some problem classes that can help narrow the set of decision aids appropriate for a problem (summarized in Table 4.4).

4.10 REFERENCES

Brown, R. V., and J. W. Ulvila (1976). "Setting Analytic Approaches for Decision Situations; A Matching of Taxonomies." McLean, VA: Decisions and Designs, Inc.

Ozernoy, V. M. (1992). Choosing the 'Best' Multiple Criteria Decision-Making Method. *INFOR* 30(2):159–171.

5

INTRODUCTION TO CASE STUDIES

The difficulty in life is the choice.
—***George Moore,*** *The Bending of the Bough*

5.1 PURPOSE OF CHAPTER

This chapter describes three case studies that involve hypothetical chemical process safety decisions. We present these cases to illustrate typical issues that arise and to demonstrate how the various decision aids can address these problems. These cases are revisited throughout the book to show that different decision aids approach the same problem in different ways, emphasize different aspects of the same problem, and lead to different types of results.

Cases Two and Three were initially presented in Chapter 8 of the *Guidelines for Chemical Process Quantitative Risk Analysis* (AIChE/CCPS 1989). In the *CPQRA Guidelines,* the cases were designed to aid in understanding and applying the risk analysis techniques covered in that text. We have modified the presentation below to provide a decision-oriented context for each case. The application of a decision aid in the problem is motivated by the need to decide on a course of action. The stimulus for action could occur either before or after the risk analysis is complete. Before any risk analysis is undertaken, the organization might recognize that a potential hazard exists, and need to decide on what action if any to take. The process of analyzing the decision might identify the need or eliminate the need for a quantitative risk analysis, or identify whether a more thorough risk analysis will aid in the decision process. Alternatively, the analysis of the decision might follow an initial risk analysis: the magnitude of the problem is initially estimated, and then the organization must decide how to proceed from there. Ideally, the decision-making process should determine whether and to what depth to undertake the risk analysis, but in practical instances the risk analysis may be commissioned first and thus shape the subsequent decision analysis effort.

In what follows, we have eliminated some details of the quantitative risk analysis. Instead we emphasize the approach taken in the risk analysis, the results obtained, and how the results are used in the analysis of the decision. For the second and third cases below, the reader should refer to the *CPRQA Guidelines* for further details on the risk analysis.

Case One involves a relatively simple problem of how to deal with an aging underground pipeline that carries a flammable liquid. Case Two involves how to respond to the risks from a chlorine rail tank car loading facility due to potential exposure to toxic vapors. Case Three involves alternatives for reducing the risk to a residential community from a distillation column containing a flammable material. Case One is fairly simple in terms of the magnitude of the risk, the number and complexity of aspects affecting the risk, the size of possible investments toward reducing the risk, the number of available alternatives, and the sensitivity of the decision to various stakeholders. Cases Two and Three introduce successively greater complexity with respect to one or more of these aspects.

Disclaimer: The reader should keep in mind that the examples presented here and worked elsewhere in the book are intended to illustrate risk analysis and decision aid methodologies, **NOT** to provide real world" answers. These cases are fabricated examples rather than actual decision problems, and where data have been used in these examples, these data are also fabricated unless otherwise noted. In particular, the cases were specifically designed so that risk analysis calculations could be done with only a hand calculator, and only simple risk models were used. The reader is cautioned not to assume that the specific approach, data, or results for any of the cases are transferable to any real problem.

5.2 CASE ONE: UNDERGROUND PIPELINE

Recently a leak was discovered in an underground pipeline at the processing facility of a large chemical manufacturer. A cleanup of the premises was required, and this was difficult and expensive because the material involved is both flammable and a potential environmental contaminant. One concern was that the pipeline involved had a rated life of 30 years, but had developed a leak after only 20 years. The company has many other sites with pipelines of similar material construction and transporting similar materials. Corporate management's concern has been expressed in a memo transmitted to the plant managers at each of the company's facilities. This memo informed the plant managers about the earlier incident, and described some of the circumstances such as the age and design of the pipeline that may have contributed to the episode. While not recommending any specific actions, the memo advised the plant managers to "determine whether the potential for similar incidents exists at your facility, and if so, to undertake appropriate actions."

The case study initially focuses on the decision to be made regarding a particular section of pipeline at one of this company's facilities. In response to the memo, the plant manager has asked the chief plant engineer if there are any pipelines at this facility with potential for problems similar to those encountered in the earlier incident. The chief plant engineer is most concerned about an aging section of underground pipeline that is two miles long and delivers a flammable liquid feedstock to the plant. Although this pipeline poses minimal risk to workers and the public, a fire could cause significant property damage and a release could cause significant cleanup costs. The plant manager has asked the engineer to do a study of this pipeline section and to recommend possible actions by the end of the week.

5.3 CASE TWO: CHLORINE RAIL TANK CAR LOADING FACILITY

This case concerns a decision on what actions to take after a quantitative risk analysis has been conducted. An existing chlorine loading facility is situated close to a populated residential area. Figure 5.1 describes the installation, and Figure 5.2 provides the layout of the facility and neighboring residences. To investigate the risk, a quantitative risk analysis was performed. This analysis provided a foundation on which to build an evaluation of possible measures that could be taken to reduce the risk.

Incidents addressed in the quantitative risk analysis included:

- A small liquid leakage of 10 minutes duration, caused by a valve leak, hose leak, or impact failure of liquid connecting pipeline
- A small vapor leakage of 20 minutes duration, caused by a valve leak, hose leak, impact failure of vapor connecting pipeline, or relief valve leak
- A large vapor release of 60 minutes duration, caused by an external fire.

For illustrative purposes this risk analysis has been highly simplified. The study was completed and the results conveyed to plant management. Figure 5.3 provides an F-N curve that describes the fatality risk estimated by this initial study.

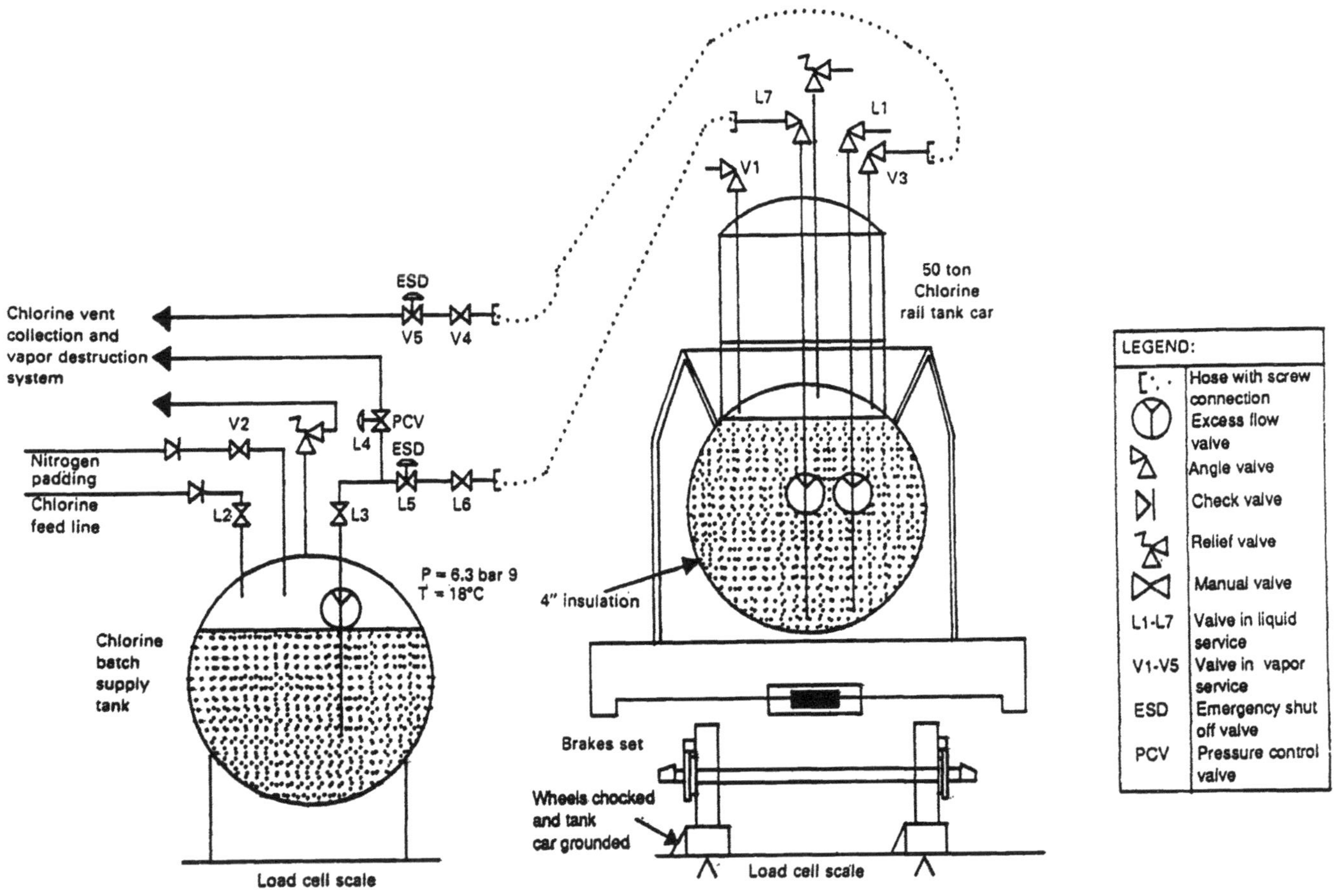

FIGURE 5.1. Diagram of liquid chlorine rail tank car loading installation. (AIChE/CCPS 1989, p. 416)

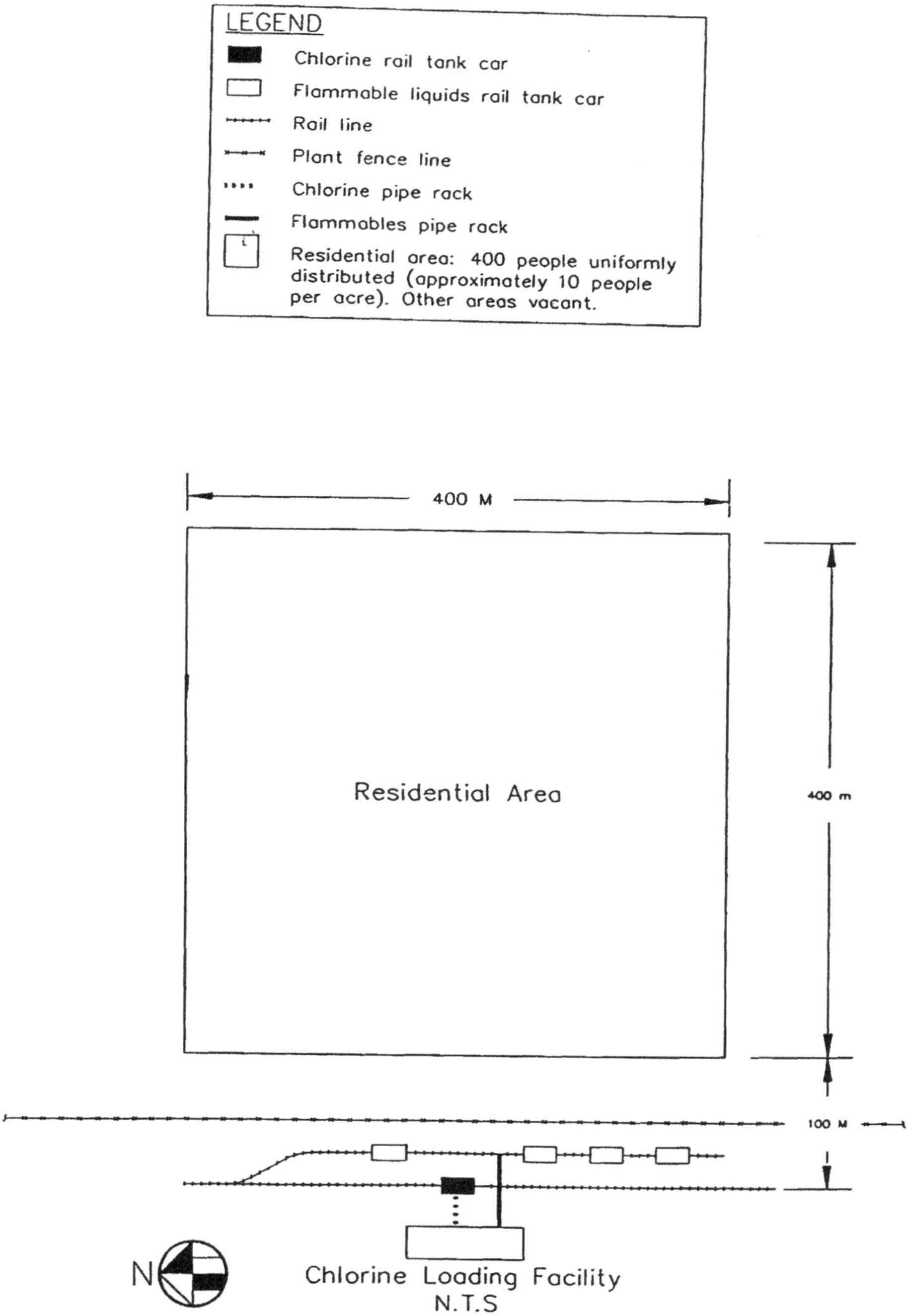

FIGURE 5.2. Layout of chlorine rail loading facility (AIChE/CCPS 1989, p. 418).

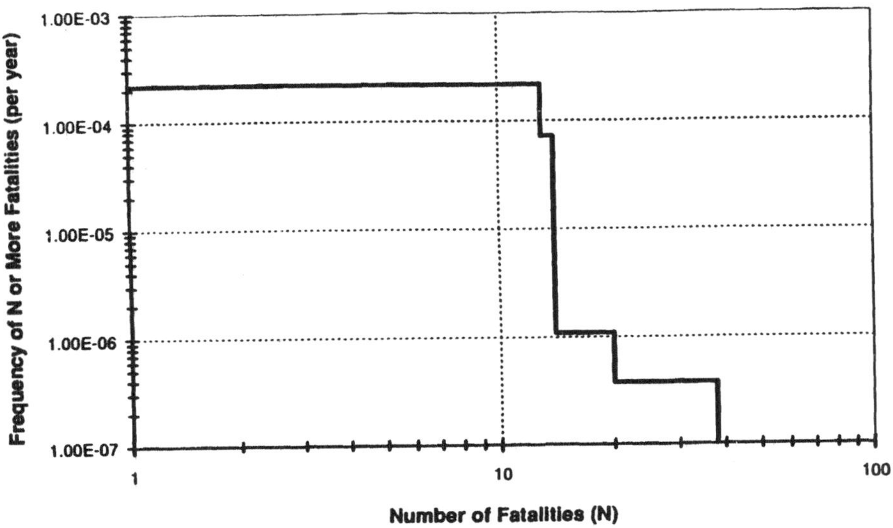

FIGURE 5.3. F-N curve showing results of risk analysis for chlorine loading facility. (AIChE/CCPS 1989, p. 440)

After the results of this study were presented, it was decided that further actions to reduce the risk of the facility should be explored. A team was formed to review the risk analysis and consider the possible next steps.

The decision maker in this case is the plant manager. To guide the plant manager's decision, an analysis team is formed to review the risk analysis, identify and refine alternatives, evaluate the alternatives, and provide recommendations on what action or actions to undertake. The team includes the risk analyst and representation from operations, maintenance, safety/health/environmental, plant process engineering, and transportation/distribution/logistics. Input can also be obtained as needed at the corporate level from the engineering and business group areas.

5.4 CASE THREE: DISTILLATION COLUMN

This case involves an existing plant that employs a distillation column to separate hexane and heptane. Close to the facility is a residential area of about 200 people. In response to management concerns, a risk analysis was undertaken. Figure 5.4 provides a schematic illustration of the system, and Figure 5.5 depicts the location of the distillation column with respect to the nearby population.

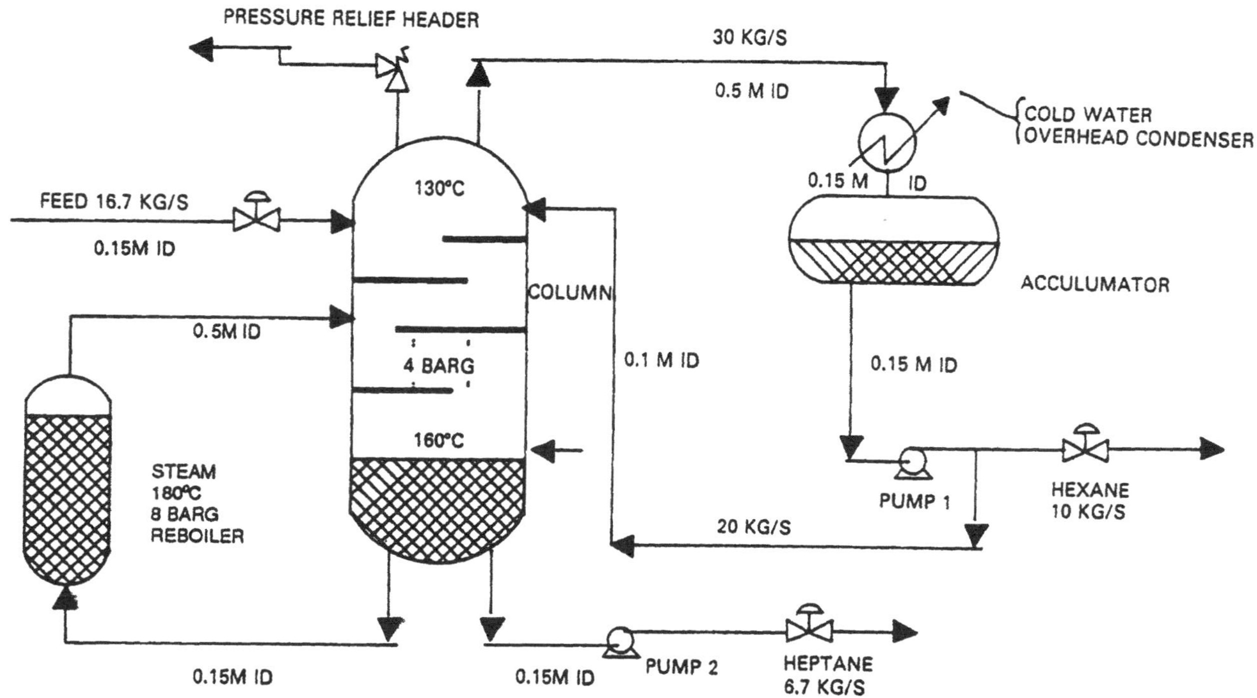

PIPING DIAMETER	TOTAL LENGTH
0.1 M	10 M
0.15 M	15 M
0.5 M	25 M

FIGURE 5.4. Diagram of distillation column system. (AIChE/CCPS 1989, p. 445)

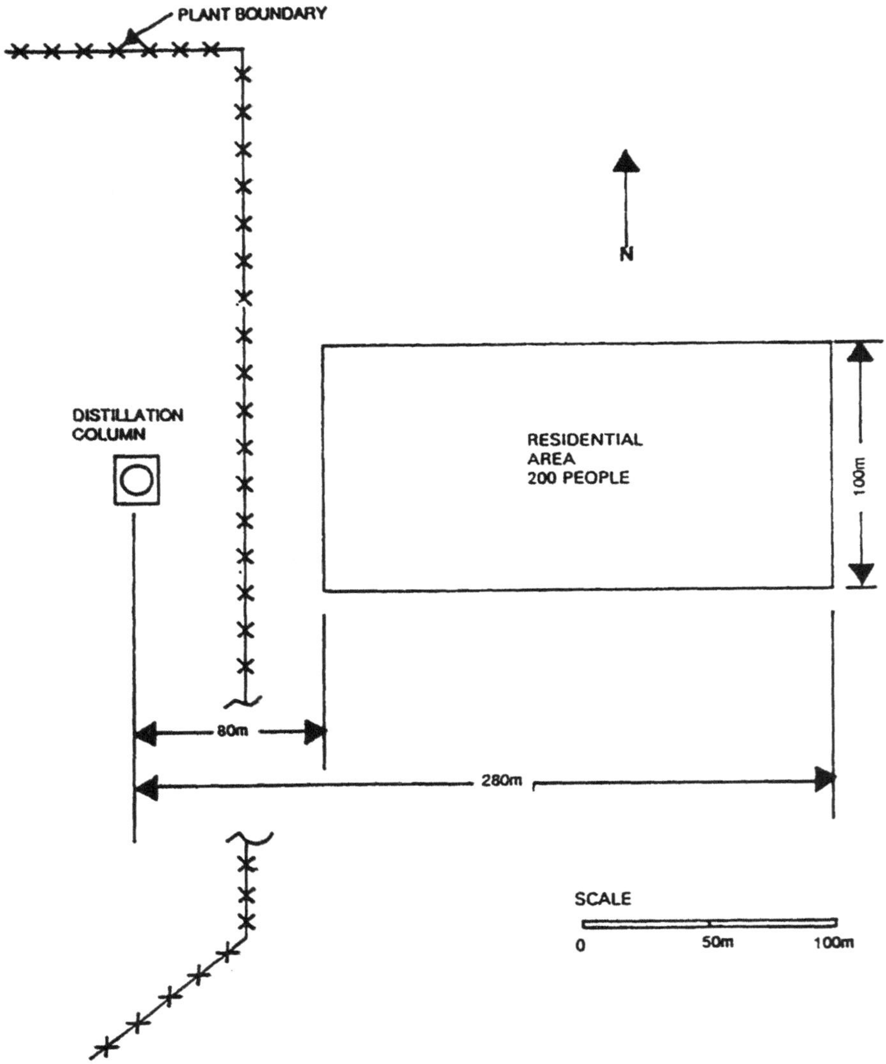

FIGURE 5.5. Layout of distillation column and nearby area. (AIChE/CCPS 1989, p. 446)

As in the preceding case, a risk estimate was prepared. This risk estimate focused on the risk to the residential community, and excluded possible concerns about worker safety that had been addressed at the time the distillation column was originally designed. Figure 5.6 depicts the types of effects considered in developing this risk estimate, and Figure 5.7 provides the

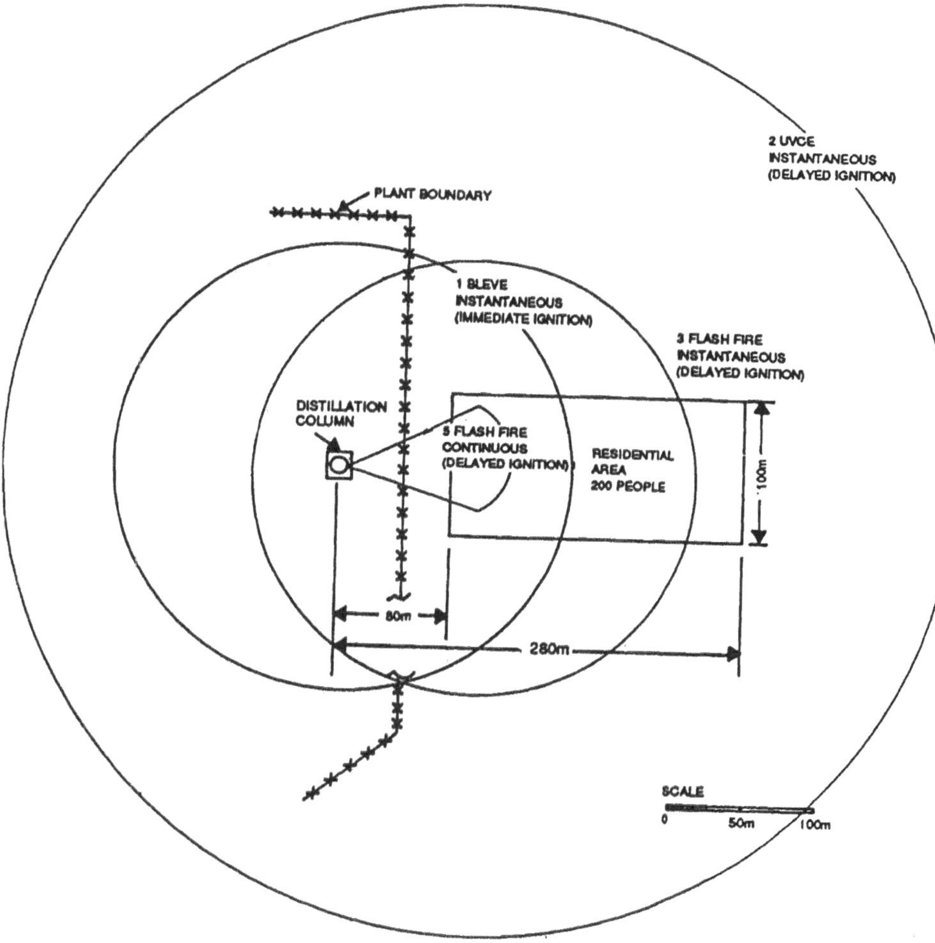

FIGURE 5.6. Effect zones addressed in risk analysis of distillation column. (AIChE/CCPS 1989, p. 466)

estimated F-N curve for the societal risk from the current operation. The effects addressed include a flash fire, a boiling liquid expanding vapor explosion (BLEVE), and an unconfined vapor cloud explosion (UVCE).

This case involves corporate-level issues due to the magnitude of the potential risk, and the importance of the distillation facility to this company's operations. An analysis team has been formed to recommend a course of action. The decision makers are a small group that includes the business director, the director of engineering, a member of the company's capital projects review body, and the operations general manager. The team will draw as necessary on the expertise of a risk analyst, a business analyst, a process engineer, a plant engineer, and safety and health professionals.

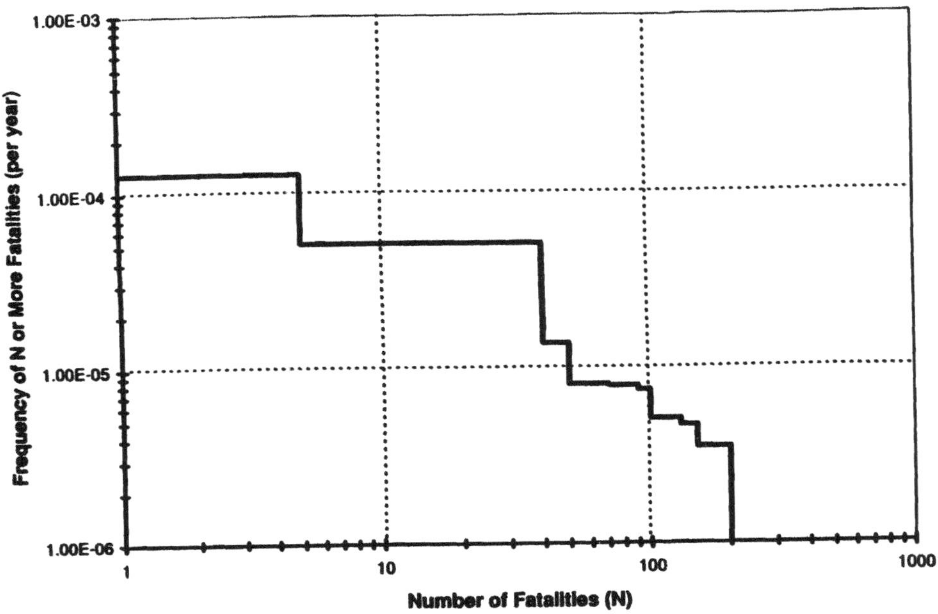

FIGURE 5.7. F-N curve showing results of risk analysis for distillation column. (AIChE/CCPS 1989, p. 476)

5.5 A ROAD MAP TO THE CASE STUDIES

Figure 5.8 provides a road map to guide the reader who wishes to follow a particular case study through the book. Following its introduction in this chapter, each case appears in two or three other chapters to illustrate the application of a specific decision aid to that problem. As necessary, the cases will be expanded and modified slightly to assist in demonstrating a particular decision aid. Finally, Chapter 13 revisits the cases by providing a summary that highlights how the decision aids vary in the approach and types of results obtained with each case.

5.6 REFERENCE

AIChE/CCPS (1989). *Guidelines for Chemical Process Quantitative Risk Analysis*. New York: American Institute of Chemical Engineers.

Discussion of Case Study	Case One Underground Pipeline	Case Two Chlorine Loading	Case Three Distillation Column
Introduction of case (Ch. 5)	✔	✔	✔
Application of decision aid to case			
Voting Methods (Ch. 6)	✔		
Weighted Scoring Methods (Ch. 7)			✔
Cost-Benefit Analysis (Ch. 8)		✔	
Mathematical Programming (Ch. 9)	✔		
Payoff Matrix Analysis (Ch. 10)		✔	
Decision Analysis (Ch. 11)	✔		
Multiattribute Utility Analysis (Ch. 12)			✔
Review of case (Ch. 13)	✔	✔	✔

FIGURE 5.8. Road map to case studies.

6

VOTING METHODS

So many men, so many opinions; every one his own way.
—***Terence***, *Phormio*

6.1 PURPOSE OF CHAPTER

Voting is by far the most common method for group decision making. Almost every business organization at some time uses some form of voting to set a course for future action. A board of directors, for example, might vote to select a new chief executive officer. A steering committee might vote to direct research and development funds. Even in the field of process safety, where the preference is to use codes, standards, and criteria for good engineering practices, a method as simple as voting has its place. Voting is a quick and economical method for turning the opinions of a group of individuals into a clear direction for future action.

In this chapter, we introduce the use of voting as a decision aid. All of us have had experience with voting, but there is much more to voting than meets the eye. As the examples in this chapter demonstrate, voting has more complexities than we often realize, particularly when we explore the implications of different voting rules. This chapter illustrates some of these complexities and shows how even simple voting systems can yield some surprising, non-intuitive results. We also describe the pros and cons of some of the more popular voting methods.

After we give an overview of voting, we introduce some criteria for judging among the different types of voting systems. We then illustrate how most of these voting systems, though intuitive and reasonable on the surface, nevertheless violate at least one of these criteria. After a brief discussion of additional topics in voting, we give an example of how we might arrive at a decision in the underground pipeline case study using some of the previously explained voting methods. This particular example illustrates how the use of different voting systems can lead to the adoption of different alternatives.

6.2 OVERVIEW OF VOTING METHODS

6.2.1 Introduction to Voting Methods

Voting is the most basic and popular method for group decision making. Groups of all sizes and types use voting, from large, representative political assemblies to business committees to informal groups of friends. It is difficult to think of a group decision-making situation in which voting is not used.

One reason why voting is so popular is its simplicity. In order to arrive at a decision through a vote, a group needs to follow only these steps:

- Each person in the group casts a vote according to his/her preference. The method of casting votes is often referred to as the *balloting method*.
- Votes are tallied according to a rule that establishes a winner. This rule is often called the *aggregation method*.

Neither the balloting method nor the aggregation method requires much time or resources. The results are arrived at quickly and easily.

Another reason why we often use voting is because we assume it is fair. Democratic traditions lend weight to voting as a just method for making decisions in group situations. From this perspective, voting not only facilitates decisions, but also empowers individuals in the decision-making process. Concepts of voting are therefore tied closely to ideals of democracy and a desire to give all participants an equal role in the decision.

Ironically, for the same reasons why voting is so popular, namely, simplicity and fairness, voting tends to be poorly understood by those who use it. Because casting and tallying votes seems so simple, we tend to believe that the results are likewise simple and transparent. Similarly, because we generally see voting as an inherently unbiased or "morally just" method for selecting among alternatives, we tend to think that all voting systems are essentially "fair" and "equal," and thus pay no attention to the particular voting method we employ. As we will see, neither of these perceptions is accurate.

The following two examples illustrate how complex and ambiguous voting can be.

6.2.1.1 Example 1: Voting by Favorite

A committee of four individuals is faced with selecting one of four alternatives to address a potential problem in process safety. For simplicity, we call the four alternatives *a*, *b*, *c*, and *d*. Preferences of the voters for the four alternatives are shown in Table 6.1.

The column under each person shows the four alternatives in order of that person's preference. This table shows, for example, that for Person 3, *b* is the favorite alternative and *a* is the least favorite alternative.

From the table, we can characterize the group's preferences as follows: While Proposal *b* is ranked first by only one person, it nonetheless has a solid base of support, with no individual ranking it lower than second place.

TABLE 6.1
Example for Voting by Favorite

	Person			
	1	2	3	4
Favorite	a	a	b	c
Second favorite	b	b	c	b
Third favorite	c	d	d	d
Least favorite	d	c	a	a

Alternative *c* has somewhat less support, and alternative *d* appears to be even less well-liked. The group's feelings about alternative *a*, however, are much more ambivalent; committee members are split as to whether this alternative is the best or worst among the set of alternatives.

Suppose the committee selects an alternative by having members vote for their favorite alternatives, a common method of voting. Using this method, alternative *a* is selected, as it receives 2 votes compared to one each for alternatives *b* and *c* and none for alternative *d*. While it is intuitively reasonable to have members select their favorite alternatives, this method leaves the committee with a chosen alternative that half of its members feel is the *worst* possible alternative.

Instead of determining a winner using only the favorite alternative of each committee member, the committee can take into account the ranking of *all* of the alternatives. A voting system that takes the complete rankings into account could somehow penalize alternative *a* for being so distasteful to some of the committee members. One method available to the committee is to select the alternative that holds the highest position *on average* across the voters. To employ this method of voting, the committee requests that voters rank the alternatives in order of preference. *Position points* are tallied for each alternative: one point for each voter who places it first, two points for each voter who places it second, and so forth. Dividing the total number of position points by the number of participants yields an average position for each alternative among the voters. The alternative with the lowest average position points is the winner. Applying this method to the above table of preferences yields a victory for alternative *b*.

This example shows how the choice of aggregation method can determine which alternative the committee chooses. Which of the sample voting methods above makes more sense? If the objective of the committee is to chose the alternative which most people feel is the single best alternative, then the first method might be better. If, on the other hand, the committee wants its vote to reflect a consensus about the alternative selected, then the second method might be more prudent. Without knowing the committee's objectives, the choice of which voting method to use is not at all clear.

6.2.1.2 Example 2: Voting Paradox

Consider another committee of three members voting on three possible alternatives. Table 6.2 presents their rankings of the alternatives.

The committee contemplates using the methods described above to determine the winner. The first method, that of recording each voter's favorite alternative, yields a tie, with one first-place vote for each of the competing alternatives. The method of assigning position points also results in a tie; each alternative receives a total of six position points, indicating that each alternative is ranked in second place on average across all voters. Unfortunately, neither method helps the committee determine a winner.

TABLE 6.2
Example for Voting Paradox

	Person		
	1	*2*	*3*
Favorite	a	*b*	c
Second favorite	*b*	c	a
Least favorite	c	a	*b*

Part of the problem may be that the votes are diluted because the committee is voting on three separate alternatives; one alternative may be *splitting the vote* of another (i.e., dividing votes among two similar alternatives). The committee decides to go back to the first method but use *head-to-head* voting (i.e., limit voting to only two alternatives at a time). Assuming that alternatives *a* and *b* are available (ignoring *c* for now) the committee prefers *a* to *b* by a margin of 2 votes to 1. Likewise, when *b* and *c* are the only alternatives considered, *b* is selected 2 votes to 1. These results seem to indicate that the correct ordering of the alternatives among the group is *a* then *b* then *c*, because *a* beat *b* head-to-head and *b* defeated *c* head-to-head.

The committee is inclined to believe these results until one of its members suggests that they try a vote considering only alternatives *a* and *c*. The results of that vote show that *c* is preferred to *a* by a margin of 2 votes to 1. The results of the entire series of votes finds that the committee prefers *a* to *b*, prefers *b* to *c*, but also prefers *c* to *a*! This result seems strange indeed. Each individual has distinct preferences among the alternatives, yet no matter how we vote on them, we cannot make a similar distinct ordering of preference for the group. And, no matter which alternative we select, there exists another alternative that is preferred by two out of the three voters.

This surprising result is known as the *voting paradox*. It has vexed voting theorists for over two hundred years. Much of the research on voting has been inspired by attempts to understand the nature of this paradox and develop methods for addressing it.

Suppose instead that the committee follows an agenda in which it casts votes on these proposals two-at-a-time. Two proposals are selected for a primary vote, and the winner is cast against the third proposal for the last vote. It does not take long to see that if the chairman (the agenda setter) knows the preferences of the other voters, then he can make any of the alternatives win by selecting which two alternatives participate in the initial election. For example, if the chairman (say, Person 3) wants alternative *c* to be selected, he could guarantee a victory for *c* by pitting *a* versus *b* in the first round, where

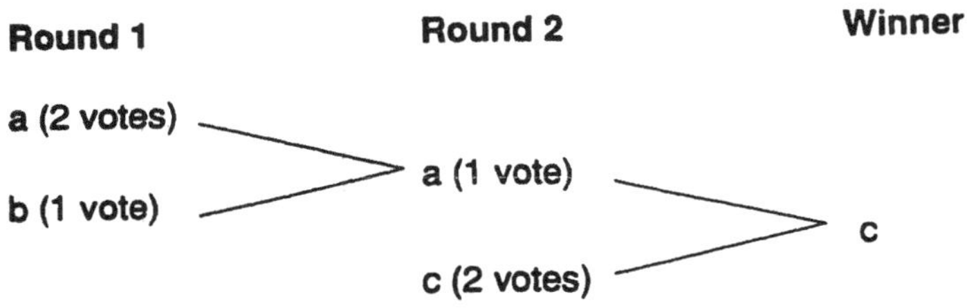

FIGURE 6.1. Voting two-at-a-time with an agenda.

a would defeat *b* two votes to one. In the second round, alternative *c* would defeat alternative *a* by the same margin. This situation is illustrated in Figure 6.1.

This state of affairs is particularly unsettling for Person 1, whose least favorite alternative wins in the end, despite the fact that his favorite alternative wins the first round. If Person 1 suspects Person 3 of manipulating the agenda in this fashion, then Person 1 might realize that by supporting alternative *b* in the first round instead of *a*, he can assure a victory by *b* in both rounds, assuming the others vote according to their preferences. The results of the two-round voting agenda is shown in Figure 6.2. Thus by strategically voting *against* his preferences (called *insincere voting*) in the first round, Person 1 can obtain a more desired result.

This example raises two more complexities of voting: first, the possibility that voters can manipulate the agenda to their benefit, and second, the possibility that some voters might be able to secure results they prefer by voting inconsistently with their preferences.

These examples serve to illustrate both the complexities of voting and the fact that different voting systems can, with the same set of voters and preferences, yield different choices. Which voting system to use is not always clear, and more importantly, not necessarily correct. In order to use the method of voting intelligently, it is important to understand the issues that surround voting as well as the relative benefits and difficulties presented by each voting system.

It is precisely because we take voting so for granted that we do not appreciate the variety of voting systems available to us, and the complexities they present. Our examples have raised a number of questions:

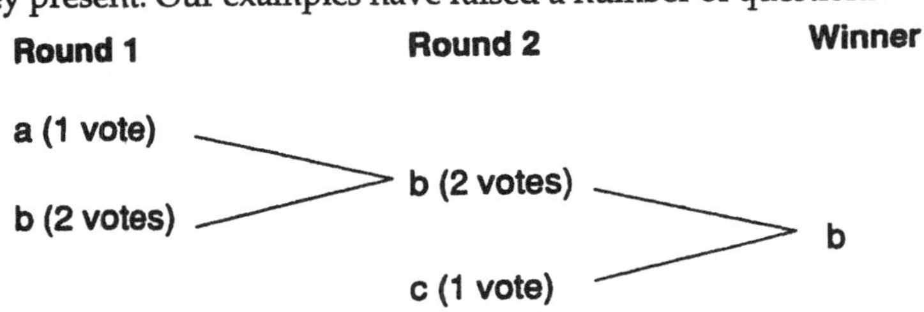

FIGURE 6.2. Alternative b wins the agenda vote when Person 1 votes insincerely in Round 1.

- ***What is the objective of voting?*** What do we wish to accomplish, vis-à-vis the preference of the group members, by voting? Is the objective to maximize the number of people who get their first choice alternative? Or perhaps to minimize the number of people who get their least favored alternative (i.e., develop a consensus)?
- ***How much do we need to know about each voter's preferences?*** How much information is contained in the vote that each person casts? Does a vote consist only of each person's favorite alternative? Or does a vote consist of all possible alternatives arranged in order of the voter's preferences? Is there additional information we could use?
- ***What method do we use to aggregate votes?*** Once we decide on the voting information to be obtained from each individual, how do we combine votes in order to yield a "decision" for the group?
- ***What is the agenda?*** Are all votes cast simultaneously or will alternatives be considered sequentially in smaller subsets, perhaps two at a time?

These questions are tied intimately with the larger question of what we want the decision process to achieve. Our lives would be simpler if there were obvious answers to these questions, and if these answers then pointed to a single voting method. Unfortunately this is not the case. There are many different voting systems, each with its own relative strengths and weaknesses.

6.2.2 History of Research on Voting

While voting itself has been used for millennia, a sophisticated study of the properties of different voting systems dates back only to the French Revolution. Two men from that era, the Marquis de Condorcet and Jean-Charles de Borda, each made significant contributions to the study of voting. Borda is credited as the first to develop a mathematical theory of elections. He saw the problems associated with our first example, where the ballot contains only the choice of most favored alternative. He proposed a method (of which the position points method is a variant) that takes into account the rankings among all alternatives. Condorcet, a contemporary of Borda, was among the first to illustrate the problems associated with our second example. These problem cases have come to be known as *cyclical majorities*.

Our own knowledge of these early pioneers in the study of voting and elections is due in part to Duncan Black, who, in 1958, published his book *Theory of Committees and Elections*. Black presented the first set of conditions that could be satisfied to prevent the problem of cyclical majorities. In addition to presenting original research, this book also serves as a reference to the works of Borda, Condorcet, and other early contributors.

Another major contributor in the twentieth century was the economist Kenneth Arrow, whose pioneering research helped define the field of social choice theory (Arrow 1963). Arrow proved an amazing result, which essentially states that there is no voting system that satisfies both certain "fairness" and "logic" conditions yet also prevents the occurrence of the voting paradox described above. His results tell us that the search for a perfect voting system is, ultimately, in vain.

Arrow's and Black's work opened a new era of research on voting. One of the more interesting results, due to Allen Gibbard (Gibbard 1973) and Mark Satterthwaite (Satterthwaite 1975), tells us that practically any system of voting is susceptible to strategic voting, as illustrated in Example 2 above. Other contributors to modern theory of voting include Amartya Sen (1970), Peter Fishburn (1973), and Robin Farquharson (1969).

6.2.3 Decisions/Situations Most Appropriate for Voting

The following are required to conduct a vote:

1. A set of alternatives for consideration
2. A group of individuals with preferences for those alternatives
3. A method for casting votes
4. A method for determining the winner from the votes cast.

You will notice that this list contains no formal information requirements for the alternatives. Voting can incorporate available information, but does not require that specific information be obtained. Consequently, voting may be most appropriate in the following situations:

- When information (or the time and resources to obtain information) is limited
- When all (or most) decision makers share a similar knowledge base or expertise level
- When the decision is sensitive to the values of the decision makers, rather than their state of information
- When there is little need to defend the logic or rigor of the decision
- When there is a desire to keep the basis for a decision confidential
- When a quick, first pass at a decision is desired to understand the feeling of the group.

In most voting methods, the preferences of all group members count equally in determining the chosen alternative. As we mentioned earlier, voting is not just a decision aid, it is a symbol of democratic institutions and reflects the sharing of power among all members of a group. Consequently, voting is also appropriate when:

- Decision-making responsibility is diffuse and shared equally among a group of individuals

- Management wants to develop a consensus about the chosen alternative
- Management wants buy-in to the decision through broader participation in the decision process.

Voting is a *group* decision-making aid that utilizes individual *preferences* of the group members. Any situation that emphasizes either group participation or the preferences of the individual group members may be appropriate for voting.

6.3 EXPLANATION OF VOTING METHODS

The earlier examples served to illustrate that even simple voting methods can lead to results that are nonintuitive or undesirable. Since Condorcet's discovery of the voting paradox (Section 6.2.1.2) in the late eighteenth century, voting experts (in most cases economists, political scientists and mathematicians) have spent considerable effort trying to understand why these situations arise, and what we can do to avoid them. They have tried to define precisely what we mean by "logical" and "intuitive" results. Is it possible to test a voting system against some standard that tells us how well the method performs?

In Section 6.3.1 we introduce a series of reasonable criteria that voting experts suggest a good voting system should obey. In Section 6.3.2 we introduce and compare a number of voting systems and show how each of these voting systems violates some of the basic criteria. From this introduction, it becomes apparent that there is no one perfect voting system. Rather, each voting system must be judged on its relative merits and failings. This brief introduction highlights some of the issues involved when comparing voting systems.

6.3.1 Key Concepts

The criteria we introduce are by no means a definitive list of all desirable properties that we may want in a voting system. They have been chosen because they are easy to explain and appear repeatedly in the voting literature.

6.3.1.1 Assumptions

To help simplify the presentation of the examples in this chapter, we make the following assumptions about each voter's preferences:

- Each voter can compare alternatives and express preferences among any alternatives presented
- Each voter has a strict preference for one of any two alternatives compared (i.e., there are no ties in the mind of an individual voter)

- Each voter's list of preferences has no *cycles* (i.e., if the voter prefers *a* to *b* and *b* to *c*, the voter *must* prefer *a* to *c*).

While these assumptions do not necessarily hold for all voting situations, they are reasonable enough approximations for our voting examples. Although there is a significant body of theory developed that addresses situations in which these assumptions do not hold, we will not address it in this book.

6.3.1.2 Some Desired Properties of Voting Methods

We will describe four desired properties of voting methods:

- Head-to-head criterion
- Monotonicity
- Independence from irrelevant alternatives
- Consistency.

HEAD-TO-HEAD CRITERION

The first criterion is the one proposed first by Condorcet and can be stated as follows. If there is an alternative which beats *every* other alternative in a two-way race, then that alternative should be selected the winner. Any alternative that defeats all other candidates in two-way races we call a *head-to-head winner* (in the voting literature this alternative is often called a *Condorcet candidate* or *Condorcet winner*). Clearly, a head-to-head winner need not exist in all voting situations; for example, there was no such head-to-head winner in the paradox discussed at the beginning of the chapter. The head-to-head criterion merely states that if such a winner exists, then the voting method must find it. This criterion seems extremely basic and natural. It may come as a surprise to the reader that many voting systems do not guarantee the choice of such a candidate.

A second part of this criterion considers the opposite type of alternative, one which loses all of the possible two-way races. It states that if such a *head-to-head loser* (or *Condorcet loser*) exists, then the voting system will *not* choose it as the desired alternative. Again, this seems intuitively reasonable.

> **HEAD-TO-HEAD CRITERION**
>
> ***Part I***
>
> If there exists an alternative which defeats every other alternative in a two-way race, then the voting system will select that alternative as the overall winner.
>
> ***Part II***
>
> If there exists an alternative which loses to every other alternative in a two-way race, then the voting system will not select that alternative as the overall winner.

MONOTONICITY CRITERION

The second criterion concerns how the intensity of support an alternative receives affects its selection. Perhaps our most basic expectation about voting is that an alternative's chance of winning should be directly related to the amount of support it receives. If an alternative receives an increase in support, then it should have a better chance of winning; if its support decreases, so too should its chance of winning. The second criterion states this expectation with respect to a winning alternative. Suppose our voting method selects alternative *a* as the winner. If a subsequent election is held in which all voters maintain their preferences except for a single group which increases its preference for alternative *a*, then alternative *a* should win again.

Like the head-to-head criterion above, this criterion also has a second part for losing alternatives: if one group decreases its preferences for a losing alternative while the others maintain their preferences, the voting system should not choose the previously losing alternative as the winner. While this criterion seems a little more complicated to state than the head-to-head criterion above, it is no less reasonable. This criterion is called *monotonicity*, because, in mathematical terms, the overall ranking of an alternative is "monotonic in" (i.e., always increases or decreases with) the amount of support it receives from the voters.

> **MONOTONICITY CRITERION**
> ***Part I***
> An increase in support for a winning alternative (keeping all other preferences the same) will not cause that alternative to lose.
> ***Part II***
> A decrease in support for a losing alternative (keeping all other preferences the same) will not cause that alternative to win.

If this criterion were not satisfied, voters might vote against their true preferences in order to secure a better result, much like what we saw in Example 2 (Section 6.2.1.2). Or worse, voters might lose confidence in both the voting system and the decision process itself. We will see, however, that one simple (and common) voting system has this very flaw.

INDEPENDENCE FROM IRRELEVANT ALTERNATIVES CRITERION

This criterion states that alternatives that do not win a vote can be ignored without changing the winner. For example, suppose *a*, *b*, and *c* are alternatives under consideration. If our voting procedure selects *a* over *b* and *c*, then re-casting votes considering *only* alternatives *a* and *b* will not cause *a* to be dethroned.

The benefit of satisfying this criterion should be clear. If a voting system did not satisfy this criterion, then astute voters could strategically introduce an irrelevant alternative (one that had little or no chance of winning) in order to make their favorite alternative win when it otherwise would not. For example, a voter could introduce a second alternative similar to the potential winner, and by splitting the vote for the potential winner, assure victory for that voter's own favorite alternative.

> **INDEPENDENCE FROM IRRELEVANT ALTERNATIVES CRITERION**
> Removing a nonwinning alternative from consideration will not change which of the other alternatives is selected as the winner.

CONSISTENCY CRITERION
This last criterion states that how we break up a committee into subcommittees should not determine which alternative we select. For example, suppose we split a large committee into subcommittees that can vote independently. If our voting method selects the same alternative in each subcommittee, then applying that method with all members voting at one time in the larger committee should result in the same alternative being chosen. If this property did not hold, then the way in which smaller groups are formed could have a decided effect on the eventual winner. In politics, an active attempt to take advantage of this situation is called *gerrymandering*.

> **CONSISTENCY CRITERION**
> Suppose a committee is divided into two groups, with each voter in exactly one group. A voting method that selects the same alternative in each group should select that alternative when that method is applied to the original committee.

6.3.2 Description of Voting Methods

In this next section we discuss some common voting systems. Our purpose in this section is to introduce some common methods of voting and to demonstrate how these seemingly reasonable voting methods violate some of the intuitive criteria described previously in Section 6.3.1. The conclusion of this section contains a table that summarizes our findings.

6.3.2.1 Plurality Method

The easiest way to conduct a vote is simply to ask for each voter's favorite alternative. This common method of voting is called the *plurality method*. Formally, the plurality method proceeds as follows:

- Each person places a single vote for that person's favorite alternative
- The alternative with the most votes wins.

When a committee is faced with exactly two alternatives, the plurality method guarantees that one of the alternatives wins a majority (or at least a tie). Unfortunately, with more than two alternatives, the plurality method cannot guarantee a majority winner; in this situation an alternative can easily win with fewer than 50% of the votes. Stated another way, the plurality method can often produce a winner that most voters do *not* prefer.

It may come as a surprise, but there may exist an alternative which beats all alternatives in head-to-head competition (a head-to-head winner), yet is not selected by the plurality method. Consider the example with five voters shown in Table 6.3. Alternative *b* beats every other alternative in a head-to-head competition, yet does not win under the plurality method because it only gets a single first-place vote. In fact, this example also exhibits that a winning alternative (*a*) can lose *all* of its head-to-head competitions! A brief look reveals the reason. While alternative *a* gets more first-place votes than any other alternative, it also gets more last-place votes as well. In contrast, *b* enjoys a relatively high level of support from all voters. The plurality method, with its concentration on first preferences, ignores this secondary support for *b* and the level of *dis*satisfaction many committee members have with alternative *a*.

Another problem with the plurality method is that it can encourage the same type of insincere voting we saw in Example 2 above in Section 6.2.1.2. If voters 4 and 5 in the previous example know ahead of time which alternatives voters 1, 2, and 3 support, they have an incentive to vote for alternative *b*, against their true preferences, to secure a victory for their second-favorite alternative instead of *a*, their least-favorite.

6.3.2.2 Borda Count Method

Borda was familiar with the problems associated with the plurality method and suggested the following procedure now called the *Borda count*, to account

TABLE 6.3
Example for Plurality Method

	Person				
	1	2	3	4	5
Favorite	a	a	b	c	d
Second favorite	b	b	c	b	b
Third favorite	c	d	d	d	c
Least favorite	d	c	a	a	a

for the entire range of voter preferences. Assuming there are N alternatives under consideration, the steps are as follows:

- Each individual ranks all alternatives.
- Each alternative is then awarded N points for each person who ranked it first, $N-1$ points for each person who ranked it second, etc., down to 1 point for each person who ranked it last.
- The alternative with the most points is the winner.

The obvious advantage of the Borda count (and its historical motivation) is that it takes into account each person's relative preferences for all of the alternatives when determining the winner. The Borda count considers more information than the plurality method does, and for that reason has a great deal of intuitive appeal. (You may recognize that this method is equivalent to the position points method described earlier.)

The Borda count is an improvement over the plurality method, but still does not satisfy two of the criteria. Although it takes into account more preference information than the plurality method does, it still cannot guarantee selecting a head-to-head winner if one exists. And because each alternative is judged on its position relative to the other alternatives, "irrelevant" alternatives can have an effect on the eventual Borda winner. The example in Tables 6.4 and 6.5 illustrates this point.

TABLE 6.4
Example for Borda Count Method

	Person				
	1	2	3	4	5
Favorite	a	a	b	c	c
Second favorite	b	e	c	e	e
Third favorite	c	b	d	b	b
Fourth favorite	e	c	e	d	d
Least favorite	d	d	a	a	a

Alternative	First	Second	Third	Fourth	Fifth	Total Points
a	2	0	0	0	3	(2 × 5) + (1 × 3) = 13
b	1	1	3	0	0	(1 × 5) + (1 × 4) + (3 × 3) = 18
c	2	1	1	1	0	(2 × 5) + (1 × 4) + (1 × 3) + (1 × 2) = **19**
d	0	0	1	2	2	(1 × 3) + (2 × 2) + (2 × 1) = **9**
e	0	3	0	2	0	(3 × 4) + (2 × 2) = **16**

TABLE 6.5
Example for Borda Count Method Showing Effect of Removing Alternative e

	Person				
	1	2	3	4	5
Favorite	a	a	b	c	c
Second favorite	b	b	c	b	b
Third favorite	c	c	d	d	d
Least favorite	d	d	a	a	a

Alternative	First	Second	Third	Fourth	Total Points
a	2	0	0	3	$(2 \times 4) + (3 \times 1) =$ **11**
b	1	4	0	0	$(1 \times 4) + (4 \times 3) =$ **16**
c	2	1	2	0	$(2 \times 4) + (1 \times 3) + (2 \times 2) =$ **15**
d	0	0	3	2	$(3 \times 2) + (2 \times 1) =$ **8**

Using the Borda count, alternative *c* wins the election narrowly over alternative *b*. Suppose, however, that alternative *e* is removed from consideration. We hope that alternative *c* would again prove victorious. Unfortunately, this is not the case. The Borda count emerging from the modified vote (Table 6.5) yields alternative *b* as the winner.

The reason for this behavior is that removing one alternative allows other alternatives to move up in the rankings of individual voters. If an alternative can move up more often than a competitor (i.e., move up in the lists of more voters), that alternative gains a relative point advantage when the point totals are recalculated. In the example above, alternative *b* moves up on three voters' lists, while alternative *c* only moves up on one voter's list. Alternative *b* gains a two (3 minus 1) point advantage over *c* by the exclusion of alternative *e*. This increase is enough to propel *b* to victory.

Thus the Borda count encourages strategic behavior by voters, not only for introducing additional alternatives, but also in revealing their preferences. Voters have an incentive to rank as last those alternatives that closely compete with their favorite alternatives, rather than giving those alternatives their true ranking.

6.3.2.3 Approval Voting Method

One problem with ranking methods in general is that they use only the *order* of preferences among alternatives and not the *strength* of preference. Consider Person 2 and Person 3 in the example above, shown together in Table 6.6 below.

TABLE 6.6
Example for Approval Voting Method

	Person	
	2	*3*
Favorite	*a*	*b*
Second favorite	*b*	*c*
Third favorite	*c*	*d*
Least favorite	*d*	*a*

These rankings tell us that alternative *c* was ranked third by Person 2 and second by Person 3. The rankings do not indicate that Person 3 actually likes alternative *c* more strongly than Person 2 does. Person 2 might be amenable to all of the alternatives, whereas Person 3 might find all of the alternatives except *b* to be repulsive. Ranking methods cannot take this type of strength-of-preference information into account.

Approval voting is a relatively new voting method that attempts to correct this deficiency. Approval voting consists of two steps:

- Each voter casts a single vote for each alternative he deems "acceptable." Voters may cast a vote for as many alternatives as they wish.
- The alternative with the most votes wins.

Approval voting allows the voter to select which alternatives get votes, though the voter is limited to giving a single vote to each alternative chosen. The winning alternative is the one that receives the largest number of "approval" votes (i.e., the alternative that the most voters approve).

The benefits of approval voting have been elucidated by its proponents, most completely by Brams and Fishburn (1983). The authors point out that approval voting, more than other procedures, encourages voters to be honest in revealing their preferences. Voters gain little or no benefit either from placing additional votes for candidates they do not like or from withholding votes from alternatives they do like. As far as our criteria are concerned, the most serious failing of approval voting is the fact that it does not necessarily select an existing "head-to-head" winner. The example in Table 6.7 illustrates. The bold italics in the table indicate the dividing line between "approved" and "unapproved" alternatives for each person. Counting up votes shows that alternative *a* wins with four total votes. Alternative *b*, which beats every other alternative in a head-to-head competition, loses the election.

TABLE 6.7
Example for Approval Voting Method Showing Violation of Head-to-Head Criterion

	Person				
	1	2	3	4	5
Favorite	a	*b*	*b*	*d*	c
Second favorite	**c**	**a**	c	a	***d***
Third favorite	*b*	c	**a**	***b***	*b*
Least favorite	*d*	*d*	*d*	c	a

It should be pointed out though, that in the framework of approval voting, such a head-to-head test may not be very meaningful. For Person 1, alternative *a* is given credit with defeating alternative *b*, while the opposite is true for Person 2. Proponents of approval voting would not consider these preferences comparable. Person 1 indicates approval of *a* and disapproval of *b*, while Person 2 approves of both, but ranks *b* higher than *a*. An approval vote consists of information that is not contained in the pure rankings of the alternatives and therefore the head-to-head criterion, based on rankings, may not be relevant.

One additional benefit of approval voting is that it is the only method described here that does satisfy the independence from irrelevant alternatives criterion. Removing a nonwinning alternative cannot take away support from a winning alternative or add support to a losing alternative and hence such a change cannot affect the alternative selected.

6.3.2.4 Copeland Method

The Copeland method thoroughly retains the spirit of the head-to-head criterion. It considers all pairwise comparisons between ranked alternatives. In the Copeland method,

- Each person ranks all alternatives.
- Results are tabulated from simulated, pairwise comparisons between each pair of alternatives. Each alternative scores a "win" when it beats another alternative in a pairwise contest, and scores a "loss" when it loses to another alternative in a pairwise contest.
- Each alternative receives a score equal to the difference between its number of wins and its number of losses. The alternative with the highest score wins.

The Copeland method is the only method introduced in this section that satisfies both parts of the head-to-head criterion. It does, however, violate the consistency criterion. Consider the two committees, Red and Blue, shown in Table 6.8. The Red Committee selects alternative *a* using the Copeland procedure. The vote of the Blue Committee results in a tie for alternatives *a*, *b*, and *c*. From these results we would expect that alternative *a* would be selected by a joint body consisting of both committees. However, Table 6.9 shows that performing the Copeland method calculations on the larger group selects alternative *b*. This directly violates the consistency criterion.

In counting "wins" the Copeland method considers only whether one alternative beats another one; it does not take into account the margin of such victory. Consequently, the Copeland method can arrive at a different winner than the Borda method, and in fact, its winner can lose its pairwise contest with the Borda winner (for an example, see Niemi and Riker 1976). According to Straffin (1980), another unfortunate aspect of the Copeland method is that it tends to produce more ties than the other methods presented here. Nevertheless it is a popular method due to the fact that it performs so well on the head-to-head criterion.

6.3.2.5 Hare (or Single Transferable Vote) Method

One of the problems identified with the plurality method is that it can lead to selection of an alternative that has the support of only a minority of voters. The *Hare method* forces a majority result to emerge from the plurality method through the following process:

- Each individual ranks all alternatives.
- Each alternative receives one vote for each ballot that places it in first place.
- If one alternative has a majority of votes, it wins.
- If no alternative has a majority of votes, the alternative with the fewest first-place votes is removed and its first-place votes transfer to the alternative ranked next on the voter's ballot.
- These last two steps are repeated until one alternative has a majority of the first-place votes.

Like the plurality method upon which it is based, the Hare method does not necessarily select a head-to-head winner if one exists. This method's primary benefit is that it will choose a candidate that has, in some sense, the support of the majority of voters. This benefit comes at a cost. The process of developing majority support introduces a violation of the monotonicity criterion. That is, an increase in support for an otherwise winning candidate can cause it to lose. Consider the example in Table 6.10 taken from Straffin (1980).

In this example, voters fall into four groups that contain individuals with similar preferences. Using the Hare method, alternative *c* is eliminated first because it has the fewest first-place votes (only 5 votes for *c* versus 6 for *a* and

TABLE 6.8
Red and Blue Committees Vote Using the Copeland Method

RED COMMITTEE

	Group of Persons		
	I	*II*	*III*
(Number of Persons in Group)	*(17)*	*(8)*	*(5)*
	a	b	c
	b	c	a
	c	a	b

Result of Pairwise Contests:
a beats *b* (22 to 8)
a beats *c* (17 to 13)
b beats *c* (25 to 5)

Copeland Scores
a = 2 – 0 = 2 so a wins
b = 1 – 1 = 0
c = 0 – 2 = –2

BLUE COMMITTEE

	Group of Persons		
	IV	*V*	*VI*
(Number of Persons in Group)	*(14)*	*(16)*	*(15)*
Favorite	a	b	c
Second favorite	c	a	b
Least favorite	b	c	a

Result of Pairwise Contests:
a beats *c* (30 to 15)
b beats *a* (31 to 14)
c beats *b* (29 to 16)

Copeland Scores
a = 1 – 1 = 0 so *a, b, and c* all win
b = 1 – 1 = 0
c = 1 – 1 = 0

TABLE 6.9
The Joint Committee Votes Using the Copeland Method

JOINT COMMITTEE						
	Group of Persons					
	I	II	III	IV	V	VI
(Number of Persons in Group)	(17)	(8)	(5)	(14)	(16)	(15)
Favorite	a	b	c	a	b	c
Second favorite	b	c	a	c	a	b
Least favorite	c	a	b	b	c	a

Result of Pairwise Contests:
- a beats c (47 to 28)
- *b* beats a (39 to 36)
- *b* beats c (41 to 34)

Copeland Scores
- a = 1 − 1 = 0
- b = 2 − 0 = 2 so *b* wins
- c = 0 − 2 = −2

6 for *b*). Removing *c* transfers five more votes to alternative *a* and it wins with a majority. Now suppose that the last group finds alternative *a* to be *more* palatable and revises its rankings so that *a* is preferred to both *b* and *c*. The revised preferences are shown in Table 6.11.

Due to the decreased support for *b*, this alternative gets eliminated in the first round instead of alternative *c*. All of *b*'s first-place votes go to alternative *c* and now alternative *c* wins with a majority of 9 first-place votes. This example clearly shows how an increase in support for a chosen alternative may cause the Hare method no longer to choose that alternative.

The Hare method seems an intuitive way to remove alternatives that are favored by few voters. Unfortunately, the Hare method violates not just the head-to-head and consistency criteria, but our other two criteria as well. It is a good example of a reasonable voting procedure that can have very unexpected consequences in practice.

6.3.2.6 Summary

None of the systems reviewed here is ideal. The plurality method considers only each person's favorite and can therefore choose an alternative that has the most first-place votes but no other support. Borda's method takes into account more information about the voters' preferences, but the way it uses

TABLE 6.10
Example for Hare Method

	Group of Persons			
	I	*II*	*III*	*IV*
(Number of persons in group)	(6)	(5)	(4)	(2)
Favorite	a	c	b	b
Second favorite	b	a	c	a
Least favorite	c	b	a	c

TABLE 6.11
Example for Hare Method Showing Violation of Monotonicity Criterion

	Group of Persons			
	I	*II*	*III*	*IV*
(Number of persons in group)	(6)	(5)	(4)	(2)
Favorite	a	c	b	a
Second favorite	b	a	c	b
Least favorite	c	b	a	c

this information introduces the possibility that alternatives with little or no chance of winning can determine the winner. The Hare modification of the plurality method has the problem that an increase in support can cause an otherwise winning candidate to lose. The Copeland method performs well on the most intuitive criteria—head-to-head and monotonicity—but it remains susceptible to violations of the consistency criterion. Approval voting is unique here in its attempt to incorporate strength of preference, yet it violates the head-to-head criterion. Table 6.12 compares the results.

6.3.3 Some Additional Issues

6.3.3.1 Single-Peaked Preferences

The previous section highlighted the theoretical shortcomings of the sample voting systems described. We showed that there was no system that was

TABLE 6.12
Summary of Voting Methods' Satisfaction of Criteria

Criterion Satisfied:	*Plurality*	*Borda*	*Approval*	*Copeland*	*Hare*
Head-to-Head	No	No	No	Yes	No
Monotonicity	Yes	Yes	Yes	Yes	No
Independence from Irrelevant Alternatives	No	No	Yes	No	No
Consistency	Yes	Yes	Yes	No	No

perfect; each had its own idiosyncrasies of which one should be aware. With our search for a perfect voting system empty, we turn our attention to some additional practical issues. Granted, some of these methods can produce unintuitive results on some carefully thought-out examples, but do they perform well in practice? Are the counter-examples we gave likely to occur in a real voting situation? Unfortunately we can never know for certain; there is no laboratory in which we test voting habits. But there are some results that are encouraging. In particular, there seems to be some evidence that the paradox of voting described earlier may not be as pervasive as we might fear. The natural question arises, "Under what conditions do we know that cyclical majorities cannot exist?" The first, and most famous result, is due to Duncan Black (1958). He showed that there exists a property of voters' preferences, called *single-peakedness*, that ensures that the voting paradox discussed earlier does not occur.

Single-peakedness is a characteristic of preferences that is easy to depict graphically. We say that preferences are single-peaked when we can arrange the alternatives in such a way that when we make a graph of the alternatives versus their rankings, the plot has only a single peak in it. The idea is best illustrated with an example. Suppose that Person 1 ranks alternatives a, b, and c in the preference order b, c, a. We can plot these alternatives against their rankings in that same order to obtain a single-peak for Person 1. We could also plot them on the x-axis in the order c, b, a to obtain a single peak. These examples are illustrated in Figure 6.3. Notice, however, that the order b, a, c, as illustrated in Figure 6.4 does not produce a single peak. It is easy to see that for any given voter, we can always find an ordering that produces a single peak. The interesting result happens when we consider a group of voters. We say that a group of voters is characterized by single-peaked preferences if there exists a single ordering of the alternatives that makes all voters' preferences have single peaks when graphed simultaneously. This situation is illustrated in Figure 6.5.

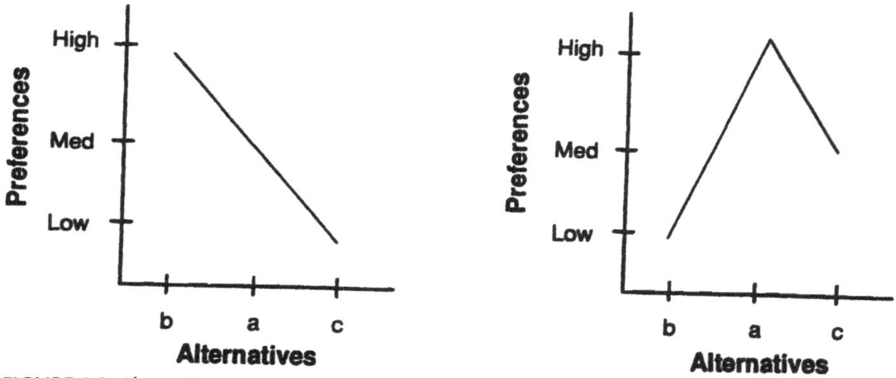

FIGURE 6.3. Alternatives arranged to yield single peaks.

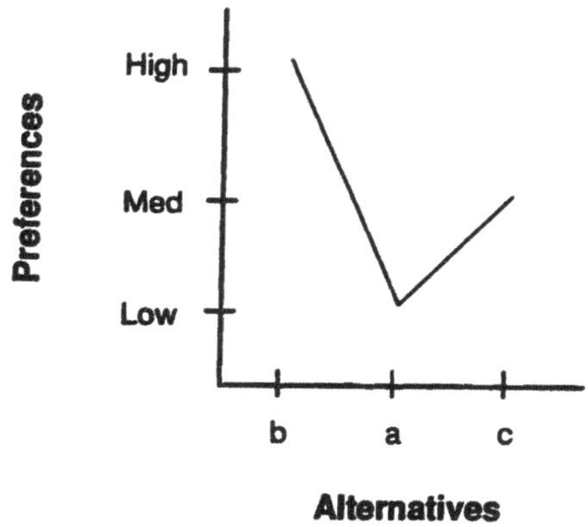

FIGURE 6.4. Arrangement that does not yield a single peak.

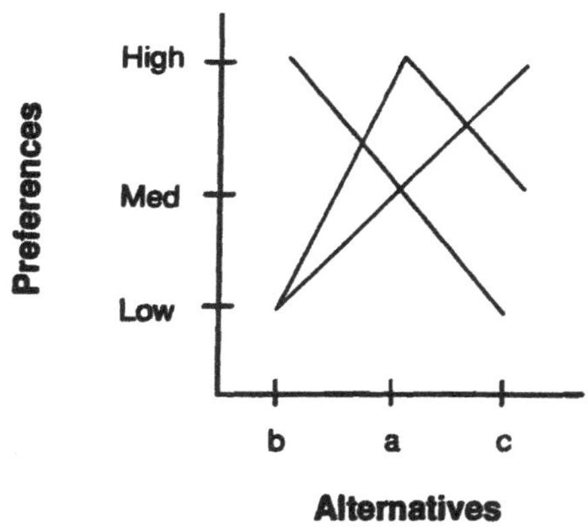

FIGURE 6.5. Single-peaked preferences for a group of three voters.

Duncan Black proved that if a group of voters is characterized by single-peaked preferences, then the aforementioned voting paradox cannot occur. The practicality of this result becomes apparent if we try to interpret the meaning of single-peakedness. Single-peakedness is an indication that the alternatives can be arranged on a scale such that each person's preferences decrease as he or she gets farther away from their preferred alternative (this applies to either direction, but independently). In some sense, it suggests that the alternatives can be placed on a single scale that is recognized as common to all voters. Each voter's preference for an alternative is equivalent to a preference for being located somewhere on the scale.

Consider, for example, a committee that votes on several alternative levels of funding for a safety-related project. It seems reasonable that while each person may have a preferred alternative, given that the funding is either higher or lower than the preferred, each would probably prefer funding levels closer to his or her favorite over funding levels farther away from that favorite. (Note that this comparison must be made only to alternatives that are on the same side of the peak.) Thus the committee members would probably have single-peaked preferences and the voting paradox would not occur in practice.

Black's result implies that if we believe in most cases of voting that alternatives can be organized along some easily recognizable single dimension, such as cost, overall risk, or even "liberal–conservative," that some of the theoretical problems with voting methods would probably disappear in practice. Black, in fact, even proved something stronger. If there is an odd number of voters, and they exhibit single-peaked preferences, then the alternative selected will be the one under the median peak.

Obviously, this result is not as good as a voting system that prevented voting paradoxes, but as we have discussed, this is not possible. Black's work on single-peakedness has inspired two major areas of voting study. One area is the simulation of voting systems. The single-peakedness result led others to question how different voting systems perform in practice. As genuine voting experiments are difficult to conduct, they settled instead on using computer simulation techniques to generate random voters and alternatives. These techniques apply statistics to see how well various voting rules perform. Another area of study concerns *spatial theories* of voting. If Black could arrange alternatives along a single dimension, could it not be possible to arrange alternatives along multiple dimensions (in "space") to get a more accurate representation of how voters make decisions? Though both areas have been fruitful, neither has yet presented results more fundamental than Black's original result.

6.3.3.2 Strategic Voting

In earlier examples, we illustrated how some of the voting systems are susceptible to *strategic voting*. In this section we summarize some information about strategic voting.

The three most common forms of strategic voting are *insincere voting, agenda manipulation* and *vote-trading*. Insincere voting is a particularly disturbing phenomenon because it strikes directly at the heart of why voting is used in the first place: to elicit true preferences on the alternatives at hand. The susceptibility to insincere voting depends on a variety of factors including the:

- Extent to which voter's preferences are known to others
- Sophistication of the voting system
- Voters' knowledge of how voting systems can be manipulated.

The more voters know about other voter's preferences, the more incentive they have to use an insincere vote. If voting is preceded by long deliberations by a committee, this information may be available to all voters. However, the usefulness of this information depends on the voting method selected. In the plurality method, a person need only know first preferences in order to understand how changing a vote might change the results. The calculations for the Copeland method, however, make insincere voting more involved. Not only does a voter need to know the entire preference structures of other voters, but a voter must also be willing to do the nontrivial calculations in an attempt to determine an insincere voting strategy. Few persons are likely to be interested in doing the requisite mental gymnastics.

The serious effects of agenda manipulation were illustrated in Example 2 of Section 6.2.1.2. The example supports the intuitive observation by Black that in a pairwise sequential voting procedure, the alternative put on the agenda last has an advantage, in that it faces fewer battles. This is equivalent to the advantage given a sports team that earns a bye (victory without competing) at an early stage of an elimination tournament. A competing alternative that may beat the preferred alternative might itself lose in an earlier round. Controlling the order of introduction of motions or alternatives can clearly be a benefit. The most notable exception to this situation is a head-to-head winner; it wins all of its pairwise contests and therefore cannot be manipulated into losing in this fashion. In some committee procedures the agenda serves to select an alternative which is then put up against the status quo. This is known as the *status quo bias*. The other problem with this type of agenda manipulation, as shown in Example 2, is that it encourages insincere voting, manipulation of a different sort. The other method for manipulating the agenda, which is an issue even when not using sequential pairwise voting, is to attempt to divert support from another alternative. Most voting systems are susceptible to this type of manipulation.

Finally, we should make a brief mention of *vote trading*. The examples in this chapter have focused on a single decision to be made by voting. In the situation where there are multiple issues to be decided, it is possible that two individuals can obtain better results for themselves if they both vote against their true preferences to help each other. The problem, of course, is that the private gain of the two vote traders may (though not necessarily) come at the expense of finding the best alternative for the entire group.

6.3.3.3 Voting Power

We earlier described the relationship between voting and democratic ideals. Voting systems can, however, reflect differences in power between individuals. Consider, for example, a committee chairman who is granted the ability to break ties. Or a steering committee that must approve alternatives before they are put up to a vote. Clearly these are examples of individuals who are more empowered than the regular voters. Is it possible to quantify in some way how powerful these positions are?

Several voting experts have developed mathematical formulas that quantify differences in voting power of individuals or voting blocs. The difficulty is making a concrete definition of what is meant by power. One method, by Shapley and Shubik, roughly ascribes power to the probability that a person's vote will be crucial in determining an outcome; that is, changing that person's vote would change the outcome. Straffin (1980) gives example calculations that reveal some surprising results. The tie-breaking chairman of a committee of four individuals, for example, has three times the voting power of the other committee members. In a similar analysis, Straffin shows that the Shapley–Shubik power figures estimate that each member of a three-member steering committee taken from a nine-member voting body has four times the power to affect the results as any of the other six members. These calculations can be useful in designing the structure of a voting body.

6.4 CASE STUDY: UNDERGROUND PIPELINE

An Application of Voting to Chemical Process Safety

This application involves Case One, the underground pipeline, which was introduced in Section 5.2; there are two additional applications of this case later in the book that illustrate the use of mathematical programming (Section 9.4) and decision analysis (Section 11.4).

This application involves a decision addressing the risks of an underground pipeline suspected of being susceptible to leaking. The company experienced a leak of a pipeline at one of their processing plants. The pipeline carries a highly flammable material that is a potential environmental contaminant. Corporate management, concerned that the same risk may exist at their other facilities, sent a memo to plant managers at all the facilities expressing its concern. The memo informed plant managers that when analyzed, the damaged section of pipeline was found to have much greater internal corrosion than anticipated. Corporate management left it to the individual plant managers to "determine whether the potential for similar incidents exists at your facility, and if so, to undertake appropriate actions."

The plant manager of one of the targeted facilities has received the memo about the risks associated with the underground pipeline. Wishing to incorporate as many perspectives as possible into the decision process, the plant manager convened a committee of senior employees in several areas of expertise: engineering, safety, and management. The committee consisted of five individuals: the business manager; the chief plant engineer; the plant environmental, health, and safety (EHS) professional; the assistant plant manager; and the plant manager. The committee was commissioned to discuss alternatives available to address the problem and then to vote to select one alternative. Each member of the committee was to be given an equal vote.

The committee meeting began with a briefing by the chief plant engineer. There was no current leakage, so there appeared to be no emergency safety concern. Most of the pipelines were relatively new. However, one section of pipeline was only five years younger than the section of pipeline that leaked at the other facility. It was unclear whether internal corrosion alone—as opposed to other factors—contributed to the previous leak. The chief plant engineer outlined two approaches available to the company. One approach was to replace the pipeline. Doing this would remove risk due to the previous pipeline, but at a significant cost for purchase, installation, and downtime as the pipeline is replaced. Also, there was a slight risk associated with the process of replacing the pipeline.

A second approach was to insert automated shutoff valves around the suspect section of pipeline to provide an additional measure of security. The valves automatically stop the flow of liquid whenever the pressure in the pipeline decreases precipitously. The valves would minimize damage from a large-size leak or rupture in the pipeline, but would have little effect on a small, extended leak. However, continuous monitoring of the area surrounding the pipeline would alert the company to any chemical release occurring over an extended time. This approach had a cost of about 5% of the cost to replace the entire segment of pipeline and also had a slight risk associated with excavation and installation.

Besides these approaches, the chief plant engineer explained, the committee could consider performing some additional tests. However, even conducting testing would not be inexpensive or without some risk. The least expensive method of gaining additional information would be to test a section of the pipeline with ultrasound to determine its approximate thickness. This test could be performed without any downtime at the plant. Alternatively, a metallurgical analysis could be performed on a removed section of the pipeline. This test would provide better information, but with some minor downtime during removal of the pipeline section. During these tests, of course, the surrounding community would be at some unquantified risk.

After some discussion the committee agreed to consider four alternatives:

- *Replace pipeline:* Replacing the pipeline is the most expensive option, but would result in the greatest long-term reduction of risk.
- *Insert Valves:* Installing automatic shutoff valves could be done at about 5% of the cost of pipeline replacement, but would probably be redundant when the pipe is eventually replaced.
- *Metallurgy Test:* A short section of pipeline would be removed, necessitating a brief plant shutdown.
- *Ultrasound Test:* A nondestructive, in-line analysis could be done with an ultrasound test, at the least cost of the options considered.

At the end of the discussions and before voting, the plant manager asked each committee member to summarize his or her position:

- *Business Manager:* The business manager was concerned about disruption of business during any required down time. No drastic action should be taken without performing additional tests, preferably with the plant still in operation. The likelihood that a similar problem would occur in the immediate future seemed remote, especially considering that the pipeline in question was a full 5 years younger than the original pipeline that leaked. If something must be done, the Business Manager preferred to insert the valves. This option would provide short-term security and allow them to delay a more costly replacement of the pipeline.
- *Process Safety Professional.* The process safety professional warned the other committee members that the pipeline was an unknown hazard and that it posed a potential threat to the surrounding community. The position of the business manager was too focused on immediate financial concerns. If a major leak were to occur from this pipeline and people were injured, the fact that the company had not acted on information from a previous accident would be much more damaging. A further concern was that the possible environmental damage from a slow leak that went undetected could itself result in a significant risk. The process safety professional favored replacing the pipeline, or at the very least performing a structural integrity test on the pipeline. If the plant must remain in operation, then they should at least perform a nondestructive test of the pipeline thickness.
- *Chief Plant Engineer.* The chief plant engineer felt that the most serious issue was reducing the short-term risk of a *major* leakage and resulting fire. The automatic shutoff valves would drastically reduce this possibility. Installation of the valves would limit any problem over the next few years to a very minor one and would allow the plant to maintain operations. Alternatively, the Engineer favored replacing the pipeline without an extended study. The pipeline would approach the end of its useful life in the next five years anyway.

- *Assistant Plant Manager.* The assistant plant manager believed that the pipeline should be replaced even if it were not currently damaged, given the reasonably short expected future life of the pipeline. The company could avoid much extra effort and expense by replacing the pipeline now. If performing additional tests would convince people of the need for replacement, then those tests should be done. Throwing away money by installing new shutoff valves on an aging pipeline would not solve the essential problem: the pipeline would be unlikely to last its rated life.
- *The Plant Manager.* The plant manager concurred with the chief engineer. The top priority is to immediately take steps to prevent the worst possible scenario, while imposing only minor disruption of plant operations.

The committee members' preferences can be summarized as shown in Table 6.13. The committee decided to vote using the plurality method, described in Section 6.3.2.1. When the members voted for their preferred alternatives, the result was a tie between Insert Valves and Replace Pipeline, as shown in Table 6.14.

To break the tie, the committee deleted the options that received the fewest first-place votes—Ultrasound Test and Metallurgy Test—and voted again. The second round of voting gave the most votes to Install the Shutoff Valve, as the business manager switched a vote from Ultrasound Test to Insert Valve. The voting results are summarized in Table 6.15. It is interesting to note that three of the five committee members listed the winning alternative either third or fourth.

TABLE 6.13
Committee Member Preferences for Pipeline Alternatives

	Committee Members				
Order of Preference	*Business Manager*	*Plant Process Safety Professional*	*Chief Plant Engineer*	*Assistant Plant Manager*	*Plant Manager*
First	Ultrasound Test	Replace Pipeline	Insert Valves	Replace Pipeline	Insert Valves
Second	Metallurgy Test	Metallurgy Test	Replace Pipeline	Ultrasound Test	Replace Pipeline
Third	Insert Valves	Ultrasound Test	Metallurgy Test	Metallurgy Test	Metallurgy Test
Fourth	Replace Pipeline	Insert Valves	Ultrasound Test	Insert Valves	Ultrasound Test

TABLE 6.14
Voting Results for Pipeline Alternatives Using Plurality Method

Option	*Total Votes*
Replace Pipeline	2
Insert Valves	2
Metallurgy Test	0
Ultrasound Test	1

TABLE 6.15
Second Round Voting Results for Pipeline Alternatives Using Plurality Method

Option	*Total Votes*
Replace Pipeline	2
Insert Valves	3

If the committee instead had used the Borda count method (Section 6.3.2.2), the result still would have been close between the Replace Pipeline and Insert Valve options, but the replacement option would have won. Recall from earlier in the chapter that the Borda count method would assign 4 points to an alternative each time it was listed as someone's first choice, 3 points each time it was listed as someone's second choice, and so forth, down to 1 point for each time it was listed as someone's last choice. The Borda count tally is illustrated in Table 6.16.

Suppose, however, that the business manager, being fairly sure how each person would rank the alternatives, anticipated that either of the two voting schemes would result at best in selecting his third-choice alternative. The business manager can influence the outcome in his favor by slightly changing

TABLE 6.16
Voting Results for Pipeline Alternatives Using Borda Method

Option	*First*	*Second*	*Third*	*Fourth*	*Total points*
Replace Pipeline	2	2	0	1	$(2 \times 4) + (2 \times 3) + (1 \times 1) =$ **15**
Insert Valve	2	0	1	2	$(2 \times 4) + (1 \times 2) + (2 \times 1) =$ **12**
Metallurgy Test	0	2	3	0	$(2 \times 3) + (3 \times 2) =$ **12**
Ultrasound Test	1	1	1	2	$(1 \times 4) + (1 \times 3) + (1 \times 2) + (2 \times 1) =$ **11**

the agenda. Recognizing that the Replace Pipeline option would probably not have enough first-place votes to be selected on the first ballot, the business manager could persuade the committee to consider the problem in two parts: first, whether the pipeline should be replaced, and then, if not, what should be done. According to this strategy, the option of replacing the pipeline would lose 3 to 2 against the option of not replacing the pipeline. The preferences of the committee members show clearly that once the alternative Replace Pipeline option is eliminated, both the Metallurgy Test and the Ultrasound Test look more favorable. Without the Replace Pipeline option, the Metallurgy Test wins with the Borda method, and the Ultrasound Test wins using the same Plurality procedure outlined above. By forcing a vote on the Replacement option, the business manager can ensure victory for one of his first two choices.

6.5 EXTENSIONS OF VOTING METHODS

The voting methods described above are only a representative sample of the voting methods available. Some of the methods not presented here involve refinements to these methods to attempt to make them more "fair." For example, the *plurality runoff* method is an extension of the plurality procedure in which the top two vote-getters are paired off for a second vote that determines the winner. This method, like the Hare system, tries to improve the plurality method by forcing a majority decision from the voters.

A more adventurous extension of voting has been the attempt to develop methods that take into account the strength of preference a voter has for the various alternatives. Just because each voter can rank four alternatives, does not necessarily imply that each voter prefers the first alternative ranked to the second or third by the same margin. One voter, for example, may only slightly favor the first-ranked alternative over the alternatives below it, while another voter may feel emphatic that he strongly prefers his first alternative over the others. Approval voting addresses this in only a rudimentary way, by allowing the voter to pick one of two levels of preference ("approved" or "not approved"). A better method might take into account the entire range of preference.

The desire to incorporate strength of preference raises two important questions. First, how does one elicit the intensity of preference? Methods for measuring preference, or *utility*, in a rigorous manner are discussed in Chapters 11 and 12. Other methods that employ utilities in voting are discussed in Nurmi (1987) and Riker (1982). However, these voting methods are resource-intensive and probably not practical. Other methods (such as *point allocation*) are related to the weighted scoring methods discussed in Chapter 7, but these methods have their own drawbacks as described in that chapter. The second question is whether it even makes sense to compare intensities of preference across individuals. At the very least, the range of one person's preference intensity could be significantly different from that of another person. Approval voting may partly address this problem. Though the range of preference that a voter can express with approval voting is small, the concept of "approval" may be one that can be appropriately compared across individuals. However, other methods that incorporate preferences are prone to different interpretations of the preference scale of different individuals.

A practical extension of voting can be made to the problem of a single decision maker evaluating a series of alternatives characterized by many attributes. If the decision maker ranks all of the alternatives separately on each attribute, then these separate rankings can be aggregated using one of the voting procedures above. The decision maker simply treats the ranking presented by each attribute as a vote cast by that attribute for the most appropriate alternative. This method implies that each attribute has equal weight, just as each voter has an equal weight in most voting methods. Alternatively, attributes can be weighted and the method will more closely resemble the weighted scoring methods described in Chapter 7.

6.6 IMPLEMENTATION NEEDS

Voting is among the simplest of decision aids. It has no specific information or data requirements; the process of voting is independent of the amount of knowledge possessed by the individuals involved. In fact, this can be a sore point to some members of a committee who feel that their expertise or superior knowledge warrant placing more weight on their votes. Ultimately, however, the only real information requirements are that there be:

- A group of two or more eligible voters
- Two or more alternatives under consideration
- A specification of which voting system will be used.

Voting requires little or no training. Most of the voting methods can be implemented with little more than simple arithmetic. If the vote takes place in the course of a formal meeting, the participants may wish to know how to conduct the voting procedure formally. Procedures such as those described by *Roberts' Rules of Order* may help here. Otherwise, the only training required is the ability to tabulate and count the votes according to the agreed upon aggregation method.

Depending on the size of the group, its familiarity with the alternatives under consideration, and the seriousness of purpose, the time required of the participants may be significant, especially if the group wishes to debate the alternatives. In addition, the voting process itself may also require resources. For example, if the number of voters is large, the group may need an automated voting system. Or, if the group is divided into strong factions, members may request a separate process to ensure that votes are counted fairly and accurately.

6.7 SUMMARY

6.7.1 Strengths and Weaknesses of Voting

Voting has many strengths that make it attractive as a decision aid. First, it is one of the few decision aids geared specifically for multiple decision makers. Second, it has a strong societal basis, which means that the process is generally understood by participants and the results generally accepted. Third, voting gives equal weight to all participants, encouraging committee members to buy into the decision process and its results. And finally, voting requires few resources; it is both quick and inexpensive.

Voting does have some serious shortcomings, however. First, because the votes themselves are all that may be known about the decision, the information and rationale behind the decision is often poorly documented. Of course, this could be a plus if the organization wishes to maintain the confidentiality of its decision making. Unless the participants document and explain their votes, reviewers are left with little to help them understand what affected the decision. Even with such documentation, the vote may not necessarily reflect rigorous thinking, complete information, or noble intent.

Second, the simplicity of voting can be a limitation as well as a benefit. Voting systems tend to be simple because they force people to make crude characterizations to their true preferences. We generally consider preferences to exist on some kind of continuous scale, but most voting systems require voters to state their preferences in terms of favorites or ranked lists. Even Approval Voting, the most sophisticated use of preference information discussed here, considers only two levels of preference: approve and "not approve." No common voting method utilizes the full range of voter preference. Moreover, when there is limited discussion and consensus-building, the result of the vote can cause divisiveness in the organization.

Third, the potential for manipulation of voting is another problem. Clearly the seriousness of this problem depends on which voting system is used and who participates in the vote. It is, however, at least theoretically possible to manipulate any voting system, and, with the possible exception of Approval Voting, the easier the voting system is to use, the easier it is to manipulate. In fact, once the members of a group become familiar with various voting systems, they may disagree on which voting system to use.

Finally, voting is not amenable to sensitivity analysis. One may be interested in knowing how the results would change if the information, values, or alternatives changed. Because each participant's vote results from his internal thought processes rather than from an explicit model, it is nearly impossible to answer "what if" questions about the decision.

6.7.2 Suggestions for Further Information/Study

Voting is a subject that touches many different disciplines. Significant works have been written by political scientists, economists, mathematicians, and philosophers. Each group approaches voting from a different point of view, and though a common base of terminology and research has been built, no definitive textbook on voting methods exists.

For an easy introduction, we recommend two articles from *Scientific American*, one by Niemi and Riker (1976), another containing an essay by Steen (1980). Straffin (1980) presents another introduction similar in spirit to this chapter. For a discussion of voting methods with regard to ideals of democracy, see *Liberalism against Populism* by Riker (1982). For a more classical introduction to committee voting, see Black (1958).

6.8 REFERENCES

Arrow, K. J. (1963). *Social Choice and Individual Values*. 2nd ed. New York: Wiley.

Black, D. (1958). *Theory of Committees and Elections*. Cambridge: Cambridge University Press.

Brams, S., and P. Fishburn (1983). *Approval Voting*. Boston: Birkhäuser.

Farquharson, R. (1969). *The Theory of Voting*. New Haven: Yale University Press.

Fishburn, P. (1973). *The Theory of Social Choice*. Princeton: Princeton University Press.

Gibbard, A. (1973). Manipulation of Voting Schemes: A General Result. *Econometrica* 41:587–601.

Niemi, R. G., and W. H. Riker (1976). The Choice of Voting Systems. *Scientific American* 234:6, 21–27.

Nurmi, H. (1987). *Comparing Voting Systems*. Theory and Decision Library Series A, Philosophy and Methodology of the Social Sciences, W. Leinfellner and G. Eberlein (eds.). Boston: D. Reidel.

Riker, W. H. (1982). *Liberalism against Populism: A Confrontation Between the Theory of Democracy and the Theory of Social Choice*. San Francisco: W. H. Freeman.

Satterthwaite, M. (1975). Strategy-Proofness and Arrow's Conditions. *Journal of Economic Theory* 10:187–217.

Sen, A. (1970). *Collective Choice and Social Welfare*. San Francisco: Holden-Day.

Steen, L. A. (1980). Essay in Martin Gardner's column in *Scientific American* 238:16–26B (October).

Straffin, P. (1980). *Topics in the Theory of Voting*. Boston: Birkhäuser.

7

WEIGHTED SCORING METHODS

Guess if you can, choose if you dare.
***—Pierre Corneille,** Heraclius*

7.1 PURPOSE OF CHAPTER

Weighted scoring methods range from informal "back of the envelope" calculations to formal, structured processes. Many people have used a weighted scoring method at some point in their lives, perhaps informally. In general, weighted scoring methods are easy to learn and use, and are particularly well-suited to decisions in which time or resources are limited.

We use *weighted scoring methods to refer to a decision-making technique in which alternatives are assigned numerical scores that rate each alternative on each of several decision criteria. These decision criteria are assigned weights according to their relative importance to the decision maker. A total score for each alternative is then determined by aggregating its ratings on each decision criterion, weighting the criteria appropriately. The alternative with the highest total score is considered to be the best alternative.*

This chapter provides a general description of the weighted scoring method. A number of variations of the general method exist. We focus on two of the more common variations, Kepner–Tregoe (KT) decision analysis and the analytic hierarchy process (AHP). In addition, we present a few of the popular methods available for developing weights and discuss how weighted scoring methods can be used in group decision making. Finally, we discuss the strengths and weaknesses of weighted scoring methods.

7.2 OVERVIEW OF WEIGHTED SCORING METHODS

7.2.1 Introduction to Weighted Scoring Methods

Weighted scoring methods are formal procedures for deciding among alternatives taking into account the explicit criteria on which the alternatives may differ. The decision maker assigns weights to these criteria that express their contribution towards the decision-maker's overall goal. Each alternative is assigned *scores* that characterize how well the alternative performs on each decision criterion. The scores are then *weighted* by multiplying the scores by the weight of the corresponding criterion. For each alternative, the sum of the weighted scores for each criterion gives the total score for that alternative. In general, the decision is made by choosing the alternative with the highest total score.

Although there are variations of this method, the general process used by decision aids that are categorized as weighted scoring methods is similar. Each alternative is scored on a set of decision criteria that have been weighted by relative importance. Many of us have employed such an approach, perhaps informally, to make personal decisions: we think about the key criteria relevant to selecting an alternative, and how well we think each alternative performs on these criteria. We then choose the alternative that we believe performs better on the more important criteria.

7.2.2 History of Weighted Scoring Methods

Many variations of weighted scoring methods have been developed over the years, for the most part independently of one another. Two weighted scoring methods, Kepner–Tregoe decision analysis and the analytic hierarchy process, have been discussed extensively both from academic and practical perspectives, and been applied broadly. The following sections briefly describe the history of these techniques. Another widely applied technique that can be viewed as a form of weighted scoring is multiattribute utility analysis (MUA), discussed in Chapter 12. We distinguish MUA from weighted scoring methods because MUA explicitly differentiates between the desirability of possible outcomes of a decision alternative and the likelihood of the outcomes. Unlike the weighted scoring methods discussed in this chapter, MUA is grounded in expected utility theory. In addition to MUA, there are other weighted scoring methods that are less widespread than the techniques of this chapter. Among these is the Simple Multiattribute Rating Technique (SMART), which was developed in an effort to make MUA more practical (Watson and Buede 1987) and Moody's precedence charts (Moody 1983). Moody's technique uses a chart format to elicit the extent of "precedence" that each factor has in making the decision. SMART and Moody's precedence charts are not discussed further in this book.

7.2.2.1 History of Kepner–Tregoe Decision Analysis

In the 1950s, Charles Kepner and Benjamin Tregoe were working at RAND Corporation and saw many decisions in government and industry that they considered to be of poor quality. They felt that a more logical, rigorous approach to decision making could improve the quality of decisions. They formed their own consulting firm and developed a collection of techniques that they referred to as the "rational management process." One of these techniques is KT decision analysis, which employs a weighted scoring method to select from among the alternatives under consideration (Kepner and Tregoe 1981).

Kepner–Tregoe decision analysis has been applied to a variety of decisions. A few of the areas of application include environmental, industrial safety, electric utilities, mining, and manufacturing. An application involving worker safety is given by Moore (1986).

7.2.2.2 History of Analytic Hierarchy Process

Analytic hierarchy process (AHP) was developed by Thomas Saaty at the University of Pennsylvania Wharton School in the 1970s while working on policy decisions for the U.S. government (Saaty 1980). Since its introduction, many books and papers have been written on the theory and application of AHP. Some authors attest to its practicality and proven usefulness in diverse applications. Other authors contest the rigor of the technique and claim that it can produce arbitrary results. Although it is subject to much debate and continuing research, AHP has been widely used in government, industry, and academia.

Analytic hierarchy process has been applied in several fields including business, energy, health, transportation, sociology, economics, and politics. Among the numerous applications in these fields are decisions in the areas of R&D, strategic planning/portfolio selection, marketing, finance, manufacturing and production, capital investment, and legal planning. For an extensive bibliography and a list of the numerous applications of AHP, see Golden, Wasil, and Levy (1989).

7.2.3 Decisions/Situations Most Appropriate for Weighted Scoring Methods

Weighted scoring methods are particularly suited for two types of decision situations. The first of these situations is where formality or explicitness is required. Weighted scoring methods involve a formal process in which the value judgments can be assessed, incorporated into the analysis, and presented explicitly. The second type of situation to which weighted scoring methods are well-suited is where the resources available to analyze the decision (time, money, analytical expertise, etc.) are constrained. As a result of their relative simplicity, weighted scoring methods can often be performed at a low cost, in a short time frame, and without the need for sophisticated analytical skills.

Weighted scoring methods are less appropriate for situations in which there is significant uncertainty and a formal process is needed to account for the risks. Since these methods do not distinguish between the values for the outcomes under each alternative and the likelihood that various outcomes occur, the results obtained with weighted scoring methods may lead to inconsistent or erroneous decisions. See Chapter 2 for a more thorough explanation of uncertainty in decision making.

Another important drawback of weighted scoring methods is that they may fail to account for relevant complexities in the preference structure of the decision maker. For example, the importance of cost as a decision criterion may depend greatly on the magnitude of the cost; weighted scoring methods generally constrain the analysis to employ a single weight per unit of cost regardless of the total cost. By making simplifications of this type, weighted scoring methods may not reflect the decision-maker's preferences well.

7.3 EXPLANATION OF WEIGHTED SCORING METHODS

In this section, we first discuss the generic process for weighted scoring methods and the key concepts used in the process. Then, we describe in more detail Kepner–Tregoe decision analysis and the analytic hierarchy process and how they vary from the generic process.

7.3.1 Key Concepts of Weighted Scoring Methods

7.3.1.1 The Generic Process

Although there are many variations of weighted scoring methods, they generally follow a similar process. This process includes the following steps (we will clarify some of the terms in the remainder of the section):

1. Define the decision
2. Create the decision criteria
3. Weight the decision criteria
4. Create the decision alternatives
5. Score the alternatives on each criterion
6. Determine a total weighted score for each alternative
7. Choose an alternative.

The key feature that weighted scoring methods have in common is that the end product of the analysis is a total weighted score for each alternative. The weighted score is intended to measure the degree to which an alternative fulfills the desired decision criteria. Typically, the decision is made by choosing the alternative that has the highest total weighted score.

7.3.1.2 Decision Criteria

As mentioned earlier, in using weighted scoring methods one must first define decision criteria upon which the alternatives are scored. Decision criteria can be framed as either *objectives* or *attributes*. Objectives have an object and a direction (e.g., maximize profit, minimize environmental impact, minimize risk, minimize cost, maximize satisfaction). Attributes are characteristics with which to quantify how well an alternative fulfills an objective. For example, if *maximize profit* is the objective, an attribute that can be used to quantify how well an alternative fulfills this objective is profit in dollars. Another objective may be *maximize job satisfaction*. Attributes that quantify how well an alternative maximizes job satisfaction may include location, salary, position or title, 'etc. Either attributes or objectives can serve as decision criteria. Weighted scoring methods vary as to whether they score alternatives on attributes or objectives. This distinction between the use of attributes and objectives as decision criteria does not alter the basic weighted scoring method process.

Ideally, the set of decision criteria should be:

- Comprehensive
- Nonredundant
- Independent.

In a comprehensive set of decision criteria, all objectives or attributes that the decision maker is considering are accounted for in the set. In practice, some less important criteria may have to be ignored. However, the attributes included in the analysis should be complete enough to capture the important differences among alternatives.

Criteria should also be nonredundant. If two criteria overlap one another, this would result in placing too much importance on the one aspect represented by these two criteria. For instance, in deciding what car to buy, two decision criteria could be *cost of operating the car* and *cost of gas*. However, these two criteria are redundant, since the cost of operating the car includes the cost of gas.

Finally, the decision criteria should be independent: how well an alternative performs on one decision criterion should not depend on how well the alternative performs on another decision criterion. If such a dependency exists, then it is impossible to score an alternative on each of the two related criteria independently, as required by the method. For instance, two criteria for the car decision may be *Cost of gas* and *Fuel economy*. The cost of the gas used in operating the car depends on the car's efficiency, or fuel economy. It would be impossible to score the car on *Cost of gas* without knowing its score on *Fuel economy*.

In general, one can structure the analysis carefully to ensure that the decision criteria meet these considerations. To do this, it may be necessary to divide overly broad criteria into narrower criteria that better distinguish one from another, or to combine one or more of the criteria to eliminate overlap.

7.3.1.3 Weights

Weights are measures of the relative importance of each of the decision criteria to the overall decision. If one alternative performs better than the other alternatives on all criteria, then clearly this is the best alternative regardless of the weights on the criteria. Usually the decision is not so easy, and the decision maker needs to make tradeoffs between criteria. For example, in deciding which of several products to buy, we often have to trade off the cost of the product and the quality of the product. Typically, one must pay more to obtain a higher quality product. Using weights to express the relative importance of one criteria over another, we quantify the value judgments engendered in these tradeoffs.

There are several methods commonly used to assess the weights. A few of these methods for constructing weights will be described in detail in Section 7.5, Extensions of Weighted Scoring Methods.

7.3.2 Description of Kepner–Tregoe Decision Analysis

As mentioned previously, Kepner–Tregoe (KT) decision analysis is part of the KT "rational management process" (Kepner and Tregoe 1981). The KT rational management process was developed by observing techniques and processes that seemed to work best for successful managers. Kepner and Tregoe observed that there are four basic routines that managers perform and developed four rational processes to aid managers in performing these routines:

- *Situation appraisal:* a process used to evaluate a situation and plan for future action
- *Problem analysis:* a process for identifying and solving problems
- *Decision analysis:* a process that aids in organizing and analyzing information for decision making
- *Potential problem analysis:* a process by which potential future problems, their potential causes, and preventive actions are identified.

This chapter is concerned with only one of these four routines, the decision analysis process. However, it is considered important to implement KT decision analysis within the context of the entire integrated rational management process, since many of the benefits derive from being a part of this whole process. In particular, the developers of the approach feel that the management process, of which KT decision analysis is one part, lends consistency to decision making and facilitates communication throughout the organization.

The rest of this section introduces the "nuts and bolts" of KT decision analysis. The steps, which we describe in detail, are shown below:

1. Construct a decision statement
2. List the objectives of the decision
3. Divide the objectives into MUSTS and WANTS
4. Assign weights to the WANTS
5. Evaluate the alternatives on the MUSTS and WANTS
6. Find a total weighted score for each alternative
7. Make a "tentative" choice
8. Examine the possible negative consequences
9. Make a final decision.

The remainder of this section explains these steps using an example problem.

7.3.2.1 Construct a Decision Statement

The decision maker in our example is going to build a house. He wants to determine where he should build it. We begin with the decision statement:

Where should I build my new house?

7.3.2.2 List Objectives

Once we have developed the decision statement, we must develop the list of objectives. For the example problem, the objectives include the following:

- Minimize distance to work
- Maximize quietness of neighborhood
- Maximize quality of local schools
- Minimize property costs.

The objectives are typically formulated as a direction of action (e.g., minimize) combined with an object (e.g., distance to work).

7.3.2.3 Divide Objectives into MUSTS and WANTS

After the objectives are developed, they are divided into MUSTS and WANTS. MUSTS are those objectives that must be satisfied. They are constraints. WANTS are those objectives that one desires but does not consider absolutely necessary.

MUSTS	**WANTS**
Property cost less than $80,000	Maximize quietness of neighborhood
Distance to work less than 30 miles	Maximize quality of local schools
	Minimize property cost

An alternative must fulfill each of the objectives that are classified as MUSTS in order to be considered for further analysis. Thus the list of MUSTS serves to screen possible alternatives and reduce the number considered in detail. Objectives that are classified as WANTS are desirable, but do not express performance levels that must be achieved for an alternative to be considered. Note that the issue of property cost is addressed in both the MUSTS and WANTS. Although the decision maker is willing to pay up to $80,000, he would like to minimize the property cost. Ideally, we would like to satisfy all of the WANTS to the greatest extent possible. In reality, this usually cannot be done, and we must make tradeoffs among wants that can be only partially satisfied by the available alternatives. Therefore, in order to maximize our total satisfaction, the WANTS are assigned weights that express their relative desirability.

7.3.2.4 Assign Weights to the WANTS

To construct the weights, the following steps are performed:

1. The most important objective is given a 10
2. All other objectives are given a number that reflects the relative importance in relation to the most important objective.

For instance, suppose the most important WANT in the example problem is *Quality of local schools*. It is assigned a weight of 10 as shown in Table 7.1. *Property cost* is one-fifth as important, so it is assigned a 2. The WANT, *Quietness of neighborhood*, is assigned a 5 because it is only half as important as the quality of the local schools.

7.3.2.5 Evaluate the Alternatives on the MUSTS and WANTS

Once the objectives have been defined and weights have been assigned, we evaluate the alternatives. To be considered, an alternative must fulfill all of the MUST objectives. For instance, in our example, a location 40 miles from the decision-maker's office is not a feasible alternative because it violates the MUST objective of being less than 30 miles from the office. In the location example, the decision maker identifies three alternative locations that satisfy all of the MUSTS:

TABLE 7.1
Weights for Location Decision

WANTS	*Weight*
Maximize quietness of neighborhood	5
Maximize quality of local schools	10
Minimize property cost	2

- Oakdale Village
- Riverside Estates
- Main Street.

Each alternative that meets all of the MUST objectives is now scored on each WANT objective as follows:

1. Give the alternative that comes the closest to meeting an objective a 10
2. Score the other alternatives in comparison to the one that comes the closest to meeting the objective.

The alternatives for the location example have been scored on all WANTS as shown in Table 7.2. The weight of each WANT is shown in parentheses under the WANT. As an illustration of the process, consider the scoring of the alternatives on *Quietness*. Riverside Estates is the least noisy neighborhood, so it was given a 10. The other locations were then scored in comparison with Riverside Estates. Main Street, the noisiest neighborhood, is 10 times less desirable with respect to noise than Riverside Estates and is assigned a 1.

7.3.2.6 Calculate Total Weighted Scores and Make a Tentative Choice

We now have all of the information required to calculate a total weighted score for each alternative. The analyst calculates a total weighted score for each alternative by multiplying the alternative's score on each objective by the corresponding weight for the objective and summing over all objectives. This calculation for the location decision is shown in Table 7.3. A" tentative" choice is then made based on the alternative with the highest total weighted score. The Oakdale Village location in the example has the highest total weighted score.

TABLE 7.2
Alternatives and Scores for Location Decision

Alternative	*WANTS (weight shown in parentheses)*		
	Quietness (5)	*Quality of schools* (10)	*Property cost* (2)
Oakdale Village	4	10	2
Riverside Estates	10	3	6
Main Street	1	2	10

TABLE 7.3
Calculation of Total Weighted Scores for Location Decision

	WANTS (weight shown in parentheses)			
Alternative	*Quietness (5)*	*Quality of schools (10)*	*Property cost (2)*	*Total Weighted Score*
Oakdale Village	4	10	2	(4 x 5) + (10 x 10)+(2 x 2) = **124**
Riverside Estates	10	3	6	(10 x 5) + (3 x 10)+(6 x 2) = **92**
Main Street	1	2	10	(1 x 5) + (2 x 10) + (10 x 2) = **45**

7.3.2.7 Examine the Possible Negative Consequence

The choice based on total weighted score is called "tentative" because the best alternative is not necessarily the one with the highest total weighted score. First, the risks of the alternatives must be considered. KT provides a step in which the possible negative consequences of choosing each alternative are considered. This step is intended to help clarify the risks involved. In *The New Rational Manager* by Kepner and Tregoe (1981), the authors suggest examining each alternative by asking five questions that force examination of the possible negative consequences:

- What requirements for success have we missed in the previous stages of this analysis?
- What factors within the organization, based on our experience, could harm its acceptance or its implementation?
- What kinds of changes within the organization could harm its long-range success?
- What kinds of external changes (such as competitive response or government regulations) could harm its long-range success?
- What kinds of things tend to cause problems in implementing this type of decision?

The questions are used to develop a list of potential negative consequences for each alternative. Both the probability of occurrence of these consequences and the seriousness of these consequences are assessed. The probability of occurrence of the consequences is rated with either a high–medium–low scale, or with a 1-to-10 scale. The seriousness of consequences is rated using a 1-to-10 scale in which a higher number indicates a more serious potential consequence. These numbers are not used in a quantitative analysis, but are relative measures of the potential negative consequences. Potential negative consequences for the location example are shown in Table 7.4.

TABLE 7.4
Potential Negative Consequences for Location Decision

	Probability	*Seriousness*
Oakdale Village		
Potential increase in the local property tax	High	4
Riverside Estates		
River may flood in spring and cause water damage	High	8
Main Street		
Possible decrease in value of property	Medium	5

7.3.2.8 Make a Final Decision

The decision maker makes a final decision considering both the total weighted scores and the relative measures of the potential negative consequences. Numerical ratings of the negative consequences draw attention to alternatives with high probability and high seriousness, but the process is not meant to rule out all alternatives that have potential negative consequences. Nor does the process provide an explicit method to factor in negative consequences. Instead, the decision maker mentally trades off the total weighted score against the potential negative consequences in reaching a final decision.

This explanation of KT decision analysis has presented only the basics of the process. For a more thorough description and an explanation of additional features of KT decision analysis see *The New Rational Manager* (Kepner and Tregoe 1981).

7.3.2.9 Strengths and Weaknesses of KT Decision Analysis

A key strength of KT decision analysis is its simple, structured methodology. KT decision analysis is easy to implement and avoids complicated mathematical computations. Decisions can be more easily communicated and justified by showing that a formal methodology was used. Communicating the judgments underlying the decision is straightforward since the weights on the objectives and the scores assigned to the alternatives are documented. In addition, the potential negative consequences that were considered can be examined. Of course, there may be sensitive decision situations in which the explicitness of KT may not be desirable.

One of the drawbacks of KT involves the way in which KT incorporates potential negative consequences into the analysis. Although there is a process for examining the negative consequences, there is no explicit tradeoff made between the total weighted score and the potential negative consequences. It may not be clear to the decision maker whether the risks incurred from the potential negative consequences are worth the returns represented by the total weighted score. If a formal and rigorous method of dealing with risk is desired, an analyst should consider using decision analysis (Chapter 11) or multiattribute decision analysis (Chapter 12), methods that address this issue explicitly.

Another weakness of KT is that the numbers can be easily manipulated to favor an alternative. One way this can be done is by adding more decision criteria on which the preferred alternative performs well until the favored alternative has the highest total weighted score. Similarly, less preferred alternatives can be made worse by eliminating decision criteria on which they perform well. It is important to carefully examine the selected decision criteria to ensure that they fulfill the requirements of comprehensiveness, nonredundancy, and independence.

The analyst must be aware that the total weighted scores used in KT or other weighted scoring methods may not be a meaningful measure of the desirability of the outcomes of a decision. One problem is that the 1-to-10 scale used to construct the weights and the scores is arbitrary, and may not be a natural representation of the quantity being assessed. Since the scale is not tied to a concrete performance measure, individuals will vary in how they score identical items. If a score is tied to a natural, concrete measure, decision makers may be more consistent and better able to express their true beliefs. As an example of a natural scale, the objective *minimize costs* could be measured in terms of dollars. Rating the cost of alternatives on a 1-to-10 scale is somewhat arbitrary, but using a dollar scale makes the rating concrete. In assigning scores from 1 to 10 on cost, the decision maker may confuse the desirability of the outcome with the importance of the objective; it is certainly difficult to separate the desirability of a cost of $100 versus $1000 from the importance of cost as an issue in the decision.

Another reason that the total weighted scores may lack meaning is that the weights on the objectives are constant over the entire range of the scores for the objectives. This means that the relative importance of an objective is the same, regardless of the extent to which the objective is fulfilled. Most persons exhibit decreasing marginal utility; that is, the utility or value for each additional increment of something desired declines as the amount of that something increases. For instance, the value for a 10-decibel reduction in the average noise level in a neighborhood may be more valuable if the noise level is 100 decibels rather than 30 decibels. Along with most other weighted scoring methods, KT assumes a constant marginal utility. This is realistic if the marginal utilities of the attributes over their ranges of possible outcomes in the decision is constant.

Despite its weaknesses, if used appropriately, the KT decision analysis process can provide insight into the important elements of the decision and can give the decision maker a feel for which alternatives are the best or the worst.

7.3.3 Description of the Analytic Hierarchy Process

The analytic hierarchy process is designed to structure decision problems hierarchically. The hierarchy consists of an overall objective at the top level, criteria that lead to the overall objective in the middle levels, and at the bottommost level, the set of possible alternatives. A decision hierarchy for a student choosing between graduate school at College A and College B is shown in Figure 7.1. For each level, weights are determined for each criterion that appears in that level. These weights, called "priorities," express how important each criterion in that level is in fulfilling the criterion just above it. These priorities are derived from a series of pairwise comparisons made between the elements on one level with respect to their contribution to the elements on the next highest level. At the bottommost level, priorities are determined for the alternatives themselves. These priorities express the degree to which each alternative satisfies each criterion in the next level up. The priorities of the alternatives on the bottom level are weighted by the priorities of the objectives in the other levels to find an overall priority for each alternative. This overall priority is the "total weighted score" referred to in the description of the generic weighted scoring process. Along with the total priorities, AHP provides a consistency measure that indicates whether the decision maker's judgments are consistent.

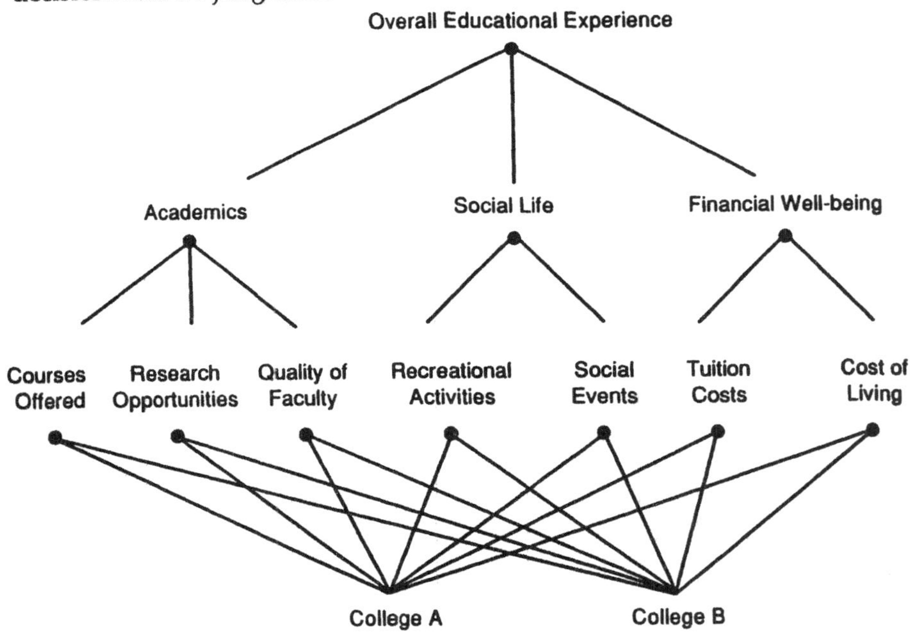

FIGURE 7.1. Decision hierarchy for college example.

As with KT decision analysis, AHP will be described using an example problem. The following is an overview of the steps that will be taken:

1. Define the decision problem
2. Structure a decision hierarchy
3. Construct pairwise comparison matrices
4. Calculate the priorities for the elements of the hierarchy
5. Find the overall priority for each alternative
6. Evaluate the results and make the decision.

7.3.3.1 Define the Decision Problem

The first step of AHP is to define the decision problem. In order to illustrate AHP, we will use a different problem involving a decision among transportation modes:

Should I fly first class or coach?

7.3.3.2 Structure a Decision Hierarchy

After the decision problem has been defined the analyst needs to structure a decision hierarchy. The overall value or focus of the decision is at the top level of a hierarchy. This is the management objective. The middle levels consist of decision criteria or subcriteria, which may be either objectives or attributes. The criteria and subcriteria contribute to, satisfy, or determine the value of the criteria above it in the next highest level. The alternatives are in the bottom-most level of the hierarchy. As much as possible, criteria should be comprehensive, nonredundant, and independent. Figure 7.2 shows a three-level hierarchy for the airline example. The overall objective is to maximize overall

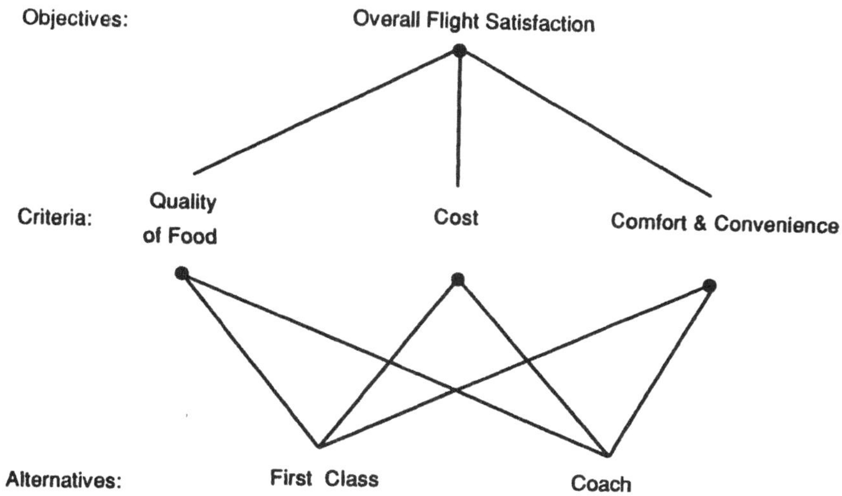

FIGURE 7.2. Decision hierarchy for airline example.

flight satisfaction. The passenger has chosen the three criteria *Quality of food, Cost,* and *Comfort & convenience* as those factors contributing to his overall flight satisfaction. The bottommost level displays the two alternatives, First Class and Coach. Lines connect each of the criteria to the overall objective as well as each alternative to each criterion.

The criteria *Quality of food, Cost,* and *Comfort & convenience* are used to determine the overall flight satisfaction. If desired, the criteria can be broken down into subcriteria. For example, subcriteria for *Comfort & convenience* are shown in Figure 7.3. The *boarding convenience*, the *comfort of the seat*, and the *quality of in-flight service* all contribute to the comfort and convenience. The number of levels to include in the hierarchy will depend on the level of detail that the analyst finds necessary and appropriate to the analysis, given the time and resources available. An attractive characteristic of AHP is that the analysis can be expanded to any desired level of detail, although expansion increases the number of required assessments. Expansion can be done subsequent to the initial analysis if desired. For our purposes, the hierarchy shown in Figure 7.2 will be sufficient.

7.3.3.3 Construct Pairwise Comparison Matrices

Once the hierarchy is structured, priorities (weights) for the elements in a level are inferred from responses to a series of pairwise comparisons between the elements of the level. These questions are posed to the decision maker or to experts familiar with that portion of the decision hierarchy. The pairwise comparisons may be expressed verbally and translated into a numerical rating using a comparison scale that Saaty developed specifically for AHP. Or, they can be expressed directly using the numerical scale. The values on the scale, shown in Table 7.5, represent the relative importance of one element (e.g., criterion) when compared to another. To use the scale, one may ask a number of comparison questions. "How much more strongly does this element contribute to or satisfy the criterion than the element to which it is being compared?" "How much more important is it?" For instance, if element A is of strong importance over element B in satisfying a criterion, the comparison

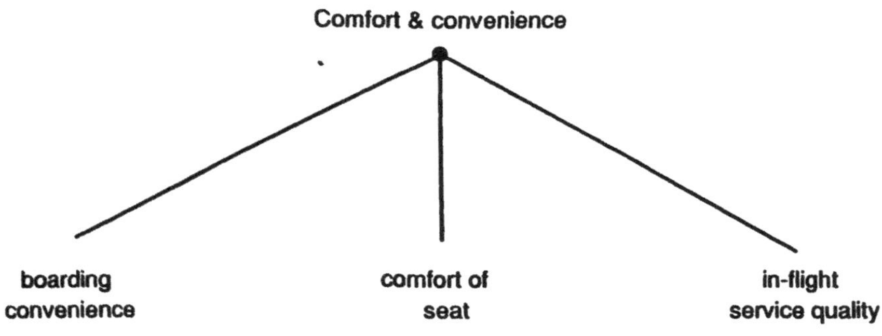

FIGURE 7.3. Possible subcriteria of *Comfort & convenience* in airline example.

receives a score of 3, meaning A is three times as important as B. Likewise, the comparison of B to A receives a score of 1/3, meaning B is one-third as important as A.

This scale is designed to be multiplicative; that is, a score of 8 implies twice the importance or influence as compared with a score of 4. Saaty reports studies demonstrating that the 1-to-9 scale may adequately represent the intensity of importance between elements on a multiplicative scale. As stated by Saaty, "experience has confirmed that a scale of nine units is reasonable and reflects the degree to which we can discriminate the intensity of relationships between elements" (Saaty 1990). However, those who have criticized the scale believe that a large discrepancy may exist between the quantitative measure that different people associate with any given qualitative statement, so that using a scale to assign a numerical value may not give results that accurately represent the decision-maker's beliefs. Caution must be used if verbal judgments are to be translated into a numerical rating using the scale. In practice, it is more common to express a numerical rating directly.

TABLE 7.5
The Pairwise Comparison Scale[a]

Intensity of Importance	*Definition*	*Explanation*
1	Equal importance.	Two activities contribute equally to the objective.
3	Moderate importance of one over another.	Experience and judgment slightly favor one activity over another.
5	Essential or strong importance.	Experience and judgment strongly favor one activity over another.
7	Demonstrated importance.	An activity is strongly favored and its dominance is demonstrated in practice.
9	Extreme importance.	The evidence favoring one activity over another is of the highest possible order of affirmation.
2, 4, 6, 8	Intermediate values between two adjacent judgments.	When compromise is needed.
Reciprocals of above nonzero numbers.	If an activity has one of the above numbers (e.g., 3) compared with a second activity, then the second activity has the reciprocal value (i.e., 1/3) when compared to the first.	

[a]Reproduced with permission from Analytical Planning: The Organization of Systems (1991), RWS Publications, Pittsburgh

The numerical comparisons are put into matrices that are manipulated mathematically to construct the priorities. The entries in the matrix represent comparisons between two criteria or alternatives at one level with respect to satisfying a particular criterion in the next highest level. An example of a pairwise comparison matrix for our airline ticket example is shown in Table 7.6. This is the matrix used by the decision maker for assigning priorities to the elements of the second level of the hierarchy with respect to overall flight satisfaction. Elements represent the number given to the item in the column to the left of the matrix when compared with the item in the row along the top of the matrix. For instance, in the airline example, the decision maker feels that Cost has strong importance over the *Quality of the food* so it is given a score of 5. A "5" is then placed in the second row of the first column of the matrix. The reciprocal, 1/5, is assigned to the *Quality of the food* as compared to Cost. Reciprocals are entered for half of the elements of the matrix. Note that elements along the matrix diagonal are 1, indicating equal importance. This is because the elements along the diagonal represent each criterion as compared to itself.

We need to determine the relative importance of the elements on a given level in satisfying each objective, criterion, or subcriterion to which they contribute on the next highest level. Therefore, a matrix will be assessed for each element that has other elements below it in the hierarchy. For the airline example, one matrix (Table 7.6) would be used to calculate the priorities of the quality of food, cost, and comfort with respect to overall flight satisfaction. For each criterion in the second level, one matrix would be used to get the priorities of the alternatives with respect to that criterion: one matrix would compare first class to coach with respect to quality of food, one would compare the alternatives with respect to cost, and a third would compare them with respect to comfort and convenience.

An important concern in the AHP methodology is whether the comparisons expressed by the decision maker are consistent. This can be determined from the consistency of the judgment matrix. A matrix is consistent when the pairwise comparisons reflect consistent preferences. For instance, imagine three decision criteria: A, B, and C. If A is preferred to B by 2 : 1 and B is

TABLE 7.6
Pairwise Comparison Matrix for Airline Example

	Quality of food	*Cost*	*Comfort & convenience*
Quality of food	1	1/5	2
Cost	5	1	4
Comfort & convenience	1/2	1/4	1

preferred to C by 3 : 1, then A will be preferred to C by 6 : 1 if the comparisons are perfectly consistent with each other. In another example, if one feels D is twice as important as E, then E must be half as important as D in order to be perfectly consistent. AHP measures the consistency of the decision-maker's judgments in a matrix through the calculation of a consistency ratio that ranges from 0 (perfect consistency) to 1 (highly inconsistent). The consistency ratio is a measure of the deviation of the matrix from one that contains entirely random judgments. It is not necessary, and may be extremely difficult, to obtain perfectly consistent judgments when using AHP. By examining the consistency ratio the analyst can determine whether it is appropriate to reassess some of the pairwise comparisons in order to obtain greater consistency. The reader is referred to *Decision Making for Leaders* (Saaty 1990) for technical details on calculating consistency ratios.

7.3.3.4 Calculate Priorities for the Elements of the Hierarchy

Once the pairwise comparison matrices have been constructed, the analyst can solve for the set of weights (or "priorities") that best match the comparison ratios expressed in each matrix. The priorities are on a multiplicative scale, and the priorities sum to 1 among criteria emanating from a single criterion on the next level above. For example, if there are two criteria, A and B, and A is 4 times as important as B, then A will get a priority of 0.8 and B will get a priority of 0.2. In the example, the priorities for *Quality of food*, *Cost*, and *Comfort & convenience* will sum to 1. The priorities for *First Class* and *Coach* with respect to any one of the three criteria above them will sum to 1.

When there are more than two criteria, the numerical priorities are not obvious from the comparison ratings, but can be calculated fairly easily. The approach used to determine the priorities makes use of all of the matrix assessments, including redundant ones; the results are a best fit to the assessments.[1] Mathematical axioms on which this is based have been developed. Refer to the paper "Axiomatic Foundation of the Analytic Hierarchy Process" (Saaty 1986) for the axioms and mathematical derivation. The airline ticket example will be used to demonstrate one method of calculating the approximate numerical priorities without performing complex vector manipulations. Using this approximation method, the resulting priority weights are very close to the true priority vector if the matrix judgments are fairly consistent, and they are exact if the matrix is perfectly consistent.

We will illustrate the construction of priorities for the criteria in the second level of airline hierarchy. The priorities for the second level of the hierarchy will give us an estimate of the relative importances of the criteria with respect to overall flight satisfaction. To construct the priorities using an approximation method, we first add the values in each column as shown in Table 7.7.

1 Mathematically, the priorities obtained are the elements of the eigenvector associated with the largest eigenvalue of the matrix.

TABLE 7.7
Column Totals for Pairwise Comparison Matrix for Airline Example

	Quality of food	*Cost*	*Comfort & convenience*
Quality of food	1	1/5	2
Cost	5	1	4
Comfort & convenience	1/2	1/4	1
Column Totals	6.50	*1.45*	*7.00*

TABLE 7.8
Normalized Pairwise Comparison Matrix for Airline Example

	Quality of food	*Cost*	*Comfort & convenience*
Quality of food	$\frac{1}{6.5} = .15$	$\frac{.2}{1.45} = .14$	$\frac{2}{7} = .29$
Cost	$\frac{5}{6.5} = .77$	$\frac{1}{1.45} = .69$	$\frac{4}{7} = .57$
Comfort & convenience	$\frac{.5}{6.5} = .08$	$\frac{.25}{1.45} = .17$	$\frac{1}{7} = .14$
Column Totals	*1.00*	*1.00*	1.00

Next we normalize each column so that the elements sum to 1. This is done by dividing each element in the column by the column total. The normalized matrix is shown in Table 7.8.

By normalizing, we have put the relative judgments in each column on the same 0 to 1 scale. We then calculate the individual priorities by averaging the elements in each row of the normalized matrix. For instance, the priority for *Quality of food* is calculated as follows:

$$\frac{.15 + .14 + .29}{3} = .19$$

The priorities of the criteria are shown in Table 7.9. The priorities indicate the importance of each criterion relative to the others on the same level. The larger the priority, the higher its relative importance. Priorities are analogous to the weights used in Kepner–Tregoe or other weighted scoring methods. In order to choose the best alternative, we are interested in the overall priorities of the alternatives with respect to each other. These are obtained by creating

TABLE 7.9
Priorities of Criteria in Airline Example

	Quality of food	Cost	Comfort & convenience	Criterion Priority
Quality of food	.15	.14	.29	$\frac{.15+.14+.29}{3}=.19$
Cost	.77	.69	.57	$\frac{.77+.69+.57}{3}=.68$
Comfort & convenience	.08	.17	.14	$\frac{.08+.17+.14}{3}=.13$

priorities for the alternatives with respect to each criterion on the next highest level and then weighting these by the priorities of the criteria. This is equivalent to finding a total weighted score for each alternative.

The priorities of the alternatives with respect to a given criterion are obtained by using the pairwise comparison process. A pairwise comparison matrix for each criterion in the airline example is shown in Table 7.10. The numbers in these matrices represent the judgments formed by the decision maker. For instance, the decision maker judged First Class to be four times as desirable as Coach with respect to the quality of food. The calculated priorities are also shown. These priorities represent "scores" of each alternative on each criterion.

7.3.3.5 Calculate the Overall Priority for Each Alternative

The scores of each alternative on each of the criteria are weighted by the corresponding priorities of the criteria to calculate the overall priorities or total

TABLE 7.10
Pairwise Matrices for Comparing Airline Travel Alternatives

Quality of food	First Class	Coach	**Priority**
First Class	1	4	.80
Coach	1/4	1	.20
Cost	First Class	Coach	**Priority**
First Class	1	1/6	.14
Coach	6	1	.86
Comfort & convenience	First Class	Coach	**Priority**
First Class	1	2	.67
Coach	1/2	1	.33

weighted scores. The overall priority for each airline alternative is calculated as shown in Table 7.11.

7.3.3.6 Evaluate the Results and Make the Decision

The overall priorities indicate that the Coach alternative is preferred over First Class. At this point in the process, the decision maker can either make the decision or perform a sensitivity analysis. Sensitivity analysis enables the decision maker to examine how variations in the individual pairwise comparisons or priorities impact the overall priorities. Although sensitivity analysis is not explicitly a part of the analytic hierarchy process, it may provide additional insight before a decision is made. Sections 7.4.6 and 7.5.3 further discuss sensitivity analysis.

The process can be extended to a problem with a larger number of levels in the hierarchy. Priorities must first be obtained for each level of the hierarchy with respect to each criterion in the next highest level. Then an overall priority is found by calculating the total weighted score.

7.3.3.7 Strengths and Weaknesses of AHP

As mentioned previously, some controversy surrounds the theory and techniques of AHP. Proponents of AHP like the structure it provides, the ease of obtaining judgments in the form of pairwise comparisons, the mathematical elegance, and the computational speed. However, others criticize AHP for its lack of a sound theoretical foundation and the arbitrariness with which alternatives are ranked according to their overall priorities.

Although AHP has axioms underlying the hierarchical structure and pairwise comparison (Saaty 1986), critics claim that it is not logically sound. These critics believe that it produces results that do not necessarily reflect the preferences of the decision maker. The ranking of alternatives by their overall priorities is considered to be arbitrary for reasons described in the following paragraphs.

The analytic hierarchy process can exhibit the phenomenon of *rank reversal*. Rank reversal occurs when the addition of an alternative that is identical to another alternative causes the relative ranking of two or more of the other

TABLE 7.11
Calculation of Overall Priorities

Alternative	*Quality of food (0.19)*		*Cost (0.68)*		*Comfort & convenience (0.13)*		*Overall Priority*
First Class	.8 × .19	+	.14 × .68	+	.67 × .13	=	.33
Coach	.2 × .19	+	.86 × .68	+	.33 × .13	=	.67

alternatives to change. For instance, a decision maker may prefer A to B and B to C from among three alternatives. However, when alternative D, which is identical to A, is added to the decision-maker's choices, AHP may rank the alternatives in the order B, A and D, C. The ranks of B and A have been reversed. A related concern is that the results of AHP, and of weighted scoring methods in general, can depend on whether or not the analysis includes an option that is clearly dominated by the other options (Belton and Gear 1983). Many people prefer to make their decisions in a manner in which the attractiveness of an alternative does not depend on the other alternatives that are being considered. (This property is called *Independence of Irrelevant Alternatives* in the behavioral mathematics literature.) For a lively discussion of theoretical criticisms and counterarguments the reader is referred to Winkler (1990).

The pairwise comparison process that is part of AHP also has critics and supporters. Many like the fact that it is easy to get numerical rankings from verbal statements by using the pairwise comparison scale. Others believe that this method replaces the task of expressing difficult value tradeoffs with a process whereby the decision maker merely accepts an assigned number. For those who feel uncomfortable with the verbal judgment scale, the pairwise comparisons can be expressed quantitatively without using the verbal scale. However, some critics view the numerical scale as arbitrary and difficult to relate to the physical characteristics (e.g., cost) of the alternatives being evaluated. Part of this difficulty is that the scale incorporates both the relative performance of the two aspects being compared and the decision-maker's value for the difference in performance.

The number of pairwise comparisons required lends both pros and cons to AHP. As a hierarchy gets larger and more complex, more pairwise comparisons are necessary. The number of comparisons needed at each level of the hierarchy, is equal to $[n(n-1)]/2$, where n is equal to the number of elements on the level. More complex problems may involve a large time commitment and the assessments of the pairwise comparisons may become tedious. However, the number of comparisons required by AHP allows the decision maker to check the consistency of his judgments, since each element is compared to all others. Informally, consistency is encouraged because the decision maker is forced to think about the relationship between all of the elements. A formal check is provided mathematically through the calculation of a consistency ratio. One refinement of AHP is to assess only a predefined subset of the comparisons, or to assess only as many as are required until a sufficient degree of consistency is obtained.

The analytic hierarchy process has the same weaknesses that were mentioned for KT decision analysis including the lack of a rigorous treatment of the uncertainties in the decision, the assumption of constant marginal utility, and the ease of manipulating results in favor of an alternative.

7.4 CASE STUDY: DISTILLATION COLUMN

An Application of the Analytic Hierarchy Process Involving Chemical Process Safety

This application involves Case Three, the distillation column, which was introduced in Section 5.4. A second application of this case illustrates the use of multiattribute utility analysis (Section 12.4).

We use the case to illustrate the key steps of the analytic hierarchy process (AHP). The details of the mathematical manipulations are not presented. For specifics on the mechanics, please refer to Sections 7.3.3.4 and 7.3.3.5. The following discussion assumes that the reader has read the description of AHP in Section 7.3.3.

The case involves a plant at which a distillation column is used to separate hexane and heptane from a feed stream consisting of a mix of the two. A residential neighborhood with about 200 residents is close to the facility. In response to management concerns, a quantitative risk analysis was undertaken. The risk analysis focused on the safety of the residential community.

The case involves corporate-level issues due to the magnitude of the risk, and the importance of the distillation facility to this company's operations. A corporate team was formed that included the business director, the director of Engineering, a member of the company's capital projects review body, and the operations general manager. The corporate team came up with some alternatives to address the issues at hand and wanted quick analysis to examine the relative attractiveness of the alternatives. The corporate team was to have a management meeting in three weeks and needed to present the preliminary analysis of the alternatives at that time. Supporting the corporate team was an analysis team composed of a risk analyst, business analyst, and process and plant engineers. The analysis team considered the short time frame and the budget constraints for the initial analysis of the alternatives and chose to use the analytic hierarchy process. In the remainder of this section, we describe the key steps taken in the analysis.

7.4.1 Define the Decision Problem

The corporate team presented the analysis team with the decision statement "What action should be taken to deal with the potential risk to the public from the distillation column?" They asked that the analysis team evaluate the following alternatives:

- Do nothing
- Replace with two smaller distillation columns
- Install a water curtain.

7.4.2 Structure a Decision Hierarchy

The analysis team met with the corporate team to discuss the overall objective and the decision criteria. They agreed that the overall objective was to satisfy, to the greatest extent possible, the corporate objectives outlined in the company's mission statement. A list of criteria contributing to the achievement of the corporate objectives was developed. This list included all of the criteria that the team members considered to be relevant to the decision. Excluded from this list were criteria on which all alternatives perform equally well. The overall objective, the criteria, and the alternatives were used to construct the decision hierarchy shown in Figure 7.4. The overall objective, to maximize the relevant corporate strategic objectives, appears at the top of the hierarchy. The criteria that contribute to this overall objective appear on the second level. *Cost* was chosen as a criterion to satisfy the corporate goal of maximizing financial return. It is defined as the cost of the action taken, including the cost of disrupting plant operations. *Risk to community* was chosen to account for the risk of public injuries and fatalities. *Community goodwill* was also included as a criterion because the company wants to be viewed as a responsible, good neighbor. One of the team members expressed concern over a possible dependency between *Risk to the community* and *Community goodwill*. However, the team decided that *Risk to the community* can be assessed independently of *Community goodwill*. *Risk to the community* refers to the actual risk of public injuries or fatalities whereas *Community goodwill* depends on the actions the company takes to reduce the risk, but not on the level of risk resulting from these actions.

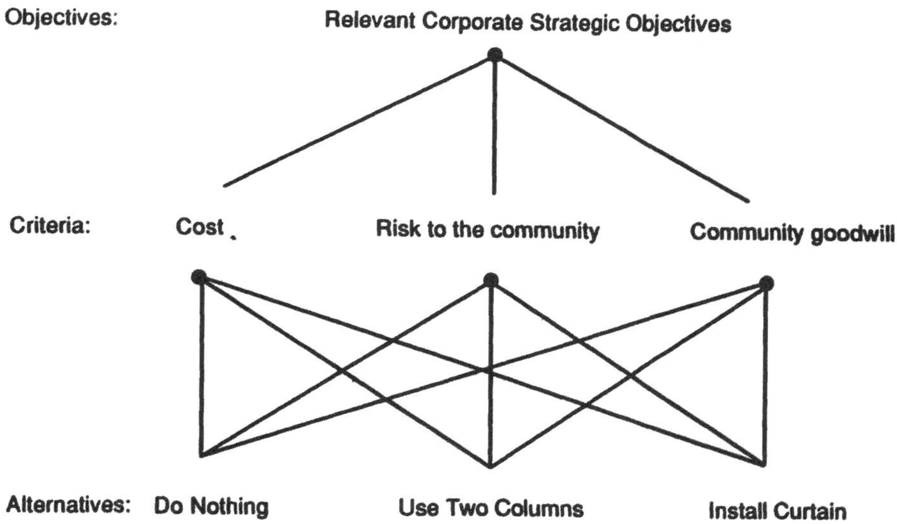

FIGURE 7.4. Decision hierarchy for distillation column example.

7.4.3 Construct Pairwise Comparison Matrices

To begin the mathematical analysis, the analysis team constructed a pairwise comparison matrix to compare the second-level criteria to one another. The corporate team was asked to make the comparisons between the three decision criteria on their relative importance to the overall corporate goals. For instance, the corporate team first compared the importance of *Cost* vs. *Risk to the community* with respect to the overall corporate objectives. They felt that the risk to the community is twice as important as the financial implications, so a "2" is placed in the appropriate element of the matrix as shown in Table 7.12. Note that the element in the first column of the second row, which represents the comparison of *Risk to the community* vs. *cost*, is a 2. The reciprocal, 1/2, has been placed in the second column of the first row as the *Cost* vs. *Risk to the community*. This approach is used to compare all criteria to each other and complete the matrix.

After studying the alternatives, the analysis team made the pairwise comparisons for the alternatives with respect to each of the criteria. These comparisons are presented in the matrices of Table 7.13. The cost comparisons were made by examining the costs of replacing the column with two smaller columns and installing a water curtain. The cost of doing nothing is $0. Note that although tangible numerical measures (in dollars) are available for the cost criteria, they cannot always be used directly in the AHP process to calculate the pairwise comparisons. For example, if one alternative costs twice as much as another, one cannot necessarily use the ratio 2 for an accurate pairwise comparison. This calculation does not recognize that each additional dollar may not add as much value as the previous dollar. In addition, ratios cannot be calculated when a $0 cost is involved as in our problem. Instead of using the costs directly, the team assigns comparisons based on the relative desirabilities of the costs given the company's financial position.

The team studied the F-N curve estimated in the risk analysis (See Chapter 5, Figure 5.7) before making the pairwise comparisons for the *Risk to the community* criterion. In order to make the comparisons, they considered the reduction in risk achieved by switching to two smaller columns and by

TABLE 7.12
Pairwise Comparison Matrix for the Distillation Column Decision Criteria

	Cost	*Risk to the Community*	*Community goodwill*
Cost	1	1/2	3
Risk to the community	2	1	5
Community goodwill	1/3	1/5	1

TABLE 7.13
Pairwise Comparisons of the Alternatives on the Decision Criteria Cost Do Nothing Use Two Columns Install Curtain

Cost	Do Nothing	Use Two Columns	Install Curtain
Do Nothing	1	9	3
Use Two Columns	1/9	1	1/3
Install Curtain	1/3	3	1
Risk to the community	Do Nothing	Use Two Columns	Install Curtain
Do Nothing	1	1/4	1/2
Use Two Columns	4	1	2
Install Curtain	2	1/2	1
Community goodwill	Do Nothing	Use Two Columns	Install Curtain
Do Nothing	1	1/5	1/5
Use Two Columns	5	1	1
Install Curtain	5	1	1

installing a water curtain. The team estimated that switching to two columns would reduce the risk by 4 and installing a water curtain would reduce the risk by 2. Therefore, when compared with Use Two Columns, Do Nothing was assigned a 1/4 (row 1, column 2) and when compared with Install Curtain, Do Nothing was assigned a 1/2 (row 1, column 3).

The comparisons for *Community goodwill* were made using the pairwise comparison scale presented earlier in Table 7.5. Since management strongly favored taking an action over doing nothing with respect to the resulting community reaction, Install Water Curtain and Use Two Columns were each assigned a 5 when compared to Do Nothing. The two actions were expected to be viewed equally by the community.

After completing the pairwise comparisons, the analysis team examined the consistency of the comparisons by calculating a consistency ratio. This is an advanced technique that we do not present here. (See *Decision Making for Leaders* [Saaty 1990] for the technical details.) For our purposes, we assume that the level of consistency in the judgments was found to be acceptable and no further assessments for the comparison matrices were needed.

7.4.4 Calculate Priorities for the Elements of the Hierarchy

Once the numerical comparisons have been assessed, the priorities for the criteria on the second level of the hierarchy are calculated. In this example, we use the approximation technique described in the preceding section. First, the elements of the columns in the matrix of Table 7.12 are summed as shown in

TABLE 7.14
Pairwise Comparison Matrix for Distillation Column Decision Criteria

	Cost	*Risk to the community*	*Community goodwill*
Cost	1	1/2	3
Risk to the community	2	1	5
Community Goodwill	1/3	1/5	1
Column Totals	3.33	1.70	9.00

TABLE 7.15
Normalized Pairwise Comparison Matrix

	Cost	*Risk to the community*	*Community goodwill*
Cost	$\frac{1}{3.33}=.30$	$\frac{.50}{1.70}=.29$	$\frac{2}{9}=.33$
Risk to the community	$\frac{2}{3.33}=.60$	$\frac{1}{1.70}=.59$	$\frac{5}{9}=.56$
Community goodwill	$\frac{.33}{3.33}=.10$	$\frac{.20}{1.70}=.12$	$\frac{1}{9}=.11$
Column Totals	*1.00*	*1.00*	1.00

Table 7.14. Then, the matrix is normalized so that each column sums to 1 as shown in Table 7.15. The analysis team calculates the priorities by averaging the normalized elements in each row of Table 7.15. The calculations and the resulting priorities are presented in the last column of Table 7.16. These priorities represent the relative importances of each of the criteria in meeting the corporate objectives. *Risk to the community* (priority = .58) is the most important criteria; community goodwill (priority = .11) has the lowest priority and is least important to achieving the overall objective.

Using the pairwise comparisons in the matrices of Table 7.13, the analysis team calculated the priorities for the alternatives with respect to the criteria in the same manner as it calculated the priorities of the criteria. Each matrix is normalized and the rows are averaged to obtain the priorities. The resulting priorities of the alternatives with respect to the decision criteria are displayed in Table 7.17. The priorities of a single criterion can be interpreted as the relative desirability of the alternatives with respect to that criterion. For

TABLE 7.16
Priorities of Criteria in Distillation Column Example

	Cost	Risk to the community	Community goodwill	Criterion Priority
Cost	.30	.29	.33	$\frac{.30+.29+.33}{3}=.31$
Risk to the community	.60	.59	.56	$\frac{.60+.59+.56}{3}=.58$
Community goodwill	.10	.12	.11	$\frac{.10+.12+.11}{3}=.11$

instance, Do Nothing is the most attractive of the alternatives with respect to *Cost* (i.e., the least expensive). This is confirmed by observing that its priority for *Cost* (.69) is greater than the priorities for the other alternatives.

7.4.5 Calculate the Overall Priority for Each Alternative

The analysis team was now ready to calculate the overall priorities, or weighted scores, of the alternatives. This is done by weighting the priorities of the alternatives by the corresponding priorities of the criteria. The method is illustrated in Table 7.18, where the priorities presented in Table 7.17 are multiplied by the priorities of the corresponding criteria.

7.4.6 Evaluate the Results and Make the Decision

Based on their overall priorities, the analysis team ranked the alternatives from best to worst as follows:

1. Use Two Columns
2. Do Nothing
3. Install Curtain

TABLE 7.17
Priorities for the Alternatives in the Distillation Column Example

Alternative	*Cost*	*Risk to the community*	*Community goodwill*
Do Nothing	.69	.14	.10
Use Two Columns	.08	.57	.45
Install Curtain	.23	.29	.45

TABLE 7.18
Calculation of Overall Priorities

Alternative	*Cost* (.31)		*Risk to the community* (.58)		*Community goodwill* (.11)		*Overall Priority of Alternatives*
Do Nothing	.69 × .31	+	.14 × .58	+	.10 × .11	=	.31
Use Two Columns	.08 × .31	+	.57 × .58	+	.45 × .11	=	.40
Install Curtain	.23 × .31	+	.29 × .58	+	.45 × .11	=	.29

The team would want to perform sensitivity analysis prior to the meeting, time permitting. They would conduct a sensitivity analysis to examine how variations in the individual pairwise comparisons and priorities affect the overall priorities. For example, Figure 7.5 shows the results of a sensitivity analysis involving the priorities for *Cost* and for *Risk to the community*. If the worst alternative, Install Curtain, were to rank lowest in all of the scenarios in the sensitivity analyses, the team could recommend that it be dropped from further consideration. In Figure 7.5 we see that it does rank the lowest for all ratios examined except those less than 0.35. In addition they could further

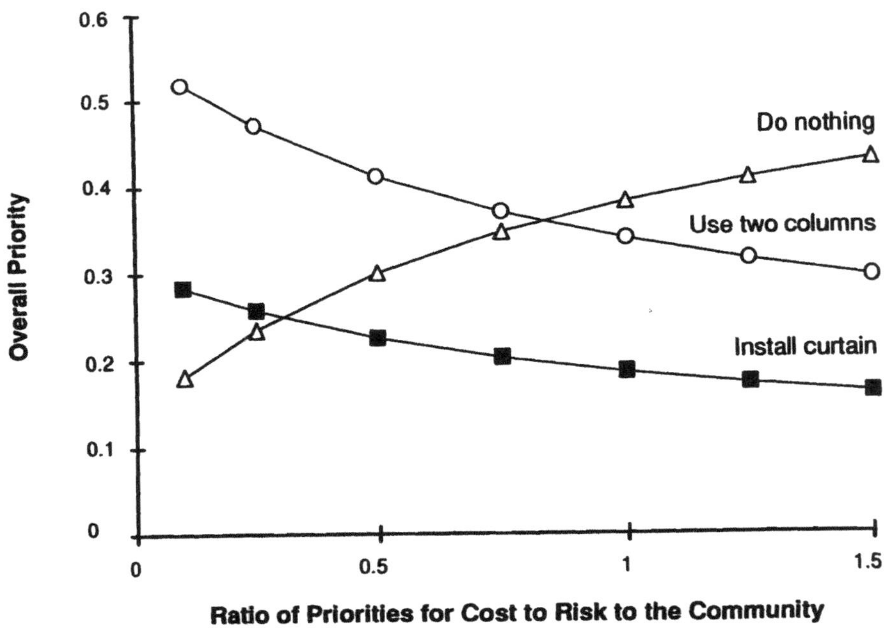

FIGURE 7.5. Sensitivity to priority for *Cost* versus *Risk to the community*.

examine the issue of dependency between *Risk to the community* and *Community goodwill.*

The results were presented along with the pairwise comparisons to the corporate team. The corporate team could request further analysis. For instance, they could be interested in seeing the implications of using different pairwise comparisons. The team would be prepared to answer this if a sensitivity analysis has been performed. Typically, a decision will not be made at this point. A second-pass analysis will be performed first in which the team may do sensitivity analysis, refine pairwise comparisons and question original assumptions and simplifications. For example, Chapter 11 presents a more detailed analysis of this problem using decision analysis.

7.5 EXTENSIONS OF WEIGHTED SCORING METHODS

This section explores three extensions of weighted scoring methods: techniques for constructing the weights on the decision criteria, approaches to group decision making, and sensitivity analysis.

7.5.1 Weight Elicitation Techniques

A variety of techniques exist to help the decision maker assess weights on the decision criteria. All of these methods are based on relative comparisons among criteria, but the elicitation method and the standard for comparison vary. Two of these methods, swing weights and ratio judgment weighting, will be described in detail below. Although the methods will be described in terms of weighting attributes, they can also be used to weight objectives.

7.5.1.1 Swing Weights

Swing weights are derived by exploring "swings" in the value of alternatives as the numerical quantity for each of the attributes changes. In order to obtain the swing weights, decision attributes must be defined and ranked in order of importance. Attributes for the airline example have been defined and ranked in Table 7.19. *Cost* is the most important; *Comfort & convenience* is the least

TABLE 7.19
Attribute Ranking for Airline Example

Attribute	*Rank*
Quality of food	2
Cost	1
Comfort & convenience	3

important. Each attribute can be thought of as having various levels. For instance, the attribute *cost* may have levels $20, $30, and $50. Levels do not have to be in quantitative terms. For instance, the levels for the attribute *Quality of food* could range from "poor" to "superior."

The first step in creating swing weights is to imagine an alternative that has the worst level of each attribute and an alternative that has the best level of each attribute. Considering just the alternative that has the worst level of each attribute, imagine that the numerical quantity for the highest ranked attribute were switched from its lowest level to its highest level, while all other attributes remain at their lowest level. The difference in value of the alternatives created by "swinging" an attribute from its lowest level to its highest level is arbitrarily assigned a swing score of 100. Similarly, imagine all attributes at their lowest level, and swing the second most important attribute from its lowest level to its highest level. The swing for that attribute is the change in overall value caused by the swing in relation to the value caused by the swing in the most important attribute. For example, if the change in value occurring under this swing is one-fourth the change that occurred when the highest-ranked attribute was swung from its lowest to its highest level, the attribute receives a swing score of 25 (i.e., 25% of the maximum swing). This process is repeated for all attributes. Attributes that cause no change in the overall value of the alternative receive a swing score of 0. The swings assigned to the attributes of the airline example are shown in Table 7.20.

Once all of the swing scores for the attributes have been obtained, the swing scores are normalized to obtain a set of weights that add to 1. Normalizing provides a consistency check for the person creating the weights. It allows the person to see what percent or fraction of the total weight each attribute has been assigned. Percent may be a more intuitive measure and allows the analyst to reexamine the scores.

Suppose that in the airline example we want to assess attribute importances using the swing weights method. We first rank the attributes in order of importance as shown in the first column of Table 7.20. We then assign a "swing" to each attribute using the swing weights method. The final weights shown in the third column are then calculated by normalizing the swing

TABLE 7.20
Swing Weights for Airline Example

Attribute	*Swing Rank*	*Swing Score*	*Weight*
Quality of food	2	30	21%
Cost	1	100	69%
Comfort & convenience	3	15	10%

scores. The weights are expressed as a percent of the total weight and would be used in a weighted scoring method for making the decision.

7.5.1.2 Ratio Judgment Weighting

In *ratio judgment weighting,* attributes are ranked in order of importance with the most important at the top of a column, and the least important at the bottom of a column. The bottom attribute is assigned an arbitrary weight of 10. Each attribute is then judged relative to the lowest one and given a weight. For instance, if an attribute is twice as important, it is given a weight of 20. Ten is used for the bottom attribute instead of one so as to avoid decimal places when the other attributes are given scores. As with swing weights, an intuitive check on the weights can be performed by normalizing the weights so that they add to 1. The importance of each attribute can then be expressed as a percent or fraction of the total weight. Table 7.21 shows possible ratio judgments for the airline example. In this example, the decision maker felt *Comfort & convenience* to be the least important attribute and assigned it a 10. The other attributes were assigned judgments relative to *Comfort & convenience.* The judgments were then normalized to get the weights shown.

Ratio judgment weighting provides a consistency check of the weight. To perform this check, a second column is formed, without the lowest ranked attribute. The second-lowest attribute is given a weight of 10 and the others are rated relative to this attribute. Likewise, this process is continued by forming new columns, disregarding the lowest attribute from the previous column, until only two attributes are left. The ratio between any two values assigned to two different attributes in a column should be the same as the ratio between the two values for these attributes in all other columns. This shows that the relative importance between columns does not change when the lower-ranked attributes are disregarded. Table 7.22 shows how the consistency of the airline example weights can be checked. First, *Comfort & convenience* is dropped from the list. Then, judgments are made with only the two remaining items. *Quality of food* is assigned a 10, since it is the most important attribute. Then suppose *Cost* is judged to be three times as important and given a 30. The new ratio of the *Cost* judgment to the *Quality of food* judgment (30/10 = 3) is compared to the ratio of the first judgments (100/30 = 3.33). We

TABLE 7.21
Ratio Judgment Weights for Airline Example

Attribute	*Ratio judgment*	*Weight*
Comfort & convenience	10	7%
Quality of food	30	21%
Cost	100	72%

TABLE 7.22
Ratio Judgment Consistency Check

Attribute	*Ratio judgment*
Quality of food	10
Cost	30

see that the judgments were relatively consistent. If the consistency ratios show a high degree of inconsistency, weights are adjusted and, if necessary, the process is repeated until the desired consistency is achieved.

7.5.2 Group Decision Making

Weighted scoring methods are often employed in group decision making. The same techniques are used, but weights and scores must reflect the desires of the entire group instead of one decision maker.

Practitioners disagree over the best way to obtain numerical judgments from groups. Some advocate assessing the judgments separately from each individual in the group, and then taking the arithmetic mean. The resulting average scores are used in further analysis. Saaty suggests that the geometric mean be used to aggregate scores in the Analytical Hierarchy Process, since the scores reflect a geometric scale. Others argue that an aggregated score is not appropriate because it may not reflect the desires of any one individual. Instead, the aggregated score may lie somewhere between the desires of the individuals in the group. Moreover, it may not be appropriate to give equal weight to the judgments of each of the individuals in the group.

An alternative to aggregating individual judgments is to have multiple decision makers reach consensus as a group on a single set of numerical judgments, not through a voting or mathematical aggregation scheme. Various processes have been suggested for achieving group consensus. While this approach is probably ideal from the standpoint of methodological rigor, achieving consensus may be difficult or impossible and is likely to require significant time and budget.

Saaty (1989) addresses issues of using AHP for group decision making.

7.5.3 Sensitivity Analysis

Many decision aids employ some type of sensitivity analysis in order to examine the "sensitivity" of the decision to various inputs of the analysis. The sensitivity is a measure of the impact of the input on the decision. Although it is not explicitly a part of the weighted scoring methods process, sensitivity analysis may prove to be useful. In using sensitivity analysis in weighted scoring methods, the analyst would vary the scores and weights and examine the impact of the changes on the total weighted scores. This may be especially useful in group decision making. When the group disagrees on either a weight or an individual score, the analyst can perform the analysis for several different values of the input in question. If the best decision is the same over the entire range of values for the input, then any value chosen for the input within the range would be appropriate.

7.6 IMPLEMENTATION NEEDS

7.6.1 Data/Information Requirements

The inputs required to use a weighted scoring method are generated by a decision maker or a decision-maker's representatives, with the facilitation of an analyst. Necessary inputs to any weighted scoring method analysis are:

- Objectives
- Weights on the objectives
- Decision alternatives
- Scores for the alternatives on each objective.

For the Kepner–Tregoe method, a list of negative consequences and scores on their probability and seriousness are also used as inputs to the analysis.

7.6.2 Personnel/Training Required

7.6.2.1 Kepner–Tregoe Decision Analysis

Kepner–Tregoe decision analysis requires training but is not analytically difficult. To implement the "rational management process," of which KT decision analysis is a part, organizations are encouraged to first prepare the line managers, who can then train others in their organization. The Kepner–Tregoe approach advocates full commitment to the KT process in the organization at all levels. As stated in *The New Rational Manager* (Kepner and Tregoe 1981), "...this system cannot be introduced by half-heartedly sprinkling a few ideas and suggestions among a random mix of the organization people in the hope that something good will happen." One of the benefits of KT decision analysis, as mentioned earlier, is that it is part of an entire management process. Implementing a management process encourages team-work and promotes decision making that is consistent throughout the organization.

7.6.2.2 Analytical Hierarchy Process

Although applying the AHP technique is also not difficult, one must first commit to learning the method. The mathematics behind the technique should be thoroughly understood before drawing conclusions from the results. Personnel using AHP should obtain a foundation in the theory and mechanics of AHP, as well as examining some real-world applications.

7.6.3 Other Resources Needed

Beyond trained personnel, no additional resources are required to implement weighted scoring methods. Commercial software is available for the analytic hierarchy process. Although Kepner–Tregoe does not license any vendors for software employing KT decision analysis, software is available for similar weighted scoring methods. Use of such software is not necessary to implement these methods. For a list of software packages see Appendix A.

7.7 SUMMARY

7.7.1 Strengths and Weaknesses of Weighted Scoring Methods

Although the soundness of the theoretical foundations of weighted scoring methods has been questioned, the methods can play a useful role in some decision situations. Weighted scoring methods provide a structured process and an analytic approach that is explicit and quantitative. The methods are commonly used, so it is easy to gain organizational acceptance. In addition, they are quick and easy to use once learned. The low resource requirement resulting from their simplicity may also make them attractive when resources are constrained. However, their simplicity also results in limitations which make them less attractive in certain decision situations. When weighted scoring methods are used, the analyst must be aware of these limitations in order to understand the degree of logical soundness in the results. This section describes the strengths and limitations in more detail (see also Dyer 1990).

One of the major strengths of weighted scoring methods is that they force the decision maker to be explicit and quantitative about preferences. When necessary, a high degree of explicitness facilitates communication and provides justification for the decision. On the other hand, on sensitive issues, a high degree of explicitness may be undesirable because the judgments made through scores and weights may be subject to controversy.

Although the values of the decision maker are captured through the weights and scores on the attributes, weighted scoring methods lack a means by which other aspects of the decision may be structured and quantified. For example, the decision maker must use intuition to address uncertainty, and to balance risks of positive and negative consequences. Intuition alone may be sufficient in some situations, but inadequate for complex problems.

When using weighted scoring methods, uncertainty must be dealt with intuitively because the methods do not separate the value associated with possible future outcomes from their likelihood of occurrence. In scoring alternatives on the various criteria, the likelihood of the events that affect that score must be accounted for intuitively. For instance, in deciding whether to go to Colorado or California for spring vacation, the weather may be one of the decision criteria. A single score for each alternative on this criteria must be obtained by implicitly taking into account the probability of favorable weather in the locations. This mental aggregation of uncertainty and consequence may be difficult to perform, and can lead to erroneous results. When uncertainty is significant and relevant to the decision, methods such as decision analysis (Chapter 11) and multiattribute utility analysis (Chapter 12) are more desirable, since the likelihood of events can be made explicit and handled rigorously.

Tradeoffs between positive and negative consequences in weighted scoring methods are not systematic and therefore, may not be apparent to the decision maker. These tradeoffs must be dealt with intuitively when assigning scores. Kepner–Tregoe decision analysis explicitly addresses potential negative consequences, but does not provide a way to incorporate the potential negative impacts into the overall evaluation of the alternatives. The decision maker is faced with the total weighted scores of the alternatives and a list of the potential negative consequences and their likelihood. However, there is no formal way to trade off between the total weighted scores and the potential negative consequences. In AHP, the potential negative outcomes must be addressed implicitly when judging the alternatives on the decision criteria.

In using weighted scoring methods, it is easy to manipulate results through the selection of criteria. For instance, if one favors an alternative, criteria on which the alternative performs well can be added, until the alternative has a high score. Criteria must be examined carefully so that they are comprehensive, nonredundant, and independent. Otherwise, disproportionate weight may be placed on one aspect of the problem.

The analyst must also be aware that weighted scoring methods do not reflect the commonly desired property of Independence of Irrelevant Alternatives (see Section 7.3.3.7). That is, the method's calculation of attractiveness of an alternative may depend on what other alternatives are being considered.

Finally, the analyst must be aware that weighted scoring methods may not provide a meaningful measure of the desirability of the outcomes resulting from the various alternatives. With an approach like KT decision analysis or AHP that involves scoring alternatives based on the decision-maker's strength of preference, all of the scores are subjective. Since the scale is not tied to concrete measures, individuals may vary in how they interpret and apply the scale. These scales are also not expressed in terms that are natural measures for decision makers, and make it hard to maintain consistency in judgments and to represent feelings accurately. In addition, the use of a single set of weights over the entire range of the attributes assumes a constant marginal utility. Methods that assume a constant marginal utility do ignore situations where the additional utility or benefit received may decrease (or increase) as one gets more of something.

Despite their drawbacks, weighted scoring methods may have a role in decisions that require a quick analysis without the need for sophisticated techniques. They can also have a role as a first-cut analysis in more complex decisions.

7.7.2 Suggestions for Further Information/Study

If one chooses to use a weighted scoring method, the particular method should be selected so as to fit well with the decision situation. This can involve using a combination of various techniques, depending on the decision situation and the decision-maker's preference. By becoming familiar with AHP, KT decision analysis, and other methods, one can decide the most appropriate method for a given problem

A good source for further information on KT decision analysis is the book *The New Rational Manager* (Kepner and Tregoe 1981). For further information on AHP, we recommend *The Analytic Hierarchy Process* (Saaty 1980) or *Decision Making for Leaders* (Saaty 1990). The first of Saaty's books offers a complete technical and theoretical explanation of AHP, while the second book provides more of a practical slant. Articles on the theory and applications of AHP can be found in journals such as *Management Science* and *Mathematical Modeling.* An annotated bibliography of applications is provided by Golden, Wasil, and Levy (1989).

7.8 REFERENCES

Belton, V., and T. Gear (1983). On a Shortcoming of Saaty's Method of Analytic Hierarchies. *Omega* 11(3):228–230.

Dyer, J. S. (1990). Remarks on the Analytic Hierarchy Process. *Management Science* 36(3):249–258.

Golden, B. L., E. A. Wasil, and D. E. Levy (1989). Applications of the analytic hierarchy process, a categorized, annotated bibliography. In *The Analytic Hierarchy Process: Applications and Studies,* B. L. Golden, E. A. Wasil, and P. T. Harker (eds.), pp. 37–58. New York: Springer.

Kepner, C. H., and B. B. Tregoe (1981). *The New Rational Manager.* London: John Martin Publishing Ltd.

Moody, P. E. (1983). *Decision Making: Proven Methods for Better Decisions.* New York: McGraw-Hill.

Moore, L. R. (1986). Kepner–Tregoe As a Safety Methodology. *Professional Safety* June:37–41.

Saaty, T. L. (1980). *The Analytic Hierarchy Process.* New York: McGraw Hill.

Saaty, T. L. (1986). Axiomatic Foundation of the Analytic Hierarchy Process. *Management Science,* 32(7):841–855.

Saaty, T. L. (1989). Group Decision Making and the AHP. In *The Analytic Hierarchy Process: Applications and Studies*, B. L. Golden, E. A. Wasil, and P. T. Harker (eds.), pp. 59–67. New York: Springer.

Saaty, T. L. (1990). *Decision Making for Leaders*. Belmont, CA: Lifetime Learning Publications.

Watson, S. R., and D. M. Buede (1987). *Decision Synthesis: The Principles and Practice of Decision Analysis*. Cambridge: Cambridge University Press.

Winkler, R. L. (1990). Decision Modeling and Rational Choice: AHP and Utility Theory. *Management Science* 36(3):247–248.

8

COST–BENEFIT ANALYSIS

A fundamental problem in implementing any cost benefit strategy, of course, is that those who receive the benefits rarely pay the costs.
—**H. W. Lewis**, *Technological Risk*

8.1 PURPOSE OF CHAPTER

Cost–benefit analysis (CBA) originated as a tool to address governmental decisions, but the technique can also address decisions of the private sector. It is helpful to understand its use in public-sector decisions in order to understand the philosophy and approach of CBA. A public agency may need to decide whether to invest public funds in a proposed venture, or must select from among alternatives for achieving an objective. The agency will usually need to justify the basis for its decision. Decisions by public agencies are expected to be made consistently with the preferences of the persons or constituency they serve.

Because an agency's objectives may not be explicitly stated, making these decisions is often difficult. How does the agency know the preferences of its constituents, and how does it deal with diverse preferences among different individuals?

Using cost–benefit analysis, the agency decides among alternatives by choosing the one with the highest net benefit to society. By net benefit, we mean the benefits of the alternative minus its costs. The costs and the benefits can include the direct effects of the project as well as indirect and intangible effects such as dislocation of jobs and traffic congestion.

A core distinction of cost–benefit analysis from other decision aids is in how the total costs and benefits are computed: where society bears costs or benefits, these are estimated by aggregating the costs and benefits across individuals impacted by the decision. A second distinction of CBA is in the basis used to estimate the costs and benefits to each individual. Specifically, the benefits and costs are estimated by determining how much individuals are willing to pay to receive the benefits or avoid the costs of the decision. While maximizing net benefits is an objective shared by other decision-making approaches, using the aggregate of individuals' willingness-to-pay as the basis for estimating costs and benefits distinguishes cost–benefit analysis from other approaches. There are practical issues in how best to obtain these inputs, as we will discuss later in this chapter.

Cost-effectiveness analysis (CEA) is a variation of cost–benefit analysis, and is discussed in Section 8.5.1. Cost-effectiveness analysis overcomes a key difficulty in CBA, the quantification of values for the intangible effects of a decision. For example, a cost–benefit analysis might assign a dollar value to the reduced fatality risk (e.g., expected number of fatalities avoided) under each of a set of safety programs. By doing this, the results can be expressed as a ratio of dollars of benefit to dollars of cost. A cost-effectiveness analysis will instead compare the programs based on the cost per unit of benefit (e.g., dollar cost per fatality avoided). This avoids the need to express a value per fatality. Although the decision rule used to compare alternatives differs between CEA and CBA, the philosophy and approach are otherwise similar.

Although the roots of CBA are in the public sector, the approach can also be used by private organizations, particularly if they wish to make decisions in a manner consistent with the public interest, or to perform an analysis in support of a request or recommendation to a public agency. If an organization wishes to maximize net benefits and to let affected individuals' preferences determine how those net benefits are to be quantified, then cost–benefit analysis may be appropriate.

Aside from practical concerns with how to obtain the inputs required in a cost–benefit analysis, there are philosophical and theoretical concerns. Among these, a key concern is the distributional issue: those in society who benefit from a decision may be different from those who bear the costs. These and other objections to CBA are addressed in this chapter.

After a brief overview of CBA, we provide a short history of the technique and a discussion of the decisions and situations in which this aid is most appropriate. We then provide a more detailed explanation of the aid, including key concepts, the steps in the analysis, and the types of results obtained. We illustrate the technique with an example from the chemical process safety area. We then describe extensions and variations on CBA including uncertainty analysis, approaches for addressing distributional issues, and cost effectiveness analysis. We describe the requirements for implementing a CBA including both data needs and staffing requirements. Finally, we summarize the strengths and weaknesses of CBA and provide a brief guide to readers wishing to learn more about the aid.

We should note that even when CBA is not the decision aid used, alternatives are often compared by evaluating costs and benefits. Decision aids differ in how they quantify costs and benefits, and in how they deal with uncertainties and value tradeoffs involving these costs and benefits. This chapter focuses on the core of cost–benefit analysis, which we define by its inclusion of external (societal) impacts of the decision, and by its use of individuals' values rather than those of the organization to quantify the costs and benefits.

8.2 OVERVIEW OF COST–BENEFIT ANALYSIS

8.2.1 Introduction to Cost–Benefit Analysis

The key purpose of cost–benefit analysis is to enable society, through governmental or private organizations acting in its behalf, to choose among alternative actions. Generally the possible alternatives will incur different costs and will accrue different benefits, and these costs and benefits may be distributed in different ways over members of society or perhaps over time. How should society decide among the alternatives?

There is no single answer. Various approaches have been suggested for choosing the most socially preferred alternative, including:

- Unanimity
- Majority rule
- Strict Pareto superiority
- Potential Pareto superiority.

We discuss each of these in turn.

Unanimity means choosing an alternative that all members of society prefer over other alternatives or from which they consistently benefit the most. While obviously a reasonable basis for selecting an alternative, unanimity rarely occurs in practice. Few decisions are so obvious that everyone in society agrees on the alternative they like best. Other, more severe barriers to unanimity are the practical problems of ensuring that each person expresses their preferences, and of efficiently collecting and managing this information. The closest society has come to this thus far is voting, and the participation rates and depth of information obtained through voting are well short of what is required in typical decisions. For realistic situations, we need a more practical approach than unanimity.

Majority rule means choosing an alternative that more than half of the individuals in society favor. In a risk decision, this might involve polling members of the community impacted by the decision. While appealing on democratic principles, there are several problems with using majority rule as a basis for such decisions. These problems include the possibility that:

- A poll of preferences is impractical
- Not everyone chooses to participate
- Individuals are unaware or misinformed
- No one alternative gains a majority vote
- The majority choice inflicts undue harm on a minority of individuals.

Strict Pareto superiority means choosing an alternative under which at least one person benefits more than under the other alternatives, and under which no one person benefits less than under the other alternatives. Again, while the choice is clear in this case, decisions are rarely so simple.

We have indicated that each of the approaches is limited in its ability to identify a preferred alternative, for both practical and theoretical reasons. We need a decision basis that helps overcome these limitations. One approach, which lies at the heart of cost–benefit analysis, is to choose the alternative that has the highest net benefit to society, where the net benefit is determined by adding up the net benefits to each individual affected by the decision. This approach has a democratic flavor, since the preferences of all members of society enter the decision. However, it goes one step further. The strength of preference, as indicated by the level of benefit to an individual, enters the evaluation, rather than simply an either–or vote.

As a basis for decision making, maximizing net benefits to society differs fundamentally from other decision-making approaches. These other approaches often use the preferences of a decision maker, not those of individual members of society, to compare alternatives.

The decision basis used by CBA is referred to as *potential Pareto superiority*. This principle relaxes strict Pareto superiority defined just above. The strict requirement that no individual be worse off under the chosen alternative can rarely be met in practice. Potential Pareto superiority says that society should choose an alternative over others if that alternative provides the greatest gains, in the aggregate. Potentially, those who fare less well under this alternative could be compensated by those who gain. If those who benefit gave some of their benefits back to those who do not, then everyone is potentially better off. In choosing among alternatives, we seek the one for which this potential improvement is greatest. This is equivalent to choosing the alternative with the highest aggregate net benefits.

While this approach has some appeal, there are philosophical problems associated with it. A key concern is that when a governmental or private organization takes actions affecting society, the organizations rarely, in fact, provide compensation for those who do not gain. Thus the potential superiority is truly only a potential one, and rarely realized in practice. Numerous government programs, for example, are paid for by a larger group of persons than those who benefit. Other drawbacks and concerns about the philosophical basis for CBA are described later in this chapter.

Putting aside concerns about the basis used to compare alternatives, how does CBA quantify the net benefits under the alternatives? These benefits are determined by the willingness of individuals to pay to receive benefits or to avoid the costs or negative impacts under the alternative. However, quantifying benefits like improved safety or enhanced aesthetics is difficult. CBA is particularly focused on estimating the costs and benefits of the decision alternatives when no market exists for these benefits and therefore market prices cannot be used to estimate people's values.

Although CBA is oriented toward public-sector decisions, we can make a case for its use by private organizations that wish to adopt a social or public-sector perspective. In particular, an organization may wish to adopt such a perspective when addressing a decision that has impacts beyond the boundaries of the corporation and its stockholders.

Traditional methods used by private companies to evaluate alternative investments, such as comparing the discounted cash flow under each alternative, ignore the external effects of the decision. For example, while the firm may benefit most under one alternative, that alternative may impose costs and negative impacts on outside stakeholders such as local residents and governmental agencies.

Frost (1975) expressed well why there may be a rationale for private entities to use this publicly oriented decision aid for some of their decisions:

> Even quite normal business decisions are likely to have effects on such factors as noise, pollution, visual amenity, and so on, and it is likely that businessmen will have to take these into account more and more. If they do not, they will be met with apparently arbitrary restrictions, which could sometimes have been avoided if the situation had been analyzed from the community's point of view in the first place. (p.9)

In light of increasing regulation of industry activities affecting public safety during the last decade, Frost's advice seems prophetic. Some of these restrictions on industry activities may stem in part from the lack of evidence of industry's recognition of public concerns in its decisions. Industry's use of what is essentially a decision aid for public-sector problems may provide a means to explicitly recognize societal values in its decisions, and thereby help foreclose the perceived need for expanded regulation.

8.2.2 History of Cost–Benefit Analysis

James Dupuit, a French economist and engineer, is considered the originator of the theory underlying CBA. His writings discussed the need for a way to formally evaluate the merits of public expenditures. In 1844 Dupuit wrote an essay titled, "On the Measurement of the Utility of Public Works." This essay introduced the concept of consumer surplus, although this term was actually created later by Alfred Marshall.

Consumer surplus refers to the extra value consumers have for a product over what they pay for it. Dupuit argued that, as a minimum, the benefit of a project to society is the quantity of project output consumed, times the price paid for this output. For example, the benefits of an energy or water supply are at least equal to the number of units of energy or water purchased, times the price paid per unit. But a project offers benefits in excess of this minimum, since surely there are some individuals who are willing to pay more than the price charged for the good. Consumer surplus is the extra value due to consumers' willingness to pay more than the price charged, to receive the benefits from the project. For a water supply, the consumer surplus is the extra amount consumers would pay per gallon for the water they buy, times the number of gallons they buy.

To evaluate a public project, it makes sense to include the consumer surplus for the project, not just the price paid for the output of the project. This is particularly true if the prices are set artificially, or where there is no competitive supplier of the goods provided by the project. Dupuit's essay launched a period of thinking and writing about this issue.

Although the standards for applying CBA are continuing to evolve, it is safe to say that CBA has become the most broadly accepted and applied formal decision aid used in the government sector. The Rivers and Harbors Act of 1902 is generally recognized as the first expression of CBA principles by the federal government. This legislation stipulated that in its project evaluations, the U.S. Corps of Engineers must factor in both the benefits to commerce and the estimated costs. More formal application of CBA began in 1936 with the U.S. Flood Control Act. Projects addressed by this legislation were to provide benefits in excess of their costs; benefits should be "to whomsoever they may accrue." Thus, inequities between who pays and who gains were acceptable so long as the net benefits were positive.

From the 1940s and onward, many different agencies including the Soil Conservation Service and the Bureau of Reclamation implemented the principle first adopted by the Corps of Engineers with the Flood Control Act. Different agencies implemented CBA in somewhat different ways. The last few decades have seen efforts within the government to standardize how CBA is applied. In particular, the Federal Interagency River Basin Committee, the Bureau of the Budget (later Office of Management and Budget), and the Water Resources Council have been key players in this effort to standardize CBA.

Many published reviews of public-sector CBA studies exist. Some representative examples address air pollution control (Merkhofer 1987), disease screening (Walker and Whynes 1991), earthquake safety (Sarin 1983), environmental protection (EPA 1987), medical treatments (Torrance, Boyle, and Horwood 1982), nuclear safety (Ulvila and Thompson 1988), product safety (Rodgers and Rubin 1989), waste management (Siegel 1991), water pollution control (U.S. Government Printing Office 1975), and wildlife protection (Boyle and Bishop 1987).

Some published applications addressing industrial health and safety include Anderson (1993), Birkner and Salzman (1986), Fleishman and Hogh (1989), and Goyal (1986). Application of CBA to nuclear plant safety is addressed by Bohn (1992).

Academic work on theoretical and practical issues of cost–benefit analysis has continued along with its widespread application. Since the 1950s, research has focused on developing a firm theoretical basis for CBA, and on refining the tools and techniques for its use. Some key contributors to this work include references at the end of this chapter and in the Topical Bibliography of this book.

8.2.3 Decisions/Situations Most Appropriate for Cost–Benefit Analysis

Cost–benefit analysis is well suited to public decisions where the objective is to maximize the aggregate benefits to those served by the agency. The breadth and depth of a cost–benefit analysis can vary with the magnitude of the decision and the resources available for the analysis. However, CBA tends to be relatively intensive in the time, effort, and data required for the analysis. Thus CBA may be particularly appropriate for decisions that involve significant economic costs, economic or non-economic benefits, and/or negative impacts, and particularly where the decision maker must be able to document a rational basis for the decision. Published CBA studies typically address programs costing millions to billions of dollars. In addition, there are some public policy decisions for which the use of CBA or a variation on CBA is mandated by law or by agency regulation.

Private-sector decisions for which CBA is most appropriate include:

- Decisions on what stand a company should take, if any, on a public policy issue that affects the company. In such a situation, the company may wish to perform its own CBA in order to base its positions on an analytic approach that is consistent with the approach used often in the public sector.
- Decisions about the company's own projects or investments that may have spillovers to the public. CBA may be particularly appropriate for decisions that will be reviewed or regulated by public agencies.

In these contexts, a private organization's use of CBA may result in a greater emphasis on societal goals and concerns over the organization's own objectives. There may be sound reasons for this, both philosophical (e.g., corporate good will) and practical (e.g., societal scrutiny of decisions). However, because this societal emphasis is implicit in CBA, the organization that uses CBA should recognize that decisions resulting from CBA may not be the same as those that would result from using a decision aid that emphasizes the organization's own preferences.

Cost–benefit analysis may also be appropriate for situations in which there is an opportunity for a private organization to share some of the costs of a project with the public. For example, suppose a company is evaluating the merits of road improvements near its plant to facilitate evacuation in the event of an accident. This company might use a cost–benefit analysis to support its appeal that the local government assume part of the cost of these improvements, since a better road may provide additional benefits to the local economy.

8.3 EXPLANATION OF COST–BENEFIT ANALYSIS

8.3.1 Key Concepts of Cost–Benefit Analysis

Cost–benefit analysis is an approach for estimating and evaluating the net benefits associated with alternatives that impose costs and benefits on society. There is no one procedure for CBA, nor is there unanimous agreement on the proper steps of the process, how to estimate the inputs to the analysis, or what decision rule to use to choose among alternatives. In what follows, we describe the basic steps of CBA as it is usually applied, and try to sensitize the reader to some of the issues that underlie differences in the various approaches. Later in this chapter, we will describe some extensions of CBA and other approaches that are related to CBA.

A key principle of CBA is that the analysis should be driven by aggregating the preferences of the individuals in society that are affected by the decision. To determine the scope of the analysis, the required inputs, and the decision rule used to choose from among the alternatives, the analyst should keep this societal focus in mind; it helps resolve how to tackle these issues. To do this, the analyst must set aside personal opinions, including viewpoints of those who make or influence the decision. No one individual is accorded more weight in the analysis than any other individual, except to the extent that an individual bears a greater share of costs or benefits. CBA has been extended to differentially weight individuals, as described under Distributional Analysis in Section 8.5.3.

8.3.1.1 Scope of the Cost–Benefit Analysis

Let us consider first how this principle of adopting society's viewpoint determines the scope of the analysis. Scoping issues were addressed from a general standpoint in Chapter 1. The scope is defined by the time horizon, the jurisdiction or portion of society addressed, and the breadth of impacts specified for the analysis. The scope should be broad enough to incorporate all individuals potentially affected by any alternative under consideration, and the time horizon should address all potential future impacts. However, an impact can almost always be found on a population well outside the immediate community, and far into the future.

For practical reasons, we avoid addressing impacts that are too small, or too similar across the different alternatives to affect the results. If an impact is large enough to alter the choices that most individuals would make acting alone, then the analyst should consider including this impact in the scope of the analysis.

As an example, consider a decision by a chemical manufacturer between expanding an existing plant, or building a new plant in a different region of the country. In using CBA to evaluate the alternatives, the firm might want to include the potential impacts on the communities near each of the two sites. Beyond this, it might be appropriate to consider the impacts on communities affected by the transportation of materials and products to and from each possible site. Extending beyond this, there may be impacts on the larger population within each state in which the plant might locate, e.g., through increased state tax revenues stemming from new jobs created by the plant. However, incorporating this latter effect is unlikely to favor one alternative over another; if the number and type of new jobs is similar at both locations. In deciding between the alternatives, excluding state tax revenues is unlikely to alter the results. Similarly, restricting the time horizon of the analysis to 15 years should be adequate unless impacts after 15 years significantly favor an alternative.

8.3.1.2 Costs and Effects Incorporated into the Analysis

We define two categories of inputs to the analysis, costs and effects. We have purposely broadened "benefits" to include both positive and negative effects.

- *Costs* are what society pays (i.e., in dollars) to undertake the project, over and above what it would pay to do nothing. Some authors refer to this component as *economic costs*.
- *Effects* are what the project provides to society as a result of its implementation. Effects can include both positive benefits such as job creation and safety improvement, and negative impacts (or *disbenefits*) such as environmental harm and residential dislocation. Some authors refer to effects as *social costs*. In an analysis where the net effects of the project are consistently positive, effects are often called simply *benefits*.

The inputs to the analysis will consist of estimates of the costs and effects to individuals over the time horizon. In some cases it is hard to separate costs and effects; in practice this distinction is usually not important since the decision rule gives the same result regardless of how the inputs are categorized. (As described further below, this result is true for most but not all decision rules commonly used in a cost–benefit analysis.) To quantify these inputs, the guiding principle is as expressed earlier: the analysis reflects how much individuals are willing to pay for these inputs. Specifically, in addition to costs, we must estimate the willingness-to-pay to receive the positive effects and eliminate the negative effects of each alternative.

8.3.1.3 Decision Rule

Once the scope is determined and the inputs obtained, the final step is to evaluate the alternatives. The analysis must determine the overall attractiveness of the various alternatives given differences in the magnitude and timing of costs and effects. Different criteria have been suggested, but again, the guiding principle is that the analysis should reflect the preferences of affected individuals. In the past, costs and benefits of alternatives were often compared using a *benefit–cost ratio*. For each alternative, the benefit–cost ratio is calculated by taking the total benefits divided by the total costs of that alternative. These ratios are then compared among alternatives, and the alternative with the highest ratio (exceeding 1) is preferred. In the last couple of decades, the benefit–cost ratio has been criticized as a decision rule and been demonstrated to lead to undesirable results. A chief criticism is that the benefit–cost ratio can lead one to choose a smaller program when a larger program might provide greater net benefit to society. As a replacement for the benefit–cost ratio, there is currently broad acceptance of the *net benefits* rule: the alternative for which the net effect (i.e., total of positive effects less the costs and negative effects) is highest, is the best alternative. This criterion has the appeal of being consistent with rational behavior expected of individuals making personal decisions. The next section discusses in more detail various decision rules that can be used with CBA.

8.3.2 Description of Cost–Benefit Analysis

In what follows we assume that earlier steps of the decision process have already been completed, such as formulating the problem and defining the alternatives.

8.3.2.1 Key Steps

The following discussion will focus on five key steps of cost–benefit analysis:

- Decide on the scope of the analysis
- Assess the costs of each alternative
- Assess the effects of each alternative
- Apply the decision rule for choosing among the alternatives
- Perform a sensitivity analysis to key inputs and assumptions.

In what follows we discuss each of these steps in turn.

8.3.2.2 Deciding on the Scope of the Analysis

Sassone and Schaffer (1978) suggest issues to resolve early in a cost–benefit analysis:

- *Project scenario:* What costs and effects are included? For example, a decision on what type of pipeline to use may be part of a larger decision on plant modernization. The analyst must determine whether to address the pipeline material decision alone, or to also address the costs and effects of the larger decision of plant modernization. It is also generally not necessary to include costs and effects that are the same under all alternatives, since these inputs will not alter the decision. Issues excluded from the quantitative analysis may warrant qualitative treatment through a written discussion provided with the quantitative analysis. The analyst should identify such issues in advance so that they can be addressed qualitatively along with the quantitative results.
- *Baseline scenario:* Against what baseline will costs and effects be evaluated? The baseline may be a "do nothing" alternative, or it may be a status quo in which some investments or effects will occur in the absence of any of the possible alternatives. The analysis will be biased if the alternatives are burdened with costs or effects that would happen anyway. Costs and effects should be measured incrementally with respect to a well-defined baseline scenario.

- *Societal scope:* What portion of society is included? This may be defined by the governmental jurisdictions, geographical boundaries, and/or demographic characteristics of the group of individuals affected by the decision.
- *Constraints on the project:* What budgetary, legal, social, political, and institutional constraints are assumed?
- *Define/Refine alternatives:* For each alternative, what variables can be changed to optimize that alternative? For example, a decision on the type of pipeline to use may depend on the costs from various suppliers. The analysis should, if possible, first resolve this issue so that the alternative is clearly defined. Where this is not possible, it may be necessary to define additional alternatives. A related concern is that it may be possible to combine alternatives into other, attractive alternatives. If so, then these additional alternatives should be addressed also; it is erroneous to directly combine results for each separate alternative to estimate the impact of the combination.
- *Discount rate:* How are future streams of costs and effects compared to one another? Generally the analysis will compute a present value for each of these streams, employing a constant discount rate. This rate needs to be specified. See Chapter 2 of this book for a discussion of discounting.
- *Time horizon:* How far into the future will costs and effects be considered? The time frame should be long enough to incorporate future impacts of sufficient magnitude to affect the decision if they were included. Usually the project lifetime is used. Due to time discounting of future costs and effects, impacts beyond 15 years are often too small to warrant inclusion. However, this will depend on the magnitude of differences among the alternatives in the impacts that occur beyond this time frame.

In addition to these issues, it is useful to resolve practical issues up front such as what data are available, how much time, staff, and other resources are available, who is the audience for the analysis, and how this audience will expect the results to be summarized and presented. Doing this in the early stages of the analysis helps ensure that the scope is appropriate.

This list of issues suggests that to define the scope of the analysis, some crucial assumptions need to be made. As part of the cost–benefit analysis, an analysis of the sensitivity of the results to the key assumptions used in scoping the analysis should be performed. For example, the analyst might assess the impact of the time horizon and discount rate on the results. It is sometimes appropriate to report the results separately under two different sets of assumptions, particularly if the results change significantly in each case and there is disagreement on which set of assumptions is appropriate.

8.3.2.3 Assessing the Costs of Each Alternative

Costs can include the materials and labor required to undertake an alternative (construction, implementation, etc.), and the ongoing expenses (operation, maintenance, repair, etc.) Costs should include costs for study and management under each alternative, interest charges, and other transaction costs.

The guiding principle in assessing the costs of each alternative is that the costs should reflect how much the individuals in society affected by the decision are willing to pay to avoid these costs. In a private-sector decision, the costs are typically paid by a private corporation. Thus the company is the affected "society" for purposes of the cost side of the analysis. In a public-sector decision, costs are more complicated, particularly if taxes or user fees are involved, and the costs impacts may be different for different individuals.

Regardless of what set of individuals is relevant to the analysis, for each alternative the costs should reflect only the incremental effects to these individuals over what would have occurred under the baseline (do-nothing) scenario. The following discussion focuses on how these principles drive the estimation of costs, but we will see that the same principles apply to estimating effects.

An engineering or accounting estimate of costs may not properly reflect how strongly the firm or society wants to avoid these costs. For example, the cost to expand an operation into an unused portion of a facility may be small, particularly if the facility is already owned and fully paid for. However, the expansion may in fact foreclose other opportunities to use this portion of the facility more profitably, perhaps even to rent out the facility. The true worth of the unused space is probably much greater than what it currently costs for that space. This true worth is referred to as the *opportunity cost*. The analyst should strive to assess the opportunity costs of the alternative, rather than relying on cost estimates based on historical figures or engineering estimates.

The concept of opportunity cost is crucial to accurately assessing the cost inputs to a cost–benefit analysis. While the example above was for a private facility, the gap between the actual costs and the true worth of these costs can be considerable for a public project. Consider a new installation to be constructed on publicly owned property. If the government already owns the property, the land costs are essentially zero. However, the value of the land to society may be large. For such a project, a cost–benefit analysis will generally consider the opportunity cost of the land, by estimating how much society would pay to avoid having property of equivalent quality, location, and extent removed from private ownership.

Sometimes a *market price* can be used to estimate a cost. Ideally, there is a fair and open market for the product, service, or other type of benefit whose costs are to be estimated, in which case the market prices can be used directly. By "fair and open market," we mean that there are sufficiently many sellers and buyers of the product or service so that none can influence the price at which the product or service sells. A fair and open market also means that there are no artificial influences that restrict or expand supply or demand for the product or service, for example, through duties, tax incentives, or regulations. The price for a product or service in a fair and open market equals the true worth of the product or service to individuals.

For a private undertaking, the cost of land may be estimated simply from the cost of equivalent land in the (presumably fair and open) local real estate market. In other situations, especially for public-sector decisions, we need to estimate the true cost by using *shadow prices*. A shadow price is simply the price at which the product or service would sell in the market if a fair and open market existed. For example, in the example above involving a government facility constructed on public land, there is no market for public land. The shadow price of the land can be estimated, however, from the price at which equivalent property would be sold on the private market.

If no free and open market exists, we need to use some other approach to measure or infer the willingness-to-pay to avoid the costs. One such approach is inferential: identify an analogous activity for which the willingness-to-pay can be obtained. For example, the value of a flood control project might be inferred from the prices paid for private flood insurance.

A second approach to estimate the willingness-to-pay is through direct assessment: conduct surveys or interviews with individuals who will be affected by the decision, to estimate their willingness-to-pay to avoid the costs. For example, a survey might be designed to elicit an individual's preferences or choices among various alternatives, and from this, the willingness-to-pay inferred. Careful design is required to avoid strategic behavior by participants, for example, by enabling participants to exaggerate how they feel about the costs and benefits of a program. A similar approach is to assess the willingness of individuals to be compensated for the negative effects of a program. This, too, could be obtained through surveys or interviews.

There are a variety of ways to assess individuals' willingness-to-pay, and these approaches differ on their accuracy, freedom from bias, and ease of administration. The literature on consumer preference and market research addresses these issues.

To summarize, possible approaches for estimating cost inputs to a cost–benefit analysis include the following (in order from most- to least-preferred approach):

1. If possible, determine the prices that exist on a fair and open market for the product or service
2. If no such market exists, impute the cost based on a shadow price, that is, a similar or analogous market that is fair and open
3. If no fair and open market can be identified, infer the willingness-to-pay based on the price paid for an analogous activity
4. Measure the willingness-to-pay directly through surveys or interviews.

8.3.2.4 Assessing the Effects of Each Alternative

As in the preceding discussion of assessing costs, the guiding principle is that the estimates of the effects should reflect how much individuals are willing to pay to receive the beneficial effects and to avoid the negative effects. Assessments should be based on the incremental effects over what would have occurred in the absence of the project. Assessments should assume that individuals are fully informed about the effects that occur or would occur under each alternative. If an alternative imposes adverse effects, these effects should be incorporated into the analysis whether or not individuals react to them or are compensated for them.

Each alternative may have numerous and wide-ranging effects. While it is impossible to draw clear distinctions, it is helpful to categorize the effects into primary, secondary, and tertiary effects:

- *Primary effects* include the products and services provided as the primary purpose of each alternative. Products might include water and recreation, while services might include waste disposal and security protection.
- *Secondary effects* include those that are not the immediate purpose of the project, but which can add or subtract significantly from the primary effects of the project. Secondary effects can include those that are sometimes referred to as "spillovers" or "externalities." *Externalities* are effects of an activity that are borne by persons or entities other than those undertaking or benefiting from the activity. Examples are road and land improvements that result from a new chemical plant, jobs created by a plant expansion, and perceived health risks from a chemical plant.
- *Tertiary effects* include additional products and services or other effects that are stimulated by the project. For example, the immediate jobs created for constructing or operating a project can infuse the local economy by stimulating the growth of business, housing, and jobs.

Note that these categories are based on the relationship of the effects to the goals of the project, not to the magnitude of the effects.

In the literature on CBA, various terms are used to categorize the types of costs and benefits that can be addressed in the analysis. Examples are *direct* and *indirect* effects, *intermediate* and *final* effects, and *inside* and *outside* effects. The basic differentiation in effects suggested by these terms are similar, and our categorization of primary, secondary, and tertiary effects captures the important distinctions.

Another distinction made is between *tangible* and *intangible* effects. We treat "intangible" effects as simply one of many types of costs or benefits that might be included in an analysis. Intangible effects are discussed later in this section, when we describe methods for quantifying inputs to a CBA.

As mentioned in the discussion of scope, the analyst needs to use judgment to decide on what effects to include in the analysis, particularly at the tertiary level. The analysis should exclude effects that are unlikely to alter the result; that is, their magnitude is too small relative to other effects to justify the effort required to quantify them. It can also be argued that unless the project is very large, its tertiary effects are far smaller than equivalent effects (and the uncertainty about these effects) from numerous other projects that society undertakes.

The possible approaches for estimating these effects are exactly the same as the approaches described above in Section 8.3.2.3 for costs:

- If possible, determine the prices that exist on a fair and open market for the product, service, or other type of effect. For example, the value of the service provided by an overnight mail program can be determined from the prices charged by the highly competitive market for this service. The value for property repossessed from a population displaced by a chemical plant expansion can be determined from the cost of similar-quality real estate in the area (prior to any announced plans for the plant that might impact property values).
- If no such market exists, impute the value based on a shadow price, that is, the observed or expected price in a similar or analogous market that is fair and open. For example, the value for a new public campground might be inferred from the prices charged at a nearby private campground. The value for unused company-owned land that will be developed in a plant expansion is not zero, even though there is no cost to acquire the land. Instead, the value is equal to the price the land (or similar land in the area) sells for in the open market for undeveloped industrial property.
- If no fair and open market can be identified, infer the willingness-to-pay based on the costs for an analogous activity. For example, the value for fire protection can be estimated from the prices paid for fire insurance. The value for a degraded environment that a community perceives a plant expansion to pose might be estimated from differences in otherwise similar properties in communities with and without a nearby plant.
- Measure the willingness-to-pay directly through surveys or interviews. For example, the value for environmental aesthetics is difficult to assess with the preceding approaches, but might be estimated from surveys or interviews.

A common stumbling block when assessing the effects of a project is to double-count effects, particularly the benefits of the project. For example, the effects of shutting down a chemical processing plant might include the value of lost production, which could be quantified in terms of the prices of the products. We would double-count this effect if, in addition to including the lost revenues, we attempted to add lost benefits to consumers of the products; these benefits are already included in the price they pay for the products.

Even where a fair and open market exists for a product or service, the market price may not reflect the true value of that product or service to society. This is because there may be some individuals who will pay a higher price for the product or service than is charged on the market. The concept of consumer surplus is used to estimate the true value of the product or service to society (Singer 1976). This requires determining the relationship between how much of a product or service will be consumed, and the price charged. Economists express this relationship with a *demand curve* as shown in Figure 8.1. The demand curve shows the quantity of product or service purchased, as a function of the price charged. If the demand curve can be obtained, then the value of the product or service can be estimated as indicated in Figure 8.1.

Economic theory has shown that the demand curve is a good approximation to consumers' marginal value for increasing consumption. In Figure 8.1, if p_0 is the price charged for the product or service, then q_0 is the quantity that will be purchased. At that price, some consumers' values are exactly equal to

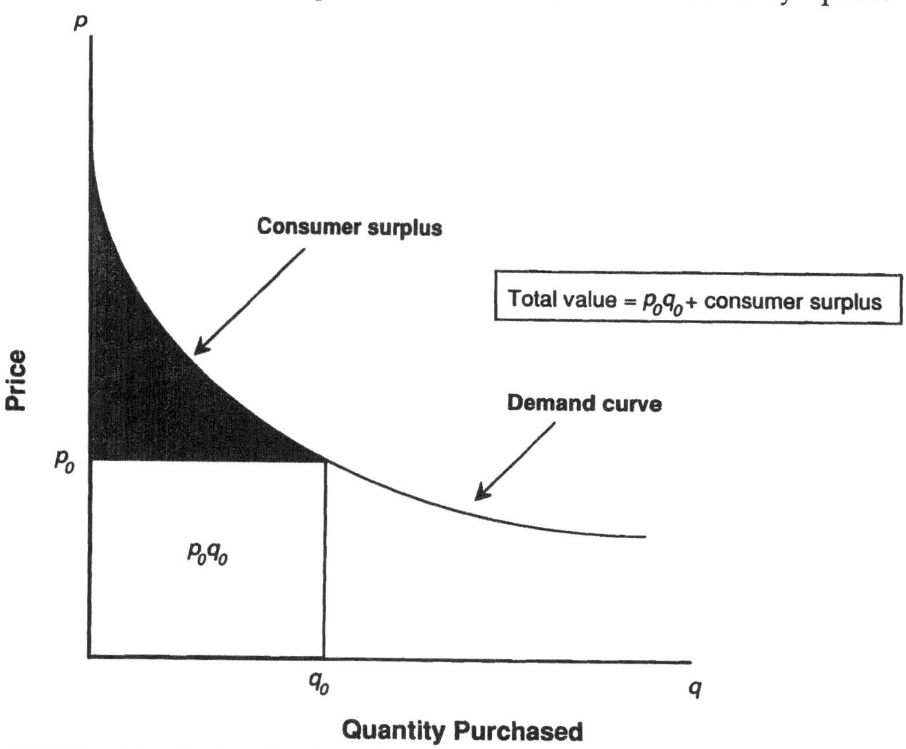

FIGURE 8.1. Estimating the value of a product or service using a demand curve.

the price. Consumers whose values for the product or service are less than p_0 will not purchase the product or service. Those whose values exceed p_0 not only purchase the product or service, but they reap extra value in excess of p_0. To determine the total value of the product or service to society, we include the shaded area in Figure 8.1. This shaded area is called the *consumer surplus*, and represents the difference between what consumers are willing to pay and what they are charged. The total value is therefore p_0q_0 plus the consumer surplus, or simply the area under the demand curve from $q = 0$ to $q = q_0$.

Unfortunately, the demand curve usually cannot be readily obtained, and thus the consumer surplus is either roughly estimated or, more often, ignored. In many CBA applications, the value of a product or service is estimated as simply p_0q_0. This tends to underestimate the benefits of the project.

Salvage value, or the worth of facilities and equipment at the end of the project lifetime, should be included in the analysis if the value is significant. Salvage value also might be relevant to an alternative involving replacement or abandonment of an existing operation before the end of its life. Salvage value should be estimated from the true market value of what is to be disposed, in its then-current condition, net of any transaction costs and tax effects. Salvage value should not be determined by the depreciated or "paper value" of the property.

For cost–benefit analysis as with any decision aid, how to address alternatives involving the risk of fatalities is a difficult issue. The *value-of-life* issue is problematic given CBA's emphasis on evaluating alternatives based on how much the costs and effects are worth to individuals in society. Applying this principle presents a need to know society's market value for reduction in the risk of fatality. Section 2.5 summarizes ways to address value-of-life issues. A useful approach in a cost–benefit analysis is to estimate what individuals are willing to pay for a statistical reduction in the risk of fatality, using one of the techniques described in Section 2.5.

Intangible effects such as visual aesthetics, noise, and prestige are also difficult to quantify. There is no agreement on how to incorporate intangible effects into the analysis. Purists argue that intangibles should be quantified in dollar terms, while pragmatists argue that it is adequate to describe and perhaps even to quantify the magnitude of the impact, but unnecessary or inappropriate to assign a dollar value (Riggs and West 1986). Another approach is to exclude intangibles from the analysis initially, unless they are a primary effect of the decision. Later, one or more intangible effects can be included in the analysis if a less preferred alternative appears to have sufficient advantages on these dimensions.

8.3.2.5 Identifying the Preferred Alternative

Different approaches have been suggested for how to combine the costs and effects to determine the recommended alternative. Chapter 2 discusses economic evaluation of projects in more detail, including alternative comparison rules.

The *net benefits rule* currently has broad acceptance and is recognized as superior to other approaches. The net benefits rule recommends selection of the alternative for which the discounted total benefits, minus the discounted total costs, is greatest.

As mentioned earlier, traditional approaches to CBA involved maximizing the *benefit–cost ratio*. That is, in evaluating each alternative, one would compute the total discounted benefits and divide by the total discounted costs, and then choose the alternative for which this ratio is greatest. The problem with using the benefit–cost ratio to compare projects is that it can lead one to choose a very small project with a high ratio; a larger project might provide attractive returns even if its benefit–cost ratio is not as high. Also, when using benefit–cost ratios, the results can change depending on whether an adverse effect is treated as a negative benefit or as a cost. Finally, the benefit–cost ratio may be less than 1 for all alternatives, in which case one would consider all projects undesirable. Despite its limitations as a basis for selecting an alternative, the benefit–cost ratio may be useful as one aspect to describe the alternatives.

Before applying the net benefits rule, it is important first to optimize each alternative. If it is possible to scale an alternative up or down (e.g., by increasing or decreasing the investment in that alternative), the alternative should be sized to the point at which net benefits are maximized for that alternative. Typically, an alternative has decreasing returns to scale, as illustrated in Figure 8.2 for a hypothetical alternative. With each extra dollar invested, there is additional benefit, but as the project grows in size, the benefit per additional dollar decreases. Assuming there is no budget constraint, the optimally sized project is the one whose marginal benefit equals its marginal cost. An extra dollar invested produces only one extra dollar in benefit, hence the net benefits do not increase by investing an additional dollar. In practice, projects usually cannot be sized continuously, but instead, only discrete alternative sizes are feasible. In this case, the analyst should evaluate the net benefits for each feasible size of the project to find the size that is optimal for that alternative. Of course, any alternatives for which the net benefits are less than zero should be eliminated from consideration.

8.3.2.6 Subjecting Key Inputs and Assumptions to Sensitivity Analysis

Key inputs to the cost–benefit analysis should be subjected to a sensitivity analysis. The purpose of sensitivity analysis is to determine how the results change under different assumptions, or under different numerical inputs. In particular, the analyst should determine and highlight where the recommended alternative changes under reasonable assumptions and inputs other than those used in the base analysis.

As an example, a key input to cost–benefit analyses of chemical process safety decisions is likely to be the assumed value for the avoided fatalities that

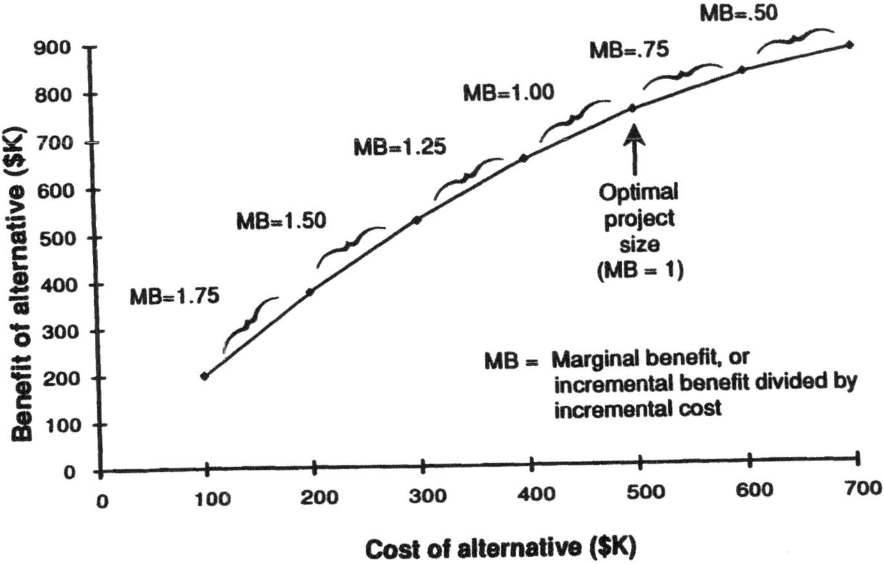

FIGURE 8.2. Decreasing returns to scale for a hypothetical alternative.

result from alternative safety improvements. There is wide disagreement as to what quantity is appropriate, and based on the approaches outlined in Chapter 2, values ranging from $100,000 to $100 million per avoided fatality may be appropriate. Sensitivity analysis can be performed by computing the results under different settings of this input, varied over this range. In reporting results, points at which the preferred alternative changes should be highlighted.

Figure 8.3 provides an example of how the results of a sensitivity analysis can be communicated.

8.3.2.7 Reporting and Presenting Results

It is important that the assumptions underlying a cost–benefit analysis be communicated along with the results. In particular, to quantify inputs often requires a combination of art and science, so it is crucial to the acceptance of the study that any such elements be documented and the results supported through sensitivity analysis.

The report and presentation should document key scoping assumptions such as the time frame and population addressed. What costs and effects are included and excluded from the CBA should be stressed. Key quantitative inputs and their sources of estimation should be provided, including reference to relevant literature or past CBA studies.

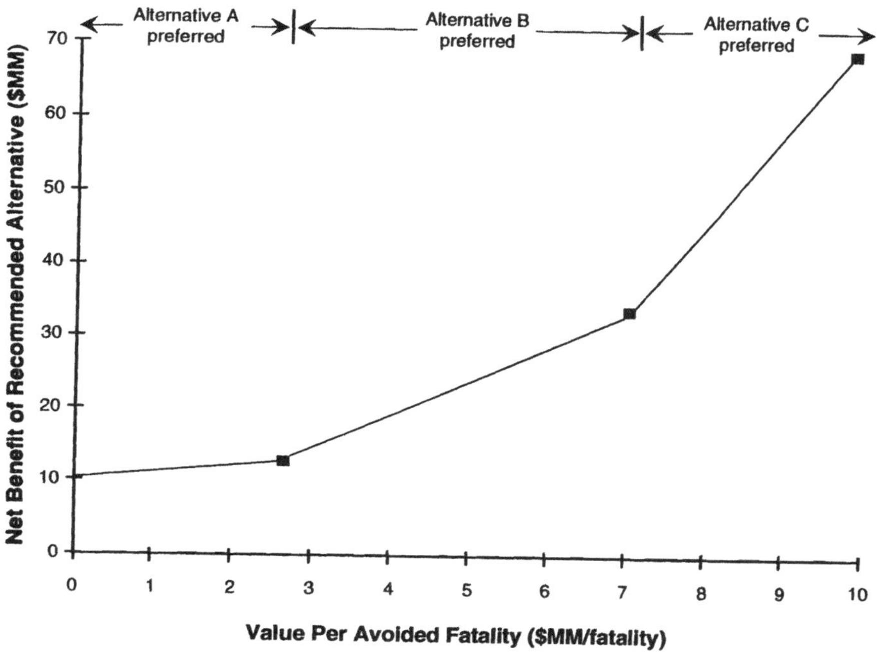

FIGURE 8.3. Sensitivity analysis to assumed value per avoided fatality (hypothetical results).

The results of the baseline analysis should be reported, and these should highlight the key elements contributing to the net effect (i.e., the key cost and effect components). The results should also emphasize alternatives that are close in value to the one preferred.

Results of sensitivity analyses should be provided, with emphasis on those inputs or assumptions to which the preferred alternative is highly sensitive (i.e., changes as that input is varied). Figure 8.3 illustrated how sensitivity analysis results can be summarized graphically. To help guide the decision maker, the analyst may want to provide a rationale for the approach used to estimate key sensitive inputs, and in some cases, provide a rationale for changing this input if there may be disagreement over the approach used. Discussion of the results can then focus on the inputs, if appropriate, rather than on the baseline results.

Finally, the report and presentation summarize the recommended alternative, including suggestions for further analysis or data-gathering if appropriate.

8.4 CASE STUDY: CHLORINE LOADING FACILITY

An Application of Cost–Benefit Analysis to Chemical Process Safety

This application involves Case One, the chlorine loading facility, which was introduced in Section 5.3. There is a second application of this case later in this book, that illustrates the use of payoff matrix analysis (Section 10.4). In this example, we intentionally keep the scope and analysis artificially simple to make the problem practical for demonstrating the decision aid.

A chlorine loading facility is situated close to a residential community. Company management prepared a quantitative risk analysis to help better understand both the frequency and impact of an accidental chlorine release at the facility.

The baseline study produced the societal risk curve shown in Figure 8.4 (this is the same curve as shown in Section 5.3 when this case was introduced.). On the basis of this estimate, the plant manager decided to allocate more resources to fully explore the options for reducing risk at the facility.

The plant manager felt that a number of simple alternatives were available to reduce the risk, but did not want to rule out the alternatives of relocating the loading facility or even shutting it down. The plant manager felt that the analysis of possible alternatives should use a societal point of view that would trade off the benefits to the company and the community against the risks of the operation and the costs of the alternatives considered.

An analysis team was formed to perform a cost–benefit analysis of the alternatives, led by a project engineer with training and experience in this

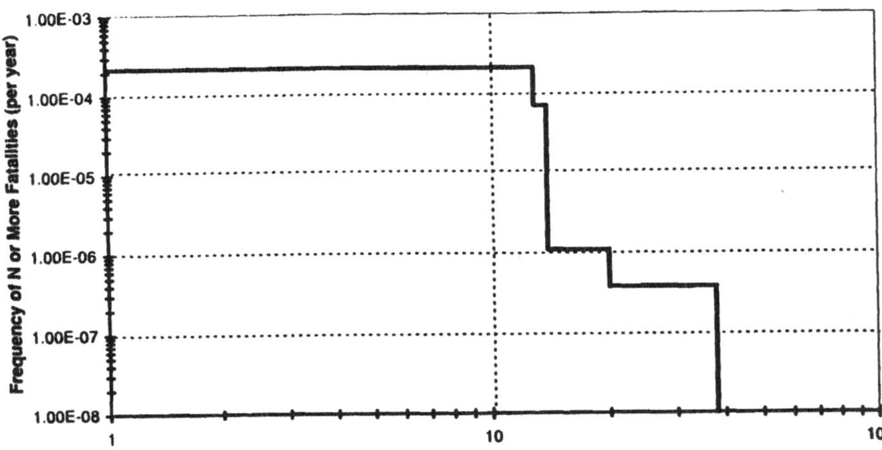

FIGURE 8.4. Societal risk curve resulting from initial quantitative risk analysis.

methodology. The project engineer was instructed to conduct the analysis using input as required from operations, maintenance, plant process engineering, logistics, and corporate-level legal, engineering, and business groups. The project engineer was asked to regularly update the plant manager on the progress of the study, to help ensure that the approach and assumptions were consistent with the plant manager's views.

In the remainder of this section, we describe the key steps of the cost–benefit analysis including developing the scope, assessing the costs of each alternative, assessing the effects of each alternative, and applying the decision rule for choosing among the alternatives.

8.4.1 Scope of Analysis

In discussions with the plant manager and brainstorming with members of the analysis team, a number of alternatives were considered. An initial list of 23 alternatives, shown in Table 8.1, was developed.

This initial list of alternatives was reduced through a preliminary screening. Specifically, the analysis team eliminated alternatives that were clearly inappropriate, could be easily and inexpensively accomplished, or were required by corporate standards. Some of these alternatives could help reduce the risk, and the cost to analyze them would likely approach or exceed the cost to implement them. These alternatives are designated by "Do anyway" in the last column of Table 8.1, and the plant manager undertook the immediate completion of these actions. The team consolidated some of the remaining alternatives into a General Upgrade option, as shown in the last column of Table 8.1. After this initial screening process, 11 alternatives remained:

- Do Nothing Further (alternative 1 in Table 8.1)
- Shut Down (2)
- Change Transportation (3)
- Move Cars (8)
- Enclose (10)
- Move Loading (11)
- Buy Greenbelt (12)
- Move Plant (13)
- Redesign Car (14)
- Refrigerate (17)
- General Upgrade (5, 7, 9, 19, 20, 23).

To scope the problem, the following assumptions were made:

- A 15-year time horizon was used.
- The analysis was restricted to the chlorine loading operation.
- The baseline against which the costs and effects were evaluated was the current operation. Thus the costs and benefits of the Do Nothing

TABLE 8.1
Initial List of Chlorine Loading Alternatives and Disposition after Preliminary Screening

Alternative	*Disposition after preliminary screening*
1. Do Nothing Further	Retain
2. Shut Down the facility (go out of business)	Retain
3. Change transport mode from rail cars to cylinders, pipeline, etc.	Retain
4. Change the form in which the chlorine leaves the plant	Not acceptable to customers
5. Reduce pipe size	Include in General Upgrade option
6. Install isolation valves on all pipelines that feed chlorine	Do anyway
7. Install scrubber on relief valve	Include in General Upgrade option
8. Move flammable tank cars away from chlorine loading site	Retain
9. Install monitoring/shutdown systems	Include in General Upgrade option
10. Enclose the chlorine loading operation	Retain
11. Move the location of the chlorine Loading to elsewhere in the plant	Retain
12. Buy Greenbelt (i.e., buy nearby houses and relocate population	Retain
13. Move the entire plant to another site	Retain
14. Redesign the tank car	Retain
15. Improve emergency response systems (e.g., fire fighting resources)	Do anyway
16. Develop community alarms and evacuation procedures	Do anyway
17. Refrigerate the chlorine	Retain
18. Improve operator training	Do anyway
19. Install spill control sump under tank car	Include in General Upgrade option
20. Install tank car stabilizers	Include in General Upgrade option
21. Install derailing devices	Do anyway
22. Install gravel end stops	Do anyway
23. Increase inspection frequency (equipment proof testing)	Include in General Upgrade option

Further alternative are both zero. This means that the results express the incremental costs and benefits over business as usual if each alternative were undertaken. The costs and benefits of Shut Down should therefore express the societal value for the current chlorine loading operation.

- The geographical scope of the analysis was restricted to the county within which the plant is located.
- Aspects included are:
  —Capital and operating costs borne by the company to implement each alternative
  —Profit from the operation
  —Risk of fatality
  —Incident-related costs of downtime, cleanup and product loss
  —Jobs supported by the operation
  —Tax revenues to the county
  —Company goodwill resulting from action taken to reduce the risk of the operation.
- Aspects excluded are:
  —Risk of injury
  —Jobs and tax revenues indirectly created to service the workers at the plant
  —The county's costs for roads and other infrastructure
  —Costs and effects to state and federal jurisdictions.
- The analysis assumed that the amount of chlorine handled by the facility would not change throughout the 15-year time horizon, nor would the costs and revenues from the operation.
- The analysis assumed that the value per avoided fatality is the same regardless of the number of avoided fatalities.
- Costs are discounted at 6% annually.

8.4.2. Assessing the Costs

The cost side of the analysis includes the discounted capital plus operating costs to implement the alternative, incremental over Do Nothing Further. Capital costs include construction, equipment, and land costs; the salvage value of any displaced facilities and equipment as well as such value remaining at the end of the 15-year time horizon; costs for severance pay, workers' compensation, and relocation costs in the event of displaced jobs; and costs to mothball the facility if it is shut down. Operating costs include changes in personnel, maintenance activities, delivery contract penalties, and costs for disruption of supply contracts (e.g., take or pay agreements).

Detailed estimates of each of these cost components for each alternative were developed by the analysis team. Table 8.2 summarizes the key elements including in the operating cost and capital cost estimates for each alternative.

TABLE 8.2
Key Components Included in Cost Estimates

Alternative	*Incremental Operating Cost*	*Incremental Capital Cost*
Do Nothing Further	None	None
Shut Down	Elimination of personnel, material, maintenance, production, transportation costs	Severance programs, delivery contract penalties, supply contract penalties, mothballing costs, future salvage value
Change Transportation Mode	Incremental staffing, transportation, and maintenance costs for alternative pipeline/cylinder supply to replace 35%/65% of chlorine loading respectively	Cost for land, compressor stations, piping, and other components of pipeline supply; cost for cylinders and bottling plant for cylinder supply
Move Cars	None	Cost to construct rail spur and loading stations
Enclose	Cost for additional staff, insurance reserve, and maintenance	Cost for building, sump, foam system, monitoring system, piping, power supply, scrubber, high stack, scrubber caustic tanks, and other required equipment including installation
Move Loading	None	Cost to construct rail spur and loading station and reroute chlorine line to this new loading area
Buy Greenbelt	Cost for landscape maintenance	Cost for property acquisition, fencing, razing, landscaping, and legal reserves, and community relations
Move Plant	Cost for training, transfers, and restaffing at another site	Cost for land, equipment, and construction of chlorine plant at another site
Redesign Car	None	Cost for design, legal reviews, and tank car fleet construction
Refrigerate	Cost for additional staff, electricity, maintenance, and spare parts	Cost to construct tankage and piping, compression and cooling facilities
General Upgrade	Costs for improved ongoing emergency response systems, improved operator training program, and increased frequency of inspections	Costs for pipeline, valves, monitoring equipment, scrubbers, sump, tank car stabilizers, and other required equipment including installation

Table 8.3 summarizes the results of the cost analysis. Note that with the Shut Down alternative, the capital cost is negative; in other words, it provides a source of revenue rather than cost. This is because of the salvage value of the plant, which is greater than other components of the incremental capital costs under this alternative. However, the substantial discounted operating costs outweigh the capital cost advantage of the Shut Down alternative.

Without further analyzing the effects of each alternative, the analysis team realized that it could eliminate some of them. These alternatives are *dominated*: dominated alternatives are those that cost more than another alternative and whose net effect is smaller. Specifically, Buy Greenbelt dominates three alternatives: Change Transportation Mode, Move Plant, and Redesign Tank Car. Buy Greenbelt costs $24.57 million and essentially eliminates all of the risk. The other three alternatives just mentioned cost 2 to 20 times more than Buy Greenbelt, result in no greater reduction in the risk, and have no other effects that differ from those of the Buy Greenbelt alternative. Thus, these three dominated alternatives were eliminated from further consideration.

As a result of this preliminary screening based on the cost estimates, eight alternatives remained to be evaluated:

TABLE 8.3
Cost Estimates for Each Alternative (Incremental over Do Nothing)

Alternative	*Discounted 15-year Incremental Operating Cost ($MM)*	*Incremental Capital Cost ($MM)*	*Discounted Incremental Total Cost ($MM)*
Do Nothing Further	0.00	0.00	0.00
Shut Down	61.90	–28.30	33.60
Change Transportation	94.86	164.00	258.86
Move Tank Cars	0.00	2.00	2.00
Enclose	0.87	3.99	4.86
Move Loading	0.00	0.90	0.90
Buy Greenbelt	0.05	24.52	24.57
Move Plant	42.56	530.00	572.56
Redesign Tank Car	0.00	144.05	144.05
Refrigerate	3.25	1.50	4.75
General Upgrade	0.17	1.23	1.40

- Do Nothing
- Shut Down
- Move Cars
- Enclose
- Move Loading
- Buy Greenbelt
- Refrigerate
- General Upgrade.

8.4.3 Assessing the Effects

The effects side of the analysis included the following potential effects from one or more alternatives, incremental over Do Nothing Further:

- Loss of profit (posttax)
- Loss of county tax revenues
- Loss of jobs
- Increase of property value
- Reduction of incident-related costs
- Reduction of risk of fatality
- Enhancement of company goodwill.

The Shut Down alternative eliminates profits and county tax revenues from the operation. Current net profits from the operation are $5.1 million per year, and county tax revenues are $178,000 per year. Discounted at 6% annually over 15 years, these effects are $49.53 million and $1.73 million, respectively. under the Buy Greenbelt option, the county loses tax revenues on the displaced housing, currently $2 million per year or $19.42 million discounted over 15 years.

There are currently 100 persons employed at the facility. Shut Down would eliminate all of these jobs, and Refrigerate Chlorine would eliminate 10% of the jobs. The value of these jobs can be estimated by the salaries paid to the workers. In this example, the average salary (including benefits) is $80,000 per year for the workers that would be affected by any alternative or by closure. The discounted present value of these salaries is $77.70 million for all 100 workers.

The analysis team estimated the impact of each alternative on property values as follows. They met with local real estate professionals to discuss the impact of the current operation on the market for nearby real estate. According to this advice, the existence of the plant, rather than the specific potential risk to the community, is the main driver of property values. A single project alone was unlikely to have a major influence on property values, but might contribute cumulatively when combined with other risk-reducing projects.

The analysis team asked the professionals to estimate the impact that eliminating the risk entirely would have on nearby property values. The professionals suggested that the effect would be to make the property values equal to those of other properties in the area located near other similar (but "non-risky") industrial operations. This presumes communication to the public about the action, and public understanding and awareness of the associated risks. The consensus was that the effect would be to increase property values by about 15%, or $3 million based on an estimated total value of $20 million for the nearby residences. The project team estimated that this would be the gain under all of the alternatives with two exceptions. under the General Upgrade alternative, the risk reduction would be substantially less than with the other alternatives, hence the plant would still be perceived as somewhat risky to the nearby population. The estimated gain in property value assuming public awareness was estimated at one-third of the amount under the other alternatives. under the Shut Down alternative, it was believed that the positive impact on property values due to a less risky nearby operation would be outweighed by the adverse impact on property values due to a depressed local economy, resulting in no net effect on property values under this alternative.

The analysis must determine the reduced risk of incidents and their associated costs and possible fatalities, as discussed in Section 8.3.2.4. The incident-related costs of downtime, product loss, and cleanup were estimated to be $15 million per incident, based on company estimates and industry-reported costs for similar types of incidents. The results were expected to be sensitive to the assumed value per avoided fatality. The analysis team agreed to use a value of $2 million per avoided fatality, and to perform sensitivity analysis using a very broad range from $100,000 to $100 million per avoided fatality, to see how this input affects the results.

The second aspect required to quantify the value of reduced incident risk with each alternative was to estimate the risks themselves. To do this, the quantitative risk assessment for the current operation was revised to reflect the impact of each alternative. For example, under the Move Tank Cars alternative, events involving the flammable tank cars were eliminated from the original risk analysis. under the Move Loading alternative, there would be no impact on the residential area, but the new location could impact a nearby office building at which 90 persons work. As examples, Figures 8.5, 8.6, and 8.7 provide the F-N curve for the Move Tank Cars, Move Loading, and General Upgrade alternatives, respectively. In each figure, the original F-N curve for the facility (Do Nothing Further) is also displayed. The shaded areas indicate the change in risk under that alternative. The alternatives to Shut Down and to Buy Greenbelt eliminate the risk entirely, within the scope of the QRA.

The value of reduced incident risk with each alternative results from calculating the expected value (mathematical expectation) of the reduction in incidents and fatalities times the dollar value per avoided fatality. See the Glossary for a definition of *expected value*, and Section 2.6.2 for details on computing mathematical expectation.

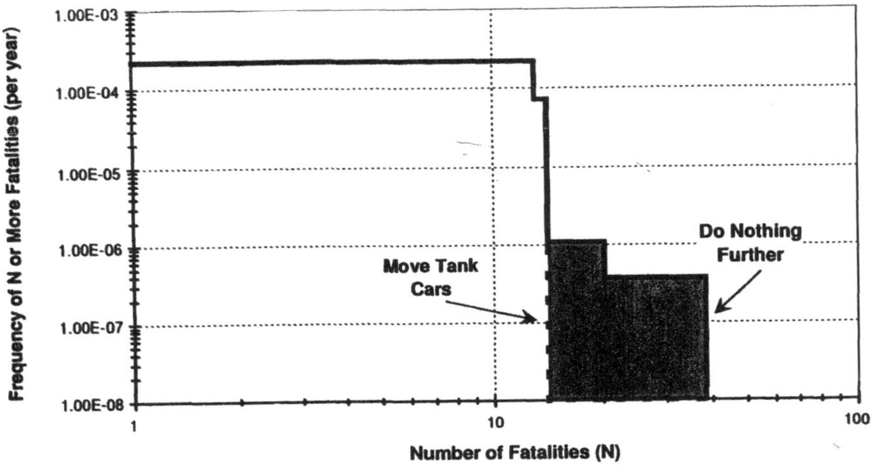

FIGURE 8.5. F-N curve for Move Tank Cars alternative.

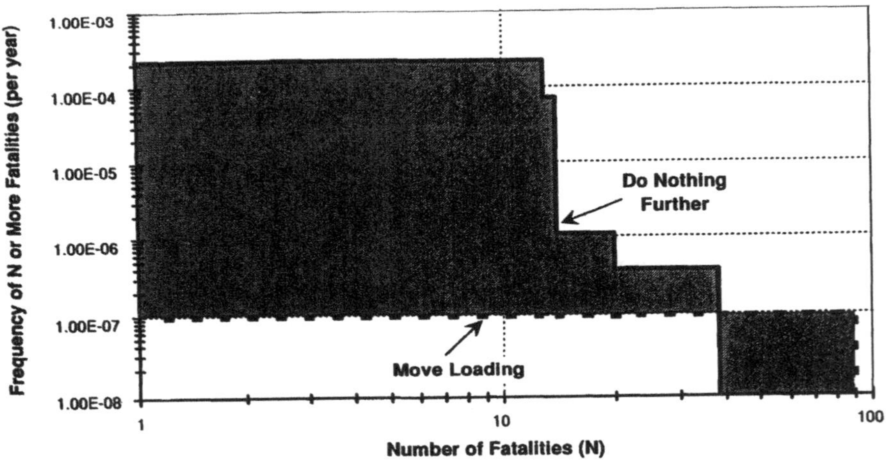

FIGURE 8.6. F-N curve for Move Loading alternative.

We assume that the reduced incident risk and fatality risk apply to each of the 15 years considered in the analysis. (This is an approximation that is appropriate for low-frequency events; to be more accurate we would need to compute all possible numbers of total fatalities that could occur in 15 years, and their corresponding probabilities.) The expected number of fatalities per year is estimated by multiplying the frequency times the number of fatalities for each possible number of fatalities that could occur, and then computing the sum of these products. This can be estimated from the F-N curve as:

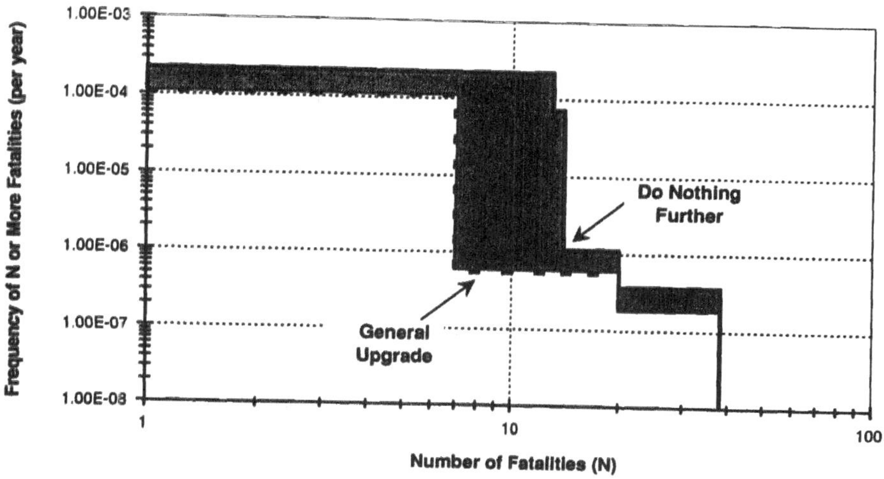

FIGURE 8.7. F-N curve for General Upgrade alternative.

$$EF = \sum_{N=1}^{\infty} N(F_N - F_{N+1})$$

where EF is the expected number of fatalities per year; N is the number of fatalities; and F_N is the frequency of N or more fatalities per year (from the F-N curve).

As an example, the expected number of fatalities per year under Do Nothing Further is computed in Table 8.4.

The result, which is the column sum for the last column of Table 8.4, is that the expected number of fatalities per year is 2.95×10^{-3} under Do Nothing Further.[1] The incident risk per year is the sum of column 3, or 2.2×10^{-4} for the Do Nothing Further alternative. Table 8.5 summarizes the reduction in incident risk and fatality risk for the alternatives addressed. The expected number of fatalities per year is 2.92×10^{-3} under the Move Tank Cars alternative. Then with Move Tank Cars, the reduction in the expected number of fatalities is $(2.95 \times 10^{-3}) - (2.92 \times 10^{-3}) = 3 \times 10^{-5}$ per year, or 4.45×10^{-4} fatalities over the 15-year horizon.

It is worth noting at this point that the value of the reduced fatality risk is quite small. For example, consider the Move Tank Cars alternative. At $2 million per avoided fatality and with an estimated 3×10^{-5} reduction in expected fatalities per year, the discounted value of the fatality reduction over 15 years is less than $1,000, as compared against a total cost of $2 million to

1 This risk is high; the numbers in this example are for illustration only.

TABLE 8.4
Computation of the Expected Number of Fatalities under "Do Nothing Further" Alternative

(1) Number of Fatalities (N)	(2) Cumulative Frequency of N or more fatalities (per year)	(3) Frequency (per year) ($F_N - F_{N+1}$)	(4) Frequency × Number of Fatalities (per year) (Col.3 × Col.1)
13	2.2×10^{-4}	1.5×10^{-4}	1.9×10^{-3}
14	7.3×10^{-5}	7.2×10^{-5}	1.0×10^{-3}
20	1.1×10^{-6}	7.2×10^{-7}	1.4×10^{-5}
38	3.8×10^{-7}	3.8×10^{-7}	1.4×10^{-5}
	Sum:	2.2×10^{-4}	2.9×10^{-3}

TABLE 8.5
Incident Risk and Fatality Risk for Chlorine Loading Alternatives

Alternative	Incident Risk (incidents per year)	Reduction in Incident Risk from Do Nothing Further Alternative (incidents per year)	Fatality Risk (fatalities per year)	Reduction in Fatality Risk from Do Nothing Further Alternative (fatalities per year)
Do Nothing Further	2.20×10^{-4}	0	2.95×10^{-3}	0
Shut Down	0	2.20×10^{-4}	0	2.95×10^{-4}
Move Tank Cars	2.19×10^{-4}	1.14×10^{-6}	2.92×10^{-3}	2.96×10^{-5}
Enclose	2.30×10^{-6}	2.18×10^{-4}	3.22×10^{-5}	2.92×10^{-3}
Move Loading	9.50×10^{-8}	2.20×10^{-4}	8.55×10^{-6}	2.94×10^{-3}
Buy Greenbelt	0	2.20×10^{-4}	0	2.95×10^{-4}
Refrigerate	1.14×10^{-6}	2.19×10^{-4}	2.96×10^{-5}	2.92×10^{-3}
General Upgrade	1.10×10^{-4}	1.10×10^{-4}	7.81×10^{-4}	2.17×10^{-3}

implement this alternative. For Shut Down, the value of the reduced fatality risk is greater by a factor of 100 over that with Move Tank Cars, but this is still small compared to the other effects. These results were somewhat surprising to the analysis team. If reduced fatality risk were the only factor considered, it would appear difficult to justify any action. However, in this analysis other benefits were also included.

Finally, the analysis team needed to assign a value to the goodwill the company might gain by taking action to reduce risks at the plant. This effect may be an important one, since some of the actions that most reduce the risk also have the most severe negative effects. For example, Shut Down is among the most costly alternatives and is also highly undesirable from the standpoint of lost profits, tax revenues, and jobs. The value of reduced fatality risk is, as noted above, quite small. However, the community would like to eliminate the perceived risk of the loading facility. Action taken by the company could enhance goodwill in the community, even though the tangible benefits of the action may be small (or even negative, when jobs and tax revenue losses are included).

Estimating the value of intangible goodwill effects is difficult. One approach considered was to survey members of the community to determine their value for eliminating the risk of the facility. This was rejected as untried, unlikely to provide useful information, and potentially harmful to the company's image. Instead, the approach used was to estimate what it would cost the company to restore the goodwill of the community, for example, by investing in public relations to explain its position and support for local programs to enhance its standing in the community. It was estimated that if no further action were taken (Do Nothing Further), roughly $500,000 per year for five years would be required, and under Shut Down and Buy Greenbelt this cost would be avoided. For the other alternatives, a cost saving of $200,000 per year for five years was used. The total present-value discounted cost savings are $2.1 million and $0.8 million, respectively, discounted over five years.

Table 8.6 summarizes the effects side of the analysis.

8.4.4 Applying the Decision Rule

The results of the analysis depend on what effects are included. Table 8.7 summarizes what effects might be included under two viewpoints, one corporate and one societal. From a corporate viewpoint, the relevant effects are the loss of profit, the reduced incident and fatality risk, and the enhanced company goodwill. From a societal viewpoint, the relevant effects are the loss of tax revenue, the loss of jobs, the reduced fatality risk, and enhanced goodwill. (Enhanced goodwill is a benefit to the company, but can be viewed as what the community would pay to eliminate the risk, over and above what they would pay for the reduced fatality risk.) The choice of effects to include

TABLE 8.6
Effects under Each Alternative (15-year Horizon, 6% Discount Rate, All Values in Millions of Dollars)

Alternative	*Loss of Profit*	*Loss of Tax Revenue*	*Loss of Jobs*	*Gain in Property Value*	*Value of Reduced Incident Risk*	*Value of Reduced Fatality Risk*	*Value of Enhanced Goodwill*
Do Nothing Further	0.00	0.00	0.00	0.00	0.00	0.00	0.00
Shut Down	–49.53	–1.73	–77.70	0.00	0.03	0.06	2.10
Move Tank Cars	0.00	0.00	0.00	3.00	0.00	0.00	0.80
Enclose	0.00	0.00	0.00	3.00	0.03	0.06	0.80
Move Loading	0.00	0.00	0.00	3.00	0.03	0.06	0.80
Buy Greenbelt	0.00	–19.42	0.00	0.00	0.03	0.06	2.10
Refrigerate	0.00	0.00	–7.77	3.00	0.03	0.06	0.80
General Upgrade	0.00	0.00	0.00	1.00	0.02	0.04	0.80

is not always clear. From a societal viewpoint, profits might arguably be included, since profits ultimately benefit society by supporting productive investments. On the cost side, we assume that the company's cost to implement an alternative can be treated as a societal cost, since it is a diversion of money from otherwise worthwhile investments. Table 8.8 summarizes the results under each point of view.

Applying an objective to maximize net benefits, the preferred alternative from the corporate viewpoint is to Do Nothing Further. It is clear from Table 8.6 that the benefits in reduced fatality risk and enhanced goodwill are negligible when compared against the costs to implement each alternative and their possible effect on profits. Moreover, Table 8.8 shows that none of the alternatives provides a positive net benefit. The closest contender is the Move Loading alternative, which has discounted net benefits (loss) to the corporation of –$11,000. This alternative costs $900,000 in discounted incremental operating and capital costs, and provides $889,000 in benefits: $32,000 in reduced incident risk, $57,000 in reduced fatality risk and $800,000 in goodwill.

From a societal viewpoint, the results are quite different. Three of the alternatives provide positive net benefits: Move Loading, Move Cars, and General Upgrade. This is because the societal viewpoint incorporates the benefit of enhanced property values not included in the company viewpoint. Table 8.8 indicates that the societal viewpoint does not consistently give a

TABLE 8.7
Effects Included in "Corporate" and "Societal" Viewpoints for Cost–Benefit Analysis

Effect	*Corporate Viewpoint*	*Societal Viewpoint*
Loss of profit	✔	
Loss of tax revenue		✔
Loss of job		✔
Increased property value		✔
Reduced incident risk	✔	
Reduced fatality rate	✔	✔
Enhanced goodwill	✔	✔

TABLE 8.8
Net Benefit of Each Alternative (Discounted Benefit Minus Costs, $MM)

Alternative	*Corporate Viewpoint*	*Societal Viewpoint*
Do Nothing Further	0.00	0.00
Shut Down	–80.94	–110.87
Move Tank Cars	–1.20	1.80
Enclose	–3.97	–1.00
Move Loading	–0.01	2.96
Buy Greenbelt	–22.39	–41.84
Refrigerate	–3.86	–8.66
General Upgrade	–0.54	0.44

greater net benefit to each alternative: the Shut Down and Buy Greenbelt alternatives, for example, are less attractive under society's point of view due to their impact on jobs and local taxes. However, property values swing other alternatives in a significantly positive direction.

From society's point of view, the preferred alternative is to Move the Loading. This has a total discounted cost of $900,000 (as before), and total discounted benefits of $3.86 million, resulting in net benefits of $2.96 million.

We noted earlier that the alternatives consistently provide negligible benefits in reduced fatality risk, assuming a value of $2 million per avoided fatality. How would the results under the corporate viewpoint change if a different value per fatality were assumed? Figure 8.8 shows the results of a sensitivity analysis to the value per avoided fatality.

In Figure 8.8, the alternatives that do not appear have a net benefit considerably less than those shown. The results of the sensitivity analysis show that the preferred alternative switches from Do Nothing Further to Move Loading when the value per avoided fatality exceeds about $3.5 million. Below this value, Do Nothing Further is preferred. From $3.5 to $100 million per avoided fatality, Move Loading is preferred.

Sensitivity analyses were performed for other key inputs, including the cost estimates, the discount rate, the time horizon, and the risk of fatality. The sensitivity analyses indicated that the results are insensitive to the discount rate, time horizon, and risk of fatality, but are somewhat sensitive to the cost estimates. To further refine and support the analysis, it would be worthwhile for the project team to explore the cost estimates more carefully before making a final recommendation.

A report was provided to the plant manager that documented the approach, assumptions, and key inputs to the analysis. The report presented the results summarized in the figures and tables above. The report made the following recommendations:

- The company should discuss possible sharing of a portion of the costs with the local government. This might be made possible through a one-time tax adjustment or other incentives. This request can be sup-

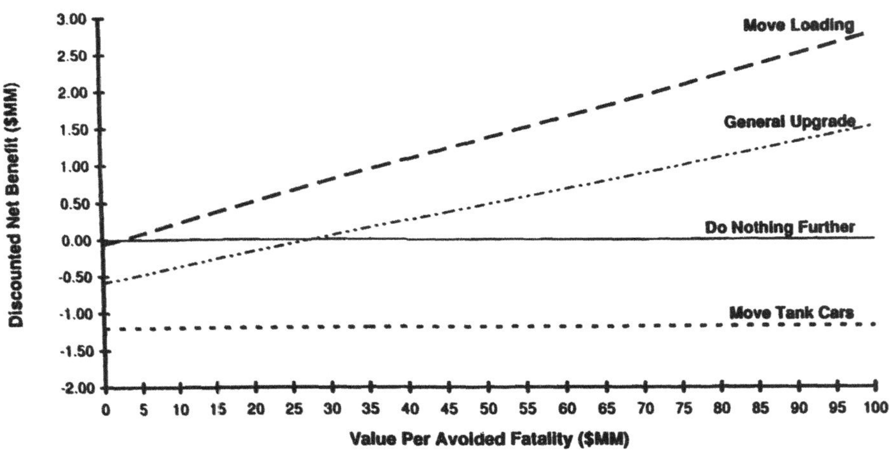

FIGURE 8.8. Sensitivity analysis for value per avoided fatality.

ported by describing the estimated societal benefits of the actions that the company might take to reduce the risks, with emphasis on substantial benefits such as enhanced property values that the company does not recoup.

- If the local government is willing to absorb some of the cost, the analysis should be revised to reflect this.
- If the local government is not willing to absorb some of the cost, the company should undertake the Move Loading alternative. Even though this represents a net loss to the company as compared with Do Nothing, the $11,000 net loss is small when compared against the uncertain consequences of taking no action. The sensitivity analysis showed that within the ranges considered for some of the uncertain inputs, Move Loading is sometimes the preferred alternative.

Before we leave this case, it is useful to consider how the same problem would be addressed by a cost–effectiveness analysis. Such an analysis would focus on the desired benefit of the action, to reduce the fatality risk, and the costs required to achieve this benefit. With a cost–effectiveness analysis, no dollar value need be assigned to the benefit, which eliminates the need to specify a value per fatality in this example. The results are provided in Table 8.9.

The cost–effectiveness analysis indicates that Move Loading is the most cost–effective alternative, at an estimated cost of about $20 million (discounted) per avoided fatality. The company would need to decide if this is within a range at which it is willing to undertake the action. It should be recognized that this approach ignores other effects (both positive and negative) of the alternatives that were addressed in the more comprehensive cost–benefit analysis of the problem. These other effects could be addressed in a cost–effectiveness analysis by treating each one separately (e.g., report the cost per dollar gain in property value) and/or by simply displaying these other effects along with the basic results of the analysis. Typically, a cost–effectiveness analysis will not combine these effects into one result for each alternative as was done in the CBA. Instead, it is usually left to the decision maker to mentally weigh these other effects.

8.5 EXTENSIONS OF COST–BENEFIT ANALYSIS

We have discussed cost–benefit analysis as it is generally applied in the majority of applications. Extensions to the basic approach address ways to incorporate:

- Intangible benefits for which the analyst prefers not to assign dollar values (cost-effectiveness analysis)
- Uncertainty in the inputs (costs and/or effects) to the analysis (uncertainty analysis)

TABLE 8.9
Results of Cost–Effectiveness Analysis

Alternative	*Discounted Incremental Total Cost ($MM)*	*Reduced 15-Year Fatality Risk*	*Cost Effectiveness ($MM per avoided fatality)*
Do Nothing Further	0.00	0	N/A
Shut Down	33.60	4.42×10^{-2}	760
Move Tank Cars	2.00	4.45×10^{-4}	4500
Enclose	4.86	4.38×10^{-2}	110
Move Loading	0.90	4.41×10^{-2}	20
Buy Greenbelt	24.57	4.42×10^{-2}	560
Refrigerate	4.75	4.38×10^{-2}	110
General Upgrade	1.40	3.25×10^{-2}	40

- Distributional issues, that is, differences among alternatives in how the costs and effects are borne by different portions of society (distributional analysis).

The remainder of this section describes these three extensions to cost–benefit analysis.

8.5.1 Cost-Effectiveness Analysis

A key difficulty in cost–benefit analysis is assessing the effects using the same scale that is used for costs. Attempting to assign a dollar value to effects that are intangible or difficult to quantify can hinder the analysis.

Cost-effectiveness analysis (CEA) eliminates the need to quantify benefits in dollar terms when comparing alternatives that have common goals or purposes. For example, if the goal is to reduce the statistical risk of fatalities, programs could be compared on their cost per unit of reduced fatality risk, without the need to assign an economic value to this effect.

Similarly, constant-cost studies assume that a fixed budget is to be spent on the project, and identify the alternative that maximizes benefits without the need to assign dollar values to the benefits. Least-cost studies assume that a fixed objective is to be achieved, and identify the alternative that achieves this objective at minimal cost.

Although cost-effectiveness analysis avoids problems of assigning economic values to benefits, the analysis may overlook important issues. For example, cost-effectiveness analysis is difficult when projects have multiple benefits that are difficult to compare on the same scale. Also, cost-effectiveness analysis is useful only when there is more than one alternative, and doing nothing is not an option.

An alternative to using cost-effectiveness analysis when effects are not estimated by dollar impacts, is to use weights to reflect the relative importance of various effects. For example, dollar costs and risk reduction might be combined by assigning a weight to each. In this case, the approach is essentially a weighted scoring method (see Chapter 7), multiattribute utility analysis (see Chapter 12), or other approach rather than what is typically viewed as a CBA. It should be noted that even though a weighting approach avoids explicitly assigning economic values to intangible effects, implicit in the weights is such an economic valuation.

8.5.2 Uncertainty Analysis

Sugden and Williams (1978) and others have advocated that uncertainty be incorporated explicitly into the cost–benefit analysis. This refinement can extend CBA in the direction of other aids that distinguish between values and uncertainties, such as payoff matrix, decision analysis, and multiattribute utility analysis, discussed in later chapters of this book.

Uncertainty can be incorporated in various ways. One approach is to define a set of possible scenarios or future states of the world. This approach pushes CBA in the direction of payoff matrix analysis, described in Chapter 10. We assess the probability that each of these possible scenarios occurs, being careful that the scenarios are mutually exclusive and collectively exhaustive (i.e., the probabilities of the various scenarios should sum to 1). under each scenario, we assess the costs and effects of each alternative, and the net benefit for each alternative as we would in a standard cost–benefit analysis. Finally, we compute the expected net benefit for each alternative, by summing the scenario probability times that alternative's net benefit, over all possible scenarios:

$$ENB_i = \sum_{j=1}^{J} p_j(B_{ij} - C_{ij})$$

where

ENB_i = expected net benefit, alternative i
p_j = probability of scenario j
B_{ij} = discounted benefit of alternative i under scenario j
C_{ij} = discountedcost of alternative i under scenario j
J = number of scenarios addressed

The alternative with the highest expected net benefit is the one chosen.

Incorporating uncertainty in this fashion may be useful if the uncertainties reflect events that are not affected by the alternatives. The approach does not, however, address uncertainties associated with the alternatives themselves, such as uncertainty in operating costs where this uncertainty varies with the alternative. In this case, it is impossible to define scenarios and probabilities that are consistent across all the alternatives.

A further refinement, then, is to incorporate uncertainty at the input level. This approach pushes CBA in the direction of decision analysis, described in Chapter 11. Under this approach, the analysis assesses uncertainty separately for each alternative, and for each cost and effect for which there is significant uncertainty. As before, alternatives are evaluated by their expected net benefit, where the expectation is computed over possible values of the inputs to the analysis. While this approach is appealing in principle, the literature provides little specific advice on how to model the uncertainties and dependencies among these uncertainties. The reader may wish to consult sources on probability theory and decision analysis for guidance (see Chapter 11).

Finally, if the analysis is to incorporate uncertainty, there is an issue as to whose uncertainties should be used in the analysis. An argument could be made that experts with knowledge about the uncertain input should provide these assessments. However, a philosophical purist might instead argue that in a cost–benefit analysis, the probability assessments should reflect the opinions and beliefs, right or wrong, of those individuals in society who are affected by the decision. In the absence of data on society's uncertainties (e.g., the risk they perceive from the operations of a chemical processing facility), the CBA may be unintentionally biased toward the beliefs of the sponsor or analyst.

8.5.3 Distributional Analysis

A key problem with traditional cost–benefit analysis is that it overlooks concerns about who pays and who gains with each alternative. It can be argued that these equity issues should have some bearing on the decision. If two alternatives are otherwise equal in net benefits, we would generally prefer the one that spreads the costs and benefits more equitably among individuals.

In practice, some alternatives may be far less equitable than others. In deciding among these alternatives, we may prefer an alternative with smaller net benefits if it provides a more equitable distribution of the costs and benefits. In particular, a decision maker may wish to choose alternatives that are believed to have more favorable distributional effects, for instance, by taking into account:

- Income considerations, for example, by giving preference to alternatives that provide greater benefits to persons at lower income levels

- Political considerations, for example, by giving preference to alternatives that give benefits to larger numbers of individuals
- Geographical considerations, for example, by giving preference to alternatives that concentrate the benefits in certain jurisdictions.

Distributional issues can be addressed in part by establishing an actual market for the goods or services provided. User fees, for example, help ensure that the beneficiaries of a program pay more of the costs. This means modifying the alternatives considered to include establishment of such a market along with implementation of that alternative.

Even if a market like the aforementioned one were established, this does not guarantee a decision that is completely "fair." Extensions to cost–benefit analysis have attempted to incorporate distributional issues explicitly into the analysis. These extensions push CBA in the direction of weighted scoring methods (Chapter 7) and multiattribute utility analysis (Chapter 12). For example, one approach is to infer the public's distributional preferences (e.g., based on the historical record of projects undertaken as compared against their benefits) and incorporate these preferences explicitly into the analysis. This can be accomplished by applying weights on the costs and effects to scale them in accord with distributional preferences. For example, an analysis might multiply the economic impacts of job loss to reflect preferences to avoid alternatives that incur greater job losses than others.

Others suggest that distributional issues be addressed qualitatively, rather than as an explicit part of the quantitative analysis. Sassone and Schaffer (1978) suggest that a qualitative "social impact analysis" be done in parallel with the quantitative analysis and the results from both analyses be included in the final presentation.

Finally, an approach called *decision-maker CBA* (Sugden and Williams 1978), can be used to penalize alternatives that the decision maker feels are less attractive for distributional reasons. This is done by incorporating quantified impacts that reflect the decision-maker's concerns about distributional issues and other impacts which seem to be left out of the analysis. To do this, the decision maker must assign dollar values to inequitable distributions or other undesirable consequences left out of a CBA. With this refinement, CBA more closely approximates the decision-maker's decision problem as addressed by decision analysis. However, in practice, analyses that use this approach often proceed without the active involvement of the decision maker in providing assumptions and assessments of the impacts.

8.6 IMPLEMENTATION NEEDS

To perform a cost–benefit analysis requires first identifying the alternatives to be considered, and defining the scope of the analysis. The alternatives must be sufficiently well defined that their costs and effects can be described and quantified. Scoping issues include the time horizon and breadth of impacts (e.g., corporate, local community, national) addressed.

Key information requirements are quantitative estimates of the costs and effects of each alternative. Often, the costs can be estimated by experts within the organization, since they typically involve capital and operating expenses of the kind the organization routinely deals with in its ongoing activities. Estimating the effects may be more difficult, particularly when intangibles are involved. In some cases the organization may draw on general literature, previous CBA studies, or outside experts. Where outside experts are used, they will often have an economics, civil engineering, industrial engineering, or management science background and/or experience in similar CBA studies.

In an organization that has not conducted CBA studies previously, the first studies should be narrow in scope and have easily defined effects. For broader studies involving difficult-to-quantify impacts, the analyst should be trained in CBA or work with experienced CBA practitioners. Although the general approach of CBA is straightforward, considerable judgment may be required to develop realistic inputs to the analysis, and experience is the best guide to these judgments. In the absence of experience with actual CBA studies, it is useful to familiarize oneself with CBA through textbooks, articles, and published CBA studies. See Section 8.7.2 for recommended materials.

A simple CBA study can be performed by one person within a few days if it involves a small enough decision, such as one involving tens of thousands of dollars. For a larger study, the time and resource requirements increase accordingly. For decisions involving tens or hundreds of millions of dollars of cost and numerous effects, several persons and two or more months of analysis may be required.

Since much of the effort and challenge in a cost–benefit analysis is in determining the numerical inputs, an organization can facilitate its use of CBA by formulating standard assumptions to be used in its CBA studies. These assumptions can include, for example, the horizon, discount rate, effects to include, and value per avoided fatality (and range used in sensitivity analysis). In addition, the organization can specify the general approach used to estimate categories of other inputs, such as how property value for displaced residents should be estimated.

General analytic software such as a spreadsheet application or analysis package can facilitate a CBA study. Software designed to address CBA is listed in Appendix A. Using a spreadsheet or other software can facilitate the numerical computations, sensitivity analysis, and development of graphical outputs that summarize the results.

8.7 SUMMARY

8.7.1 Strengths and Weaknesses of Cost–Benefit Analysis

The underlying principle of cost–benefit analysis, that decisions affecting society should account for the costs and effects borne by society's members, makes it attractive for public-sector decisions, and for private decisions that involve societal impacts. This simple guiding principle drives all aspects of the approach. With a long history of theory and application, cost–benefit analysis has perhaps the most solid track record of any formal, quantitative decision aid.

In a cost–benefit analysis, the assumptions and inputs to the decision are explicitly identified. There is a logical, documentable basis for the decision, and sensitivity analysis can show the degree to which the results depend on the assumptions. Applied properly, cost–benefit analysis is objective: the preferences and influence of any person or group of persons are accorded no more weight than of any other.

Despite its strengths and extensive use, cost–benefit analysis has theoretical and practical drawbacks. Two of the main limitations of cost–benefit analysis are its failure to address uncertainties and distributional issues; we described some possible extensions to address these issues, but their theory and application are incomplete.

The core of CBA is the principle of potential Pareto superiority. This principle holds that society should choose an alternative over others if that alternative produces the greatest aggregate gain to society, even though some members of society do not gain. Those who gain could potentially compensate those who do not gain, and all will be better off. A key objection made to cost–benefit analysis is that while the potential Pareto superiority principle appears to be fair in principle, in fact the hypothetical redistribution that would make everyone better off need not (and rarely does) occur. Moreover, the hypothetical redistribution may involve a relatively small number of those who gain, and a large number of those who lose. Some may view this as a violation of democratic principles.

Even if we accept the principle of potential Pareto superiority, we may not accept the viewpoint that all of the costs and effects can be quantified by a willingness-to-pay. This viewpoint is objectionable to those who feel that there is an intrinsic or infinite value to certain things such as environmental quality or natural resources.

Another objection is in the use of market prices or willingness-to-pay to assess costs and effects. This approach assumes that there currently exists an equitable distribution of income in society. Inequities in the income distribution, for instance, may imply that the willingness-to-pay of some individuals, or their current consumption of a good, understates their true value for the good. Often there is an implicit assumption in a CBA of full employment by members of society impacted by the decision, and this assumption may not be valid.

Cost–benefit analysis assumes linear preferences for money. This means that the value per dollar of benefit or cost is the same regardless of the magnitude of the benefit or cost. Without this assumption, one could not simply add up all the costs and benefits, since their value would vary from one individual to another. Most individuals have decreasing marginal value for money. Results could change if this more realistic valuation were incorporated into the analysis.

Although time discounting of money is an issue in most decisions, the choice of a discount rate is a particularly sensitive issue in CBA. There is disagreement on what rate reflects society's preferences. Public-sector CBAs have often been criticized as using too low a discount rate, which favors larger public investments with delayed, longer-term benefits.

Distributional issues are often a concern in CBA, particularly for public decisions. Critics have expressed concerns about the equity or fairness of an analysis that essentially treats the benefits to one segment of society and the costs to a different segment as interchangeable. Unless these concerns are addressed explicitly by modifying the analysis, the analysis may result in economically optimal decisions that are politically or socially unacceptable. This implies that important value judgments have been left out of the analysis.

The decision rule under cost–benefit analysis incorporates the collective preferences of society. A decision maker may object to this basis, particularly in decisions that are essentially private ones. A private decision maker who finds the net benefit principle attractive may prefer to use a decision-analysis approach. Decision analysis employs essentially the same decision criterion, but quantifies the costs and benefits based on the decision-maker's own preferences rather than society's preferences.

Past, high-profile CBAs have been criticized on the basis of the assumptions used to quantify intangible effects. Although this has led some to conclude that CBA is more art than science, the same objection applies to any quantitative decision aid that assigns value to intangible effects. In particular, criticism has often focused on the handling of the value-of-life issues. All decision aids face the same challenge when encountering this issue. CBA has what some consider to be a drawback that the analysis requires specification of a dollar value for these effects. Moreover, in CBA there is a high standard to achieve: the values should be those of the individuals affected, not the values of the decision maker. This significantly complicates the value-of-life issue in CBA applications.

Another criticism of CBA is that the analysis assumes that the project itself does not change the underlying societal preferences. Remember that societal preferences form the basis for quantifying the inputs to a CBA. A large project could shift these preferences sufficiently to render the assumptions underlying the analysis inappropriate. For example, a project that creates many new jobs could shift supply and demand for jobs to such an extent that wages (which may have been used in the CBA) are no longer the same as would be expected from historical experience.

Finally, on a practical level cost–benefit analysis is relatively sophisticated and time-consuming. Correct application of cost–benefit analysis usually requires a trained analyst. For some decisions, consultation with an economist may be required to determine how to appropriately quantify some of the impacts.

8.7.2 Suggestions for Further Information/Study

For a brief overview of cost–benefit analysis techniques, the reader may want to start with a basic text on engineering economics such as Riggs and West (1986), Thuesen and Fabrycky (1989), or White, Agee, and Case (1989). There are numerous texts that address cost–benefit analysis in more detail, including those by Anandarup (1984), Anderson and Settle (1977), Dasgupta (1972), Frost (1975), Gramlich (1990), Mishan (1988), Sassone and Schaffer (1978), Sugden and Williams (1978), and Thompson (1980). For reviews and critiques of CBA and related social choice approaches, see Baram (1980) and Merkhofer (1987).

8.8 REFERENCES

Anandarup, R. (1984). *Cost–Benefit Analysis: Issues and Methodologies*. Baltimore: Johns Hopkins University Press.

Anderson, E. V. (1993). Chlorine Producers Fight Back against Call for Chemical's Phaseout. *Chemical and Engineering News* 7(9):11–12.

Anderson, L. G., and R. F. Settle (1977). *Benefit-Cost Analysis: A Practical Guide*. Lexington, MA: Lexington Books.

Baram, M. S. (1980). Cost–Benefit Analysis: An Inadequate Basis for Health, Safety, and Environmental Regulatory Decision Making. *Ecology Law Quarterly* 8:473–531.

Birkner, L. R., and L. S. Salzman (1986). Assessing Exposure Control Strategy Cost–Effectiveness. *American Industrial Hygiene Association Journal* 47(1):50–54.

Bohn, M. P. (1992). "Decision Making under Uncertainty: An Investigation into the Application of Formal Decision-Making Methods to Safety Decisions" (NUREG/CR-5906). Division of Safety Issues Resolution, Office of Nuclear Regulatory Research, U.S. Nuclear Regulatory Commission.

Boyle, K. J., and R. C. Bishop (1987). Valuing Wildlife in Benefit–Cost Analysis: A Case Study Involving Endangered Species. *Water Resources Research* 23(5):943–950.

Dasgupta, A. K. (1972). *Cost–Benefit Analysis: Theory and Practice*. London: Macmillan.

EPA (1987). *EPA's Use of Benefit–Cost Analysis: 1981–1986*.

Fleishman, A. B., and M. S. Hogh (1989). *The Use of Cost Benefit Analysis in Evaluating the Acceptability of Industrial Risks: An Illustrative Case Study*. London: BP International Limited Britannic House.

Frost, M. J. (1975). *How to Use Cost Benefit Analysis in Project Appraisal*. 2nd ed. New York: Wiley.

Goyal, R. K. (1986). Probabilistic Risk Analysis—Two Case Studies from the Oil Industry. *Professional Safety* 31(7):11–19.
Gramlich, E. M. (1990). *A Guide to Benefit–Cost Analysis*. 2nd ed. Englewood Cliffs, NJ: Prentice-Hall.
Merkhofer, M. W. (1987). *Decision Science and Social Risk Management: A Comparative Evaluation of Cost–Benefit Analysis, Decision Analysis, and Other Formal Decision-Aiding Approaches*. Dordrecht, The Netherlands: D. Reidel.
Mishan, E. J. (1988). *Cost–Benefit Analysis*. 4th ed. London: Unwin Hyman.
Riggs, J. L., and T. M. West (1986). *Essentials of Engineering Economics*. 2nd ed. New York: McGraw-Hill.
Rodgers, G. B., and P. H. Rubin (1989). Cost–Benefit Analysis of All-Terrain Vehicles at the CPSC. *Risk Analysis* 9(1):63–69.
Sarin, R. K. (1983). Social Decision Analysis of the Earthquake Safety Problem: The Case of Existing Los Angeles Buildings. *Risk Analysis* 3 (1):35–50.
Sassone, P. G., and W. A. Schaffer (1978). *Cost–Benefit Analysis: A Handbook*. New York: Academic Press.
Siegel, M. L. (1991). A Spreadsheet for Benefit–Cost Analysis. *Government Finance Review* 7(6):40–43.
Singer, N. M. (1976). *Public Microeconomics: An Introduction to Government Finance*. 2nd ed. Boston: Little, Brown and Company.
Sugden, R., and A. Williams (1978). *The Principles of Practical Cost–Benefit Analysis*. Oxford: Oxford University Press.
Thompson, M. S. (1980). *Benefit–Cost Analysis for Program Evaluation*. Beverly Hills, CA: Sage Publications.
Thuesen, G. J., and W. J. Fabrycky (1989). *Engineering Economy*. 7th ed. Englewood-Cliffs, NJ: Prentice-Hall.
Torrance, G. W., M. H. Boyle, and S. P. Horwood (1982). Application of Multi-Attribute Utility Theory to Measure Social Preferences for Health States. *Operations Research* 30(6):1043–1069.
Ulvila, J. W., and B. B. Thompson (1988). A Computerized Risk Analysis Aid. *Computers and Operations Research* 15(2):123–134.
U.S. Government Printing Office (1975). *Evaluation of Techniques for Cost–Benefit Analysis of Water Pollution Control Programs and Policies*. Washington, D.C.: U.S. Govt. Printing Office.
Walker, A., and D. K. Whynes (1991). Participation and Screening Programmes for Colorectal Cancer: More Would Be Better? *Journal of Health Economics* 10(2):207–225.
White, J. A., M. H. Agee, and K. E. Case (1989). *Principles of Engineering—Economic Analysis*. New York: Wiley.

9

MATHEMATICAL PROGRAMMING

L'embarras des richesses. The more alternatives, the more difficult the choice.
—Abbé d'Allainval

9.1 PURPOSE OF CHAPTER

The decision aids discussed thus far are intended to help decision makers select from among a relatively small number of alternatives. Decision makers, however, often face a problem that involves an enormous number of alternatives. For example, a manager may wish to allocate a fixed research and development budget among five promising technologies. How should the budget be allocated? The number of alternative ways to allocate dollars among the five technologies is immense. To vote on possible allocations would require enumerating them explicitly. Even if the manager could reduce the number of alternatives—say, by considering allocations in chunks of $25,000—there are far too many combinations to consider explicitly. A decision aid that can address this type of problem is *mathematical programming*. Mathematical programming is a class of techniques geared toward helping a decision maker select among many or infinite alternatives.

Mathematical programming addresses the problem of allocating resources in an optimal fashion. By "optimal" we mean in fulfillment of a specified goal or objective. The objective included in a mathematical program typically is to maximize or minimize some quantity of interest. Common examples include "maximize profit," "minimize cost," and "minimize transport time." In solving a resource allocation problem, mathematical programming can consider an enormous number of alternatives, and identify the one that helps the decision maker best achieve an objective.

Mathematical programming is one of the most prevalent and well-established techniques in the field of management science. The word "programming" in this context refers to planning. How can a decision maker manage resources (develop a "program") to achieve objectives? The word "mathematical" refers to how the problem is expressed. The aim of mathematical programming is to describe the planning problem in mathematical terms so that it can be solved using advanced mathematical techniques. Allocating resources is often difficult: not only are the resources limited, but various factors can restrict the options available for allocating them. These factors can include requirements and limitations that constrain the decision-maker's options. Mathematical programming techniques explicitly consider the constraints and goals of the decision maker and find an allocation of resources that best fulfills the decision-maker's objectives.

This chapter introduces some of the specific techniques that comprise mathematical programming. After an overview of mathematical programming and a description of its brief history, we introduce the general form of a mathematical program and give some specific examples. We then give an example of how mathematical programming can be applied to a problem of chemical process safety. After a brief discussion of possible extensions to the basic type of mathematical programs, we conclude with a summary of the strengths and weaknesses of mathematical programming as a decision aid.

9.2 OVERVIEW OF MATHEMATICAL PROGRAMMING

9.2.1 Introduction to Mathematical Programming

Mathematical programming encompasses a class of tools that are particularly suited to problems of efficient resource allocation. The "programming" of mathematical programming refers to "planning." Mathematical programming is used extensively in industry to solve a variety of resource planning problems:

- *Transportation problems.* A company must ship products from warehouses to retail outlets. We know the demand for products at each outlet, the supply available at each warehouse, and the shipping costs from each warehouse to each outlet. How much product should be shipped from each warehouse to each of the outlets in order to satisfy demand for the product and minimize shipping costs?
- *Assignment problems.* A manager has ten workers and ten separate tasks to complete. The workers vary in their proficiency at performing individual tasks. How should the manager assign the workers to tasks, one worker to a task, to minimize the total number of man-hours spent doing the ten tasks?
- *Portfolio Planning*: A company has a series of investment options available. How much should the company invest in each option to maximize return subject to the need to diversify its expenditures?

These are just a few of the many applications of mathematical programming. Additional examples include product blending, production scheduling, and inventory planning.

Notice that these problems have three important characteristics in common: a large set of possible alternatives, a desired goal, and a set of constraints that affect the decision-maker's choice of alternatives. In mathematical programming, these characteristics of the problem are described using mathematical expressions. The statement of the problem using mathematical notation is called the mathematical "program." Mathematical programming as an activity refers to the process of developing and solving mathematical programs. The solution to a mathematical program helps the decision maker understand how to allocate resources to best satisfy the defined objective.

Exactly what is a mathematical program? A mathematical program is a mathematical *model* of the decision problem. The mathematical model captures the essential elements of the problem by expressing key aspects and their relationships mathematically.

Building a model serves many purposes. First, the model simplifies the problem and focuses attention on its most important aspects. Second, one can experiment with the model. Analyzing and testing the model can reveal insights and relationships of which the decision maker was previously unaware. And last, in solving the model new alternatives can surface that were not considered previously.

Mathematical programming models have a special form which we discuss later in the chapter. When we speak of branches of mathematical programming, such as linear programming and dynamic programming, we refer to specific classes of mathematical programming models.

9.2.2 History of Mathematical Programming

The field of mathematical programming has several roots. Most agree that its establishment can be traced to the seminal work of George Dantzig in the late 1940s. Before that time, several noted mathematicians and economists had created models to analyze planning problems. The outbreak of World War II spurred the development of techniques to manage military resources efficiently. The military brought together individuals from a wide array of academic disciplines into groups of "operations researchers" whose expertise and collaboration solved many of the problems of wartime management.

During the war, George Dantzig recognized that many important planning problems could be expressed as a series of linear inequalities. Dantzig called this the problem of "programming in a linear structure," as "program" was then a synonym for a military schedule or plan. Later, Dantzig added an objective to the general formulation to represent the explicit goals of the decision maker. In 1947, shortly after the war, Dantzig developed a solution technique called the "simplex method" that, when carried out, guaranteed the solution of these "linear programs."

Two developments spurred further progress in mathematical programming, one evolutionary, the other revolutionary. The evolutionary change was the gradual development of solution procedures for other types of planning problems that shared the general framework of the linear programming problem, that is, maximizing a mathematical function subject to a set of mathematical constraints. Later developments included the establishment of nonlinear programming (Harold Kuhn and Albert Tucker) for problems with nonlinear constraints and/or objectives; integer programming (Ralph Gomory) for problems which restrict the alternatives to integer variables; dynamic programming (Richard Bellman) for problems of multistage or multiperiod optimization subject to constraints; and goal programming (Abraham Charnes and William Cooper), for the problem of simultaneously satisfying multiple goals.

The revolutionary development was the rapid advance of the digital computer, and the influence of this development cannot be underestimated. Solution procedures for mathematical programming problems are far too complex and tedious to be done by hand for any nontrivial problem. The introduction of faster and more economical computers has made solutions to realistically complex problems increasingly accessible to industry. The rapid advance of computer technology has also influenced the development of new mathematical programming solution procedures.

Today, thousands of companies around the world use mathematical programming techniques, and in particular, linear programming, to make planning decisions. The introductory textbook by Cook and Russell (1985) gives examples of successful applications of mathematical programming by such major companies as Shell, Monsanto, and Exxon.

9.2.3 Decisions/Situations Most Appropriate for Mathematical Programming

The central task in mathematical programming is to express the decision problem in mathematical terms. Consequently, mathematical programming is most appropriate in situations where relationships between decisions and outcomes are explicit and can be captured in mathematical expressions. The requirement that relationships be described mathematically means that the decision problem must first be expressed precisely; mathematical programming is most suitable for decision situations in which goals are unambiguous and the factors that constrain the set of possible decisions are clear and explicit. For example, "the action we take must be acceptable to state regulators" must be expressed more concretely in order to be used as a constraint for mathematical programming. What is the decision to be made? What "values" can this decision take on? How can the requirement of "acceptability" be translated into a constraint on the values that the decision can take on? A possible alternative constraint might be that "there must be at least two emergency facilities within 25 miles of the proposed plant."

The goal of the decision maker is described in a mathematical program as an *objective function*. The objective function, which also must be stated explicitly, may or may not incorporate the values of the decision maker. Common applications of mathematical programming involve allocating resources in very concrete, physical situations. Often these allocations are performance oriented, with goals whose achievement can be measured objectively (such as the goals to "maximize profit" or "minimize cost"). Subjective criteria can also be incorporated into the statement of the objective function by stating explicitly how an action affects the values of the decision maker. Thus, mathematical programming is most appropriate where goals and values of the decision maker are explicit.

With the notable exception of the technique called dynamic programming, mathematical programming techniques were not originally equipped to address uncertainty in the decision-making environment. Great strides are being made toward developing probabilistic (i.e., "stochastic") versions of the traditional mathematical programming techniques and this remains an active research area. Despite these advances, few software packages are available that explicitly take into account uncertainty in mathematical programming formulations. Thus mathematical programming techniques, with the exception of dynamic programming, are less suited to decision situations characterized by much uncertainty. Sensitivity analysis, however, can partially address uncertainty in some aspects of the problem.

A practical consideration that unites mathematical programming methods is that the solution techniques for all but the most trivial problems are fairly resource intensive. Mathematical programming is best left for situations where there is significant time for analysis, where computers are readily available, and where persons trained in the required computer software are available.

9.3 EXPLANATION OF MATHEMATICAL PROGRAMMING

9.3.1 Key Concepts of Mathematical Programming

Mathematical programming is a technique of management science and, like other management science approaches, can be described by the following five-step procedure (Daellenbach, George, and McNickle 1983; Ackoff 1956):

1. Formulate the problem
2. Construct a mathematical model
3. Derive a solution to the model
4. Test sensitivity of solution
5. Implement and maintain the solution.

The specific characteristics of mathematical programming, as distinct from other management science solution methodologies, are most relevant in steps 2, 3, and 4. We describe the steps below, with particular attention to steps 2, 3, and 4.

9.3.1.1 Step 1: Formulate the Problem

Formulating a problem correctly is the first step in understanding and quantifying the problem. What is the situation at hand? Who are the affected parties? What are the objectives? What are the specific decisions to be made? How do these decisions affect the rest of the organization?

At first, questions like these may be answered in a vague or ambiguous way. "We need to operate more efficiently," or "We need to increase our competitiveness." Problem formulation is the process of taking vague, higher-level goals and processes and translating them into concrete objectives and decisions.

Consider, for example, a carpenter in the business of assembling two products, ladders and chairs. The carpenter has a limited store of raw materials he can use, in different combinations, to build his two products. The products are sold at the local market. Before the carpenter can use mathematical programming, he must formulate his problem. What decision does the carpenter need to make? What is the carpenter's goal? Assuming that the carpenter is interested in maximizing the financial return from his investment in raw materials and tools, a specific objective may be to maximize revenue from his production efforts. And how does the carpenter affect his revenues? By building and selling his ladders and chairs. The obvious decision facing the carpenter is how many ladders and chairs he should produce. With a limited amount of raw materials to work with, the carpenter will be limited in the number of finished products he can deliver. Of course, selling different combinations of chairs and ladders will generate different revenues. If the carpenter has unlimited demand for his goods, and can sell everything he makes, he might formulate his problem as the following: How many chairs and how many ladders should I make in order to maximize my revenues?

Problem formulation is an important stage of the mathematical programming process because all other steps depend upon it. An organization cannot hope to make good decisions, if the objectives, decisions, and operating constraints cannot be appropriately identified.

9.3.1.2 Step 2: Construct a Mathematical Model

The process of creating a mathematical program is often called modeling and the resulting program called a *model*. The essence of mathematical programming is the construction of a simplified representation—a model—of the decision problem. Just as an architect creates a clay model to study the relationships among the various rooms and open spaces of a proposed building, an analyst creates a mathematical model to study the relationships among important parts of a decision problem.

The success of any model depends on how well the model captures the important elements of the problem it tries to represent. No model can or should try to capture every detail. The appropriateness of a model is judged, in part, on what features it accurately represents. A particular architectural model, for example, may be designed to illustrate the use of open space in the architectural design. Such a model may include walls, floors, ceilings, and large pieces of furniture. While the final building may include carpeting, this feature may be left out of the model—quite appropriately—because carpeting may not help illuminate the fundamental features of the house. Likewise, when constructing a mathematical model we should represent important factors and illuminate the most fundamental relationships without paying excessive attention to detail. The only difference is that the object being modeled is a decision, and the materials with which the model is constructed are mathematical symbols rather than clay or wood.

Just as different architects would create different models for the same architectural plan, so too may different analysts create different mathematical models to represent the same problem. There is no one correct mathematical model. Models may illustrate different aspects of a decision and provide different insights. Nevertheless, because the modeling process has very concrete objectives, we can list some desirable properties of mathematical models:

- *Simplicity*: Does the model have few components, and are these components easily understood?
- *Completeness*: Does the model address all, or most, of the major factors that influence the decision?
- *Tractability*: Can the model be solved using available techniques?
- *Transparency*: Are the model's basic structure and assumptions easily identified?

When creating a mathematical model, the analyst strives to balance these modeling objectives along with other goals of the decision process. Not surprisingly, good modeling is as much an art as a science, requiring ingenuity and creativity in addition to training and experience.

Mathematical programming utilizes special types of mathematical models; these models are characterized by three key components. The first component is the set of alternatives available to the decision maker. In mathematical programming, an alternative is simply a number; the symbol that represents the alternative is called the *decision variable*. Making a decision corresponds to selecting a number to assign to each decision variable.

A mathematical program generally contains many decision variables because planning problems generally involve allocating resources to many different activities; each decision variable represents the level of a different activity. It is not unusual for a large, industrial application of mathematical programming to have thousands of decision variables. The purpose of solving a mathematical program is to find quantities for all decision variables that enable the decision maker to best achieve the goal of interest.

Let us return to the carpenter example. Given a supply of 600 planks and 3000 nails, how many ladders and how many chairs should the carpenter make? In this example, the decision to be made is "how many ladders and how many chairs?" A specification of these two numbers constitutes an alternative. We can represent these numbers, the alternative, with mathematical symbols. Let L represent the number of ladders made and C the number of chairs. L and C are the decision variables of the problem. "Build 30 ladders and 25 chairs" is one alternative available to the carpenter. This alternative can also be stated as "Set the decision variable L equal to 30 and the decision variable C equal to 25."

The goal of the decision maker constitutes the second key component of a mathematical program. The mathematical statement of this goal, called the *objective function*, expresses the benefit (or cost) attained under any possible quantities taken on by the decision variables. By changing the quantities of the decision variables, the decision maker changes the degree to which the goal is accomplished, as expressed by the objective function. Suppose, for example, that the carpenter wishes to maximize his revenue from his construction efforts, and is convinced that he can sell every product he makes. Moreover, suppose that the carpenter sells each ladder for \$50 and each chair for \$25. Recalling that L and C represent respectively the number of ladders and chairs made, the carpenter can express his revenue objective as a mathematical function of the decision variables: $50L + 25C$.

Making four ladders and five chairs provides the carpenter with $4 \times \$50 + 5 \times \$25 = \$325$ revenue. Solving the mathematical program is the process of finding what quantities to assign the decision variables to produce the maximum (or minimum, depending on the model) objective amount. The carpenter would like to find out what quantities he can assign to L and C to make that expression, his revenue, as large as possible.

At first glance, it appears that the objective function above implies that the carpenter can make more and more ladders and chairs to make his revenues increase without bound. The carpenter, however, has limited resources. These limited resources represent the third key component of mathematical programs—the component that characterizes the decision-making environment. Decisions are never entirely isolated; the decision maker must work within the requirements or limitations of the environment that affect the selection of alternatives, or, in other words, that constrain the quantities that the decision variables can take on. In planning, such constraints are typically resource constraints or performance requirements. In our example, the carpenter is constrained by the amount of raw materials he has previously purchased.

Like the objective function described above, the constraints of a mathematical program are mathematical expressions that involve the decision variables; each constraint describes requirements or limits on something whose use is related to the quantities of the decision variables. The carpenter has constraints on the amount of nails and wood planks available. Suppose that the construction of each ladder requires 48 nails and 12 planks, and that each chair requires 30 nails and 5 planks. The carpenter must ensure that, no matter how many ladders and chairs are made, the total number of nails used does not exceed 3000 and the total number of planks 600. The carpenter can express these requirements as the following mathematical inequalities:

$$48L + 30C \leq 3000 \quad \text{(constraint on total nails available for use)}$$

$$12L + 5C \leq 600 \quad \text{(constraint on total planks available for use)}$$

The only combinations of chairs and ladders the carpenter can make without running out of raw materials are those that satisfy *both* of the above inequalities.

Table 9.1 expresses the carpenter's problem in both plain English and as a mathematical model. Notice that there are three types of symbols in the mathematical statement of the problem. There are mathematical symbols, such as + and ≤, there are variables, such as L and C, and there are numbers. The numbers that appear in a mathematical program are called the *parameters* and play a special role. They represent assumptions that the decision maker

TABLE 9.1
Two Statements of the Carpenter's Problem

	Statement of Problem in Words	*Mathematical Statement*
Decision	Decide how many ladders and chairs to make.	Let L = number of ladders and C = number of chairs
Objective	Maximize the revenue obtained. Revenue is $50 per ladder, $25 per chair.	Maximize $50L + 25C$
Constraints	Each ladder requires 48 nails and 12 planks. Each chair requires 30 nails and 5 planks. We can use up to 3000 nails and 600 planks. We can make as few as zero ladders and zero chairs.	subject to $48L + 30C \leq 3000$ $12L + 5C \leq 600$ $L \geq 0, \quad C \geq 0$

has made or data that have been collected. In the carpenter example, we have assumed that each ladder requires 48 nails for assembly. Of course, some ladders may require more, say due to faulty nails or mistakes during production. It is important to realize that any solution derived from this mathematical program assumes that each ladder requires exactly 48 nails.

Two important characteristics of a mathematical program affect solution procedures: the nature of the decision variables and the mathematical relationships described by the objective function and the constraints. These factors are so important to the process of solving a mathematical program that mathematical programs are usually named by these characteristics.

Decision variables can represent two types of quantities: continuous quantities (like a volume of chemical or length of pipe) or discrete quantities (such as a number of nails or planks or a sequence of tasks for building a chair). Whereas continuous quantities can take on any number, subject to the constraints, discrete quantities must be represented by integer (i.e., counting) numbers. The presence of integer-valued decision variables so affects the solution process of a mathematical program that mathematical programs that contain integer-valued decision variables are given the special name *integer programs*.

The types of functions represented by the objective function and constraints are also crucial to the development of algorithms. The key distinction made here is between linear and nonlinear programs. Linear programs have constraints and objectives described by linear functions. One of the important properties of linear functions is that their magnitude is directly proportional to the quantities of the decision variables—double the quantities of the decision variables inside the linear function and the quantity of the total function is doubled as well. Linear programs are much easier to solve than nonlinear programs as a result of this property.

Regardless of how we classify a mathematical program, we can give it a general form, both in plain English and in mathematical notation. The word form is:

> Select the quantities of the *decision variables* that maximize the *objective function* subject to the *constraints.*

In the carpenter example, it was "Select the number of ladders and chairs to make that maximizes revenue subject to the limitations on nails and planks." Additional examples are:

> Select an amount to invest in each of a series of stocks that minimizes the risk in the stock portfolio subject to the constraints of achieving a desired return.

> Select the number of delivery vehicles to utilize that minimizes delivery costs subject to the constraint of delivering all items on their scheduled delivery dates.

To exhibit the general mathematical form for a mathematical programming problem we assume that there are n decisions to be made (represented by $x_1, x_2, \ldots, x_n$) and m number of constraints. The dependence of the objective function on the n decisions can be expressed, as $f(x_1, x_2, \ldots, x_n)$ indicating that it is some mathematical function of "n" decision variables. The carpenter's problem involves two decisions, and its objective function, $f(x_1, x_2)$, is $50x_1 + 25x_2$. (It makes no difference whether we call the decision variables x_1 and x_2 or L and C as we did before.)

The constraints are expressed as mathematical inequalities. One side of the inequality represents how the decisions affect the constrained quantity; for the first constraint we express this as $g_1(x_1, x_2, \ldots, x_n)$. The other side represents the actual requirement or limit involved, for example, b_1 for the first constraint. The entire first inequality can be stated $g_1(x_1, x_2, \ldots, x_n)$ b_1. In the carpenter's problem we had two constraints. The first constraint, $g_1(x_1, x_2)$, representing the limit on nails, is $48x_1 + 30x_2$. Its numerical limit, b_1, is simply 3000.

The general mathematical program, then, looks like the following:

$$
\begin{aligned}
\text{maximize} \quad & f(x_1, x_2, \ldots, x_n) \\
\text{subject to} \quad & g_1(x_1, x_2, \ldots, x_n) \leq b_1 \\
& g_2(x_1, x_2, \ldots, x_n) \leq b_2 \\
& \quad \vdots \\
& g_m(x_1, x_2, \ldots, x_n) \leq b_m
\end{aligned}
$$

where $(x_1, x_2, \ldots, x_n) \geq 0$. The last constraint reflects the fact that the decisions represent physical quantities that have no meaningful interpretation if they are negative.

All mathematical programming problems can be written in this form, though there may be times when other forms are more convenient. For example, some mathematical programs minimize, rather than maximize the objective function. Also, others may involve strict equality (=) constraints in addition to, or in place of, the inequality (≤) constraints.

9.3.1.3 Step 3: Derive a Solution to the Model

Solving the mathematical program involves finding which alternative maximizes the objective, among all alternatives that satisfy the constraints. Constraints by their very nature limit the possible alternatives from which a manager can select. In fact, there is no guarantee that there exists any decision that satisfies all of the constraints. Each constraint limits a decision-maker's choice, and it is entirely possible, after adding constraint upon constraint, that all choices are eventually eliminated. If this is the case, the mathematical program is called *infeasible*. If there exists at least one alternative that satisfies all of the constraints, then the mathematical program is called *feasible*, and such an alternative is called a *feasible solution*. Note that in mathematical programming terminology, the word "solution" is synonymous with "decision."

In the carpenter example, we can see clearly that the program itself is feasible because we can find at least one alternative—namely, making one ladder and one chair—that satisfies all constraints on the carpenter's resources. Certainly, this alternative yields little revenue for the carpenter, but its existence indicates that the carpenter does have options to choose from. The decision to make ten thousand ladders, on the other hand, is an infeasible solution because the carpenter simply does not have the resources to do so.

It should be noted that many mathematical programs contain hundreds or thousands of constraints. Unlike the carpenter's problem, it may not be obvious at first glance—or, for that matter, easy to calculate by hand at all—whether a feasible solution exists. One advantage of using mathematical programming solution techniques is that they can usually find feasible solutions if they exist, and provide such solutions to the decision maker. For complex decision problems, discovering a single feasible solution may help the decision maker immensely.

Clearly, our carpenter is not interested in just feasible solutions, he is interested in the alternative that maximizes his revenue. An alternative that best achieves the decision-maker's objective is called an *optimal solution*.

Because the carpenter's problem is relatively simple, and contains only two decision variables, we can solve it using a simple graphical technique that illustrates some of the geometry of mathematical programming problems. Each constraint in the model can be graphed so that the standard x- and y-axes represent possible values of the decision variables, in this case L and C. The shaded area of the graph (called the *feasible region*) contains all alternatives that represent combinations of L and C that satisfy both constraints. All feasible mathematical programs have a feasible region. Solving a mathematical program means finding a point in the feasible region that makes the magnitude of the objective the greatest. Figure 9.1 shows the feasible region for the carpenter's problem.[1]

Each possible revenue can be obtained by a variety of combinations of chair–ladder production levels. For example, a revenue of $1000 could be obtained by producing either 20 ladders and no chairs, or 40 chairs and no ladders. The set of all combinations of ladders and chairs yielding $1000 in revenue forms a line. We can also plot this "constant revenue line" represented by the revenue equation $50L + 25C = \$1000$. Every point of intersection between this constant revenue line and the feasible region represents a different combination of chairs and ladders which yields this constant amount of revenue and satisfies the constraints. If we increase the amount of revenue desired, the line moves toward the top right corner of the graph. We would like to find out how far we can move the line (i.e., how high we can make revenue) while still intersecting the feasible region (i.e., while making sure we

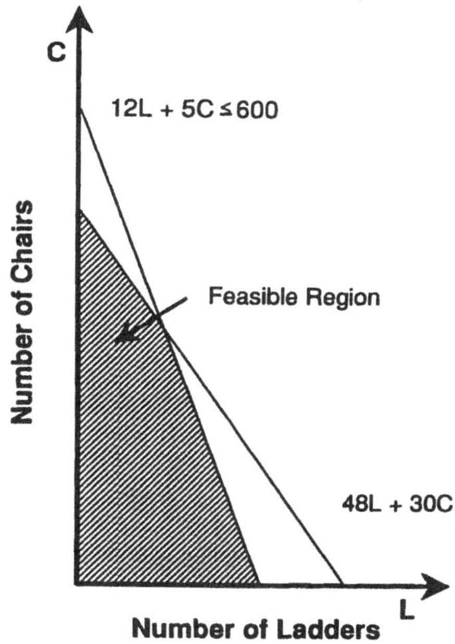

FIGURE 9.1. Feasible region of the carpenter's problem.

1 Strictly speaking, because the carpenter must make ans sell "whole" ladders and chairs, only points corresponding to integer values of L and C are feasible for the carpenter's problem. We address the issue of integrality constraints in Section 9.3.2.3.

have enough resources to achieve that revenue). Figure 9.2 illustrates that the highest revenue is achieved when the constant revenue line is moved so as to intersect the outermost corner of the feasible region, labeled "A" in the figure. The resulting decision is to produce 25 ladders and 60 chairs at a revenue of $2,750.

This graphical solution procedure illustrates some of the geometry behind mathematical programming. This procedure is not practical for larger problems (more decision variables and more constraints) because the resulting feasible region increases in dimension. A problem having three decision variables is three-dimensional. The corresponding constant revenue plots will be an entire plane, rather than a line. As the number of variables and constraints get higher, we can no longer draw a picture, or even depend on our geometric intuition about where the solution might be. More sophisticated procedures are necessary.

A procedure for solving a mathematical program is performed on a computer using an *algorithm*. An algorithm is a well-defined series of computational steps designed to solve a specific type of problem. An example is the simplex method mentioned above; it was the first algorithm available for solving linear programming problems. Many of the great strides made in the field of mathematical programming have come not by formulating new types of mathematical programs, but by developing algorithms that enable computers to solve mathematical programs that have already been formulated. Algo-

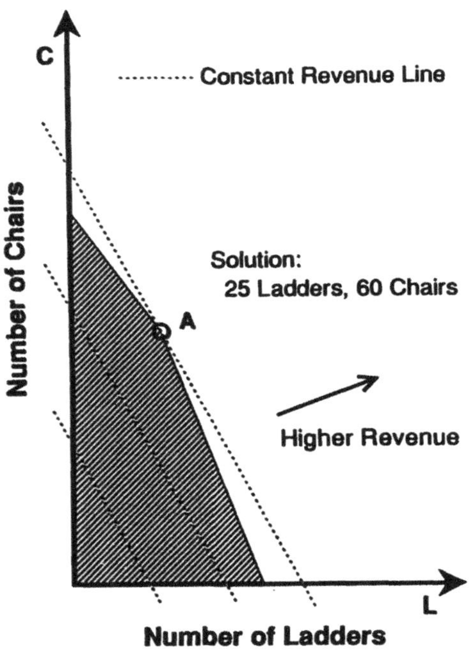

FIGURE 9.2. Solution to the carpenter's problem.

rithms exist for most mathematical programming techniques, though they vary a great deal in their efficiency and power.

9.3.1.4 Step 4: Test Sensitivity of the Solution

The decision maker often wishes to know how the solution to the problem might change if some of the assumptions that went into formulating the problem are varied. This is the purpose of sensitivity analysis. By solving his original mathematical program, the carpenter knows that producing 25 ladders and 60 chairs yields the highest revenues. This mathematical model assumed, among many things, that the carpenter had exactly 600 planks available for construction. How might the desired production of chairs and ladders change if the carpenter had a different number of planks? What if he miscounted and found only 550 planks in storage? What if his supplier sent him 650 by mistake? What is the effect on potential revenue?

Mathematical programming techniques vary in the types of sensitivity analysis they allow. Some techniques, such as linear programming, have well-defined procedures for obtaining information about the effects of changes in the input parameters. Other techniques have no such procedures and require re-solving the model for each possible sensitivity condition.

9.3.1.5 Step 5: Implement and Maintain the Solution

Once analysts solve the model and perform the appropriate sensitivity analyses, they should have a good understanding of what decisions the model supports and why. Successful implementation of the solution requires communicating the model and its results. Key decision makers need to understand what factors and relationships are represented in the model, what factors and relationships have been left out of the model, what assumptions have been made, the recommended decision, and the sensitivity of this decision to important assumptions about the problem.

9.3.2 Mathematical Programming Formulations

All mathematical programs follow the five steps described above, and can be considered variations of the same approach. What differentiates one mathematical programming technique from another is the formulation of the model and the availability of solution algorithms and computational codes. We discuss some representative mathematical programming formulations and their relative strengths and weaknesses.

9.3.2.1 Linear Programming

The first mathematical programming formulation to be developed was linear programming. It is by far the most widely used. Management science literature is replete with examples of how linear programming models have saved companies millions of dollars. Interested readers should consult past issues of the journal *Interfaces* for applications of linear programming techniques. The defining characteristic of linear programs is that both the objective function f and constraint functions g in the model are *linear* functions of the decision variables.

The assumption of linearity is so central to the success and efficiency of linear programming, that it merits further discussion. Linearity implies three important characteristics for these models. The first characteristic is *proportionality*. The amount of resources used is directly proportional to the quantities of the decision variables. In a linear program that models a chemical blending problem, for example, doubling the quantity produced of the finished product doubles the amount of raw materials required to produce that product. Halving this quantity reduces the raw material requirement by half. This proportionality rule also holds for the objective function. A chemical manufacturer wishing to maximize gross revenues from sales will double gross revenues from those sales if sales volume doubles (assuming the selling price per unit is fixed).

Clearly, not all resource allocation situations strictly follow a rule of proportionality. There can be setup costs, economies of scale, or other situations that make the relationship between the decision variables and the objective gained or resources used nonproportional. Nonetheless, in many of these situations, the computational benefits of formulating a linear model are so great, and the potential inaccuracy so relatively small, that an analyst will approximate the relationship with a linear one so that linear programming can be applied.

The second characteristic of linear functions is *additivity*. Additivity means that each decision variable contributes separately to the total magnitude of the objective and to the amount of resources used; the benefit of setting one decision variable to a quantity of 20, for example, does not depend on the quantities of any other decision variables. This does not say that the decisions themselves are independent, only their benefits and use of resources. For example, in the carpenter problem, each additional ladder sold always provides an additional $50 revenue regardless of how many chairs have been produced. Yet the decisions are not themselves independent; producing another ladder reduces the potential number of chairs that can be produced because they are made from the same limited resources.

The third characteristic of linear functions is called *divisibility*. Divisibility means that each decision variable can take on any quantity in a continuous range, whether that quantity is a whole number, fraction, or decimal (though it must be positive). In some cases, this may not be desirable. A mathematical model that suggests you build 2.54 factories is providing only limited help. Integer programming (discussed in Section 9.3.2.3) considers the problems presented when the solution is restricted to be an integer value.

One reason that linear programming is such a prevalent decision aid is that extremely efficient solution algorithms are available. Even massively large and complex problems can be solved fairly rapidly with linear programming. In 1947 George Dantzig, considered the father of linear programming, developed the *simplex method*, a practical technique for solving linear programs. The simplex method works by moving around the feasible region, from one corner to an adjacent corner, finding a corner with a better objective value at each step until it finds the optimal solution. As discussed above, linearity ensures that a solution that optimizes the objective is at a corner, and Dantzig's algorithm is an efficient procedure for finding that best corner without having to check all of them (Dantzig 1963). Figure 9.3 illustrates how the simplex method might solve the carpenter's problem. In this simple example, there are only four corners to check and there are only two possible routes from the initial solution to the optimal solution. In more realistic applications, there are many more corners to consider and the number of possible routes to the optimal solution is immense. (For example, consider how many paths there are from one corner of a cube to the opposite corner.) In large-scale applications the simplex method shows its efficiency in solving the problem.

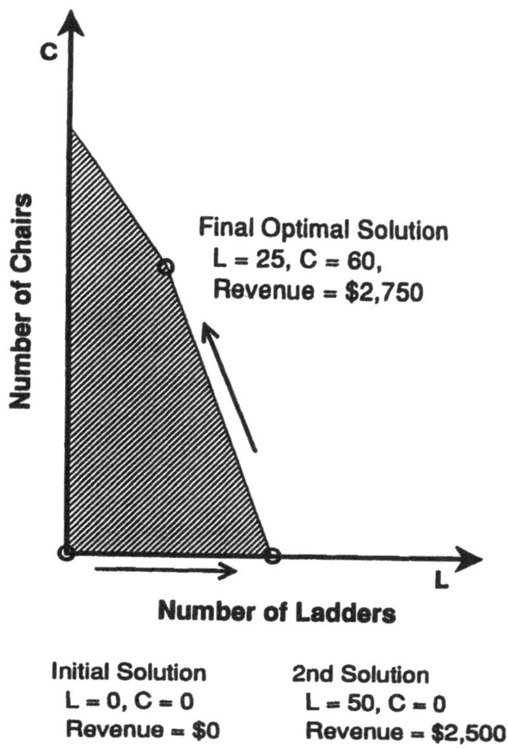

FIGURE 9.3. Solution to the carpenter's problem using the Simplex Method.

In the years following Dantzig's discovery, researchers and practitioners improved the computational aspects of the algorithm and developed standard techniques for performing sensitivity analysis. Both the simplex method and the sensitivity analysis techniques have been coded into commercial linear programming packages that are readily available on almost every computer platform. In fact, computer systems are now so fast, and software so efficient, that today's linear programming packages can solve linear programs that include up to a hundred thousand decision variables and tens of thousands of constraints.

Software codes for linear programming determine the optimal quantities of the decision variables and the resulting optimal quantity of the objective function. In addition to this basic information, most software codes also report which constraints are *binding* (or *tight*) at the optimal solution. A binding constraint is one for which the resource is entirely used up at the optimal solution. Nonbinding constraints represent resources not used to their capacity in the optimal solution. Resources corresponding to binding constraints can be viewed as valuable resources: if you had more of them, you might be able to improve the magnitude of the objective. So just knowing which constraints are binding is useful. However, standard linear programming analysis can also provide the approximate value, as expressed by the objective function, of an additional unit of that resource. Because this represents the maximum price we should be willing to pay per additional unit of the resource, this is often called the *shadow price* of the resource.[2] Shadow prices tell us how much the objective function would change with an additional unit of resource (assuming that no other constraints are binding).

Linear programming also has well-defined procedures for performing sensitivity analysis. Due to the mathematical structure of the linear programming problem and the calculations made while solving the problem through the simplex method, sensitivity analysis can be performed as an extension of the solution procedure. There are procedures for testing the sensitivity of the solution to parameters in the objective function and constraints. Solution packages vary in the amount of sensitivity analysis they provide to the user.

For most of the history of linear programming, the simplex method has been the most practical and efficient algorithm available. In 1984, Narendra Karmarkar of AT&T Bell Laboratories announced a new algorithm for solving linear programs based on a technique that moves through the interior of the feasible region, rather than traveling around it from corner to corner. This new algorithm appears to be significantly faster than the simplex method for some large-scale linear programs having thousands of constraints.

The chief advantages of linear programming versus other mathematical programming formulations are the efficiency of the solution procedure, the availability of sensitivity analysis, and the wide availability of computer

2 This use of "shadow price" slightly differs from its use by economists.

codes. The only weakness is that the linearity assumption may not be appropriate for modeling certain types of relationships, such as setup costs, nonproportionality, or integrality constraints. Consequently, other mathematical programming formulations have been developed, and some of these are described below.

9.3.2.2 Nonlinear Programming

Nonlinear programming differs from linear programming in that the objective function and constraints can be general (nonlinear) functions of the decision variables. By "nonlinear" we mean that either the proportionality assumption or the additivity assumption of linear programming does not hold. Nonlinearities can come from a variety of places: production scheduling problems with increasing or decreasing marginal costs, portfolio selection problems where the objective is to minimize risk subject to a desired return, and production problems that consider price elasticity in demand.

For example, suppose that the carpenter's chairs are a special type that appeals to collectors of fine furniture. Suppose instead of charging \$25 per chair regardless of production, he charges \$(100 – C), where C is the number of chairs he makes, under the assumption that the fewer chairs he makes, the more his customers will be willing to pay for them. The objective function for revenue then becomes

$$50L + (100 - C)C = 50L + 100C - C^2.$$

This function is a nonlinear function of C, so if the carpenter wanted to solve this problem, he would have to rely on nonlinear programming methods.

As this example illustrates, nonlinear programs allow more flexibility in specifying relationships. Using nonlinear programming the carpenter could model a very complicated relationship between production and revenue. This freedom, however, comes at a cost. Nonlinear programs are much more difficult than linear programs to solve because the mathematics is more complex and the solution procedures more sophisticated. Most algorithms that solve linear programs are variants of the basic simplex method. There is no common algorithm for nonlinear programming problems, and different algorithms are targeted for particular types of nonlinear programs.

One major problem with nonlinear programming solution techniques is that they can get caught in "locally optimal" solutions: the algorithm may appear to have found the best solution overall, but has only found a solution that is better than others close to it. This situation is analogous to that of a mountain climber who wants to ascend to the highest point on the earth. He may think he has achieved his goal after climbing Mt. McKinley in Alaska. As far as he knows, he is higher than any point around him. He cannot tell that thousands of miles away there may be mountains whose peaks are even higher. Likewise, nonlinear programming algorithms often find solutions that only seem to be the best, because they are clearly better than any solution nearby. Linear programming does not have this problem; using only linear functions ensures that there will never be more than one mountain.

Whether or not local optimality is a problem depends on the types of nonlinear functions appearing in the mathematical program and the specific type of algorithm used to solve the problem. Some nonlinear functions possess a special property (called *convexity*) that implies that they are "single-mountain" functions. When a nonlinear program is based on such functions, then there will be no problems with locally optimal solutions. In cases where local optima exist, the solution returned by the nonlinear programming algorithm can be sensitive to the initial guess that the algorithm uses to start its solution procedure. One starting point may lead the solution procedure to one local optimum, while another may lead it to a different local optimum. Often, these difficulties can be avoided if the analyst understands the characteristics of the nonlinear functions used in the problem and has some familiarity with the solution technique employed by the software package used.

Unfortunately, sensitivity analysis is more difficult for nonlinear programming than for linear programming. Computer codes can provide analysts with the same type of shadow price information available in linear programming. More sophisticated types of sensitivity analysis, involving changes to the objective function and certain constraint parameters, are usually not available. The analyst must change the problem formulation and solve the problem again to see how the solution would change.

Despite these disadvantages, nonlinear programming remains a viable tool for problems that cannot be handled adequately with linear programming.

9.3.2.3 Integer Programming

Many situations involve decisions, constraints or costs that result from distinct (or "discrete") alternatives. For example, a company may wish to decide whether to build a new plant in Topeka or Dallas. Or a manager may want to decide how much chemical to produce, subject to the requirement that production above a certain volume requires a fixed-cost investment in additional machinery. These situations violate the divisibility requirement of ordinary linear programming; there is no "in-between." The processing plant must be located either in Topeka or Dallas; it makes no sense to suggest a location between Topeka and Dallas. Likewise, the production decision will either require or not require additional production capacity; building partial capacity serves no purpose. Discrete alternatives are considered by a special branch of mathematical programming called *integer programming*. Integer programming gets its name because discrete alternatives must be represented by decision variables that take on discrete (i.e., integer) values. When all of the decision variables are integer, the technique is sometimes called *pure integer programming*. If some of the decision variables are allowed to be continuous quantities, it is called *mixed integer programming*.

Integer-valued decision variables can represent many phenomena. They can represent quantities, orderings or assignments, and yes/no decisions. Yes/no decisions are represented by using decision variables that can take on only two values (called *binary variables*), usually 0 and 1.

The use of binary variables in particular gives integer programming the flexibility to handle special constraints that cannot be handled by other mathematical programming techniques. Suppose the decision variables $x_1, x_2, \ldots, x_{10}$ represent 10 possible projects the decision maker can fund; each variable takes on the quantity 1 if the project is funded, 0 if not. Integer programming can represent, for example, the following logical constraints on the funding of projects:

1. Exactly 5 out of the 10 projects must be funded

$$x_1 + x_2 + \cdots + x_{10} = 5$$

2 Project 2 only can be funded if Project 1 is also funded

$$x_1 - x_2 \geq 0$$

In addition, by restricting a variable to a binary quantity of 0 or 1, one can automatically restrict other variables to certain quantities:

3. The number of machines purchased (x_{11}) must be 2 if Project 1 is funded, and 12 if Project 1 is not funded.

$$10x_1 + x_{11} = 12$$

Creative use of integer-valued decision variables allows the analyst to model a wide variety of logical constraints.

Unfortunately, integer programs are much more difficult to solve than ordinary linear programs. It may seem at first glance that integer programs should be easier to solve than linear programs. After all, the integrality restrictions suggest that there are far fewer possible solutions to consider when searching for the optimal solution. The flaw in this reasoning is that linear programming solution procedures do *not* have to consider all possible solutions; they rely on more efficient methods that are possible, at least in part, because the decision variables are continuous, something no longer true in integer programs. Thus a need (or desire) to test all possible solutions is a step backward in terms of solution efficiency. For example, an integer programming problem that finds the best assignment of 30 people to 30 different jobs has on the order of 10^{32} possible assignments. Even if you could evaluate 1 billion assignments in a second, it would still take longer to find the optimal assignment than the current age of the universe. Clearly, more practical solution procedures are necessary.

A naive approach would be to solve an integer program by ignoring the integer constraints (called a *relaxation* of the integer program), solving the linear program, and rounding the resulting decision variables to the nearest integers. If, by chance, solving the problem yields a decision all in integers, then no rounding is necessary and the solution is valid. (We were lucky in precisely this way with the carpenter's problem.) Unfortunately, this technique usually does not yield the optimal solution. There are two problems with this approach. First, there is no guarantee that the rounded values of the decision variables even comprise a *feasible* solution to the original problem. And second, even if the resulting solution were feasible, it may not necessarily be optimal. There may be situations where the rounding approach may be appropriate. For example, a linear programming solution to produce 5,683.68 gallons of chemical product can probably be rounded to 5,684 without much concern; the rounding is well within the accuracy of the original problem parameters. But what if solving the linear program relaxation yields a recommendation to build 2.4 processing plants and .6 distribution facilities? The decisions involved may represent large cash exposures and the risk of rounding to a nonoptimal solution could be great.

Software is available for solving integer programs. These programs utilize a variety of techniques including linear program relaxations, which are legitimate under certain situations. Another popular approach, called *branch and bound*, breaks the set of possible solutions into smaller subsets and uses special searching techniques to eliminate sets of solutions that are not optimal. Another popular technique uses *cutting planes*; cutting planes are specially formulated constraints that remove nonoptimal solutions when added to the model. Sometimes these techniques are used effectively in combination. Today's sophisticated computer codes can solve integer programs with thousands of binary and general integer variables.

Integer programming has some distinct advantages over other mathematical programming techniques. First, integer programs are designed to handle integer restrictions that cannot ordinarily be solved using standard linear or nonlinear programming techniques. Also, integer programming can model more general logical constraints through the clever use of integer, and particularly binary, variables. Unfortunately integer programming computer codes can only tackle problems that are a fraction of the size of problems solved by linear programming software. Also, sensitivity analysis is more difficult. In order to find out how the solution might change when certain parameters change, it is usually necessary to re-solve the entire problem.

9.3.2.4 Dynamic Programming

Dynamic programming differs from the other mathematical programming formulations discussed here in that it is more accurately described as an approach to solving optimization problems rather than as a technique itself. Up to now, the problems we have described involve a one-time decision.

Dynamic programming is best suited to multiperiod or multistage decision-making situations in which a sequence of related decisions must be made. The best decision alternative to chose in one period may depend on the decisions and outcomes that are likely to occur in subsequent periods. The fundamental idea of dynamic programming is that a larger problem consisting of many "stages" can be broken up into a series of smaller, linked decision problems each of which can be solved much more easily than the original problem.

Two central concepts in dynamic programming are *stage* and *state*. *Stage* refers to a point in time at which the decision maker makes a decision. The stage can be a literal point in time (such as the beginning of each month) or can refer to an event (such as the completion of Phase 1 of a project).

State refers to the decision-maker's situation at the time the decision is made. It encompasses the important information that is needed in order to make the decision. For example, suppose a retailer wished to solve the problem of how much inventory to hold of a particular item. The actual decision to make might be how much product to order. The stages might be the points in time at which the retailer must place orders, perhaps the beginning of every month. The state could be the information necessary to know how much product to order, perhaps the current inventory level and the current expected "low," medium," or "high" forecast of demand. One state, for example, could be "5 units in stock and forecast for high demand." Chances are that this would result in a different order than the state "25 units in stock and a forecast of low demand." A dynamic programming solution procedure provides a list of what decision alternative to choose in each of the possible states; this list is called the *decision policy*.

Each dynamic programming decision has two important consequences. First, some benefit or cost accrues; this is called the *return* (or *reward*) from the decision. Second, the decision itself changes the state for the next decision to be made at the next stage.

The motivating principle behind dynamic programming is that an optimal decision policy developed by dynamic programming—one that minimizes cost over *all* stages, for example—has the following property: given the current state, the optimal decision policy for the remaining stages is independent of the decisions made, and the states encountered, in previous stages. In other words, knowing the current state is *all* you need to know in order to specify the best alternative from this point onward; how you got to your current state is irrelevant. In the inventory example, it does not matter what last month's inventory level was, or its demand forecast; all the retailer needs to know for this month's order is this month's inventory and demand forecast.

We take advantage of this principle by cleverly solving the multistage decision problem as a series of nested subproblems, one at each stage. The subproblems are connected by a *recursive relationship* that relates the benefit of the decision at the current stage to the subproblem of a previous stage. The key in dynamic programming problems is to start small, with only a single stage, and build up to solve the decision problem for all stages. Initially, we solve the subproblem for the last stage; because this subproblem is only a single-stage, there are far fewer decision variables than the original problem, so it can be solved relatively quickly. Solving the single-stage subproblem produces an optimal decision alternative for each state of the last stage. We can use this solution to quickly develop the best policy for the two-stage subproblem. We save computation time because we need not consider *all* two-stage policies; having previously solved the subproblem for the last stage, we need only consider two-stage policies that contain the optimal last-stage decisions we already identified.

This procedure is best illustrated through an example. Imagine the following problem: a group of college students wants to travel from Bangor, Maine to the West Coast for a summer on the beach. They are indifferent to four possible West Coast destinations, all in California. They want to get to the West Coast by the shortest route possible, over the course of three days, with intermediate stops in the East and Midwest. Assuming they have identified a set of possible destinations for each day, if they begin their journey in Bangor, Maine, which route should they take (i.e., through which cities should they travel) in order to minimize the time spent driving?

To formulate this as a dynamic program we need to define decisions, stages, and states. Quite naturally, we can consider each morning of the trip a different stage. Stage 1 represents the morning the students start their trip. The second stage would be the second morning, Stage 3 would be the morning they leave their Midwest stopover. At each stage, the students can start in any one of a number of cities. For example, in Stage 2 they will be in one of four possible Eastern cities. At Stage 3 they will be in one of their four possible Midwest cities. The decision they make at each stage is where to drive that day. Each decision lands them at a new city at the next stage (the following day).

In Figure 9.4 we show the full decision problem in a schematic form. The stages are represented by the vertical columns of circles. (There is no column of circles for the destinations because once the destination is reached, there is no decision to make.) Each circle represents a different city (dynamic programming state) that the students visit that day. The arrows represent decisions, and illustrate the possible highway routes from cities in one stage to cities in the next stage. Note that decisions in each state are limited, i.e., it may not be possible to reach all cities in the next stage from cities in the current stage. Each arrow also has an associated time (not shown on the diagram) that represents the time it takes to drive from the city at one stage to the city at the arrowhead. The problem is to find a route—a path through all stages—that minimizes time on the road.

Figures 9.5 through 9.7 illustrate the dynamic programming approach to solving the problem, focusing on the nature of the solution, rather than the calculations required. We begin first with Stage 3. For each city in Stage 3 (the Midwest), we find the quickest route to a city on the West Coast. A possible

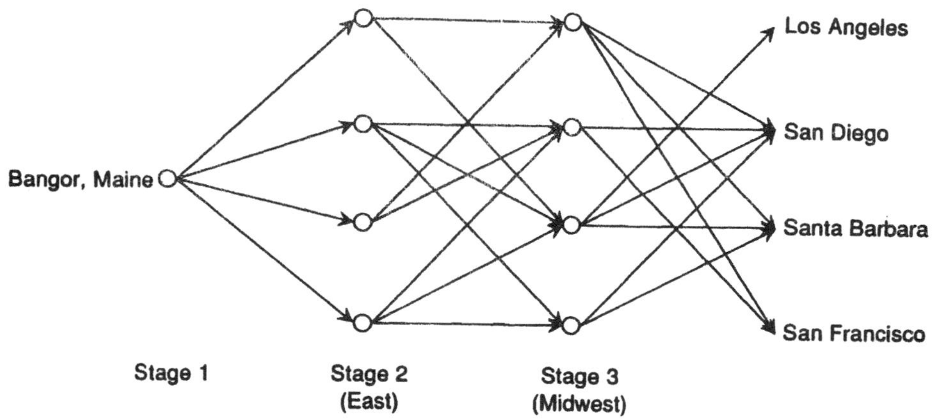

FIGURE 9.4. Dynamic program.

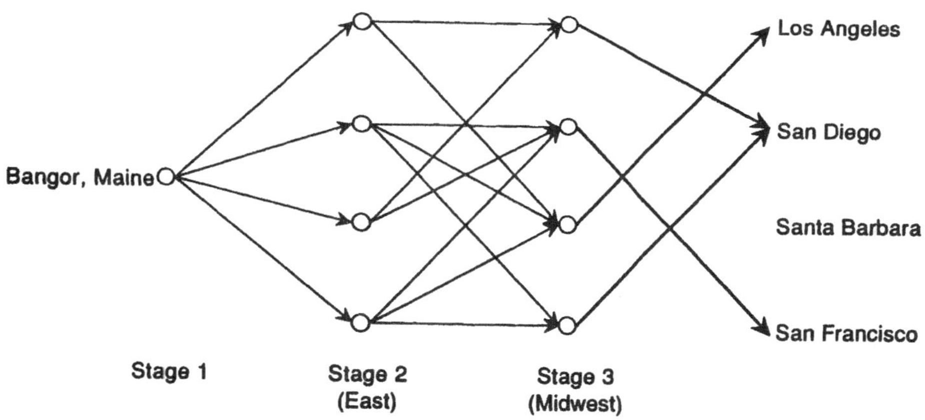

FIGURE 9.5. Dynamic program (first step of solution).

result of this one-stage optimization is shown in Figure 9.5. Each state in Stage 3 is now associated with a single path that represents the quickest route from that city to the West Coast. In other words, no matter in which midwestern city the students wake up on the third day, they know toward what destination they should drive that day. The one-stage subproblem has been optimized. Note that in solving the one-stage subproblem, we ignore all information about Stages 1 and 2. We pretend that the *only* problem at hand is how to get from the Midwest to the West Coast.

To solve the two-stage subproblem, we must figure out the quickest route from all possible states in Stage 2 (East) to the West Coast. Each decision we make at Stage 2 will leave us in *some* city in Stage 3. By the results of our

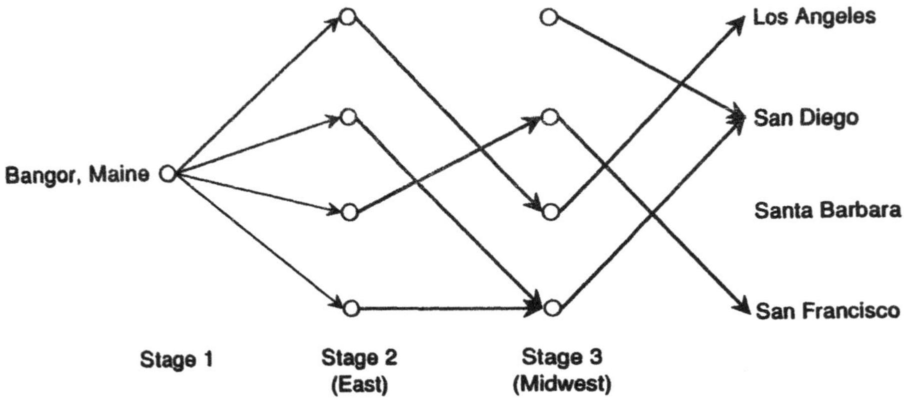

FIGURE 9.6. Dynamic program (second step of solution).

previous optimization we already know the best route from all Stage 3 cities to the West Coast. Thus once we choose a path from Stage 2 to Stage 3, we use the results of the first optimization to continue that path all the way to the West Coast. In this way we can quickly find the total driving time from any Stage 2 city to the West Coast. By using the results of the first optimization, we were able to remove many possible two-stage routes that we knew, from solving the one-stage subproblem, would not be optimal. Having the previous solution at hand means that there are fewer routes to test and we can solve the two-stage subproblem fairly quickly. This is the key to the efficiency of dynamic programming. This solution is illustrated in Figure 9.6.

We repeat this procedure at the first stage, relying on the solution of the three-stage subproblem to help us select which city the students should drive toward on their first morning. No matter what path they take from Bangor, Maine, we have already determined the fastest way to get from any second city to the West Coast. The number of combinations to test is fairly small, so we can solve the subproblem quickly. Figure 9.7 illustrates the complete policy obtained from the dynamic programming solution, the optimal route from each city to the West Coast. The students are interested in the quickest route from the first city, Bangor. This route is shown in Figure 9.8.

In any given application, the analyst must determine the appropriate specific recursion equation. Whereas linear and nonlinear programs are distinguished precisely by—in fact, named for—the type of functions used, dynamic programming is not limited to any special types of functions, nor does it impose any restrictions on the decision variables. Dynamic programming can handle all types of function and discrete or continuous variables. The key to using the dynamic program is to be able to identity stages and states, to formulate an appropriate recursion, and to solve the subproblem

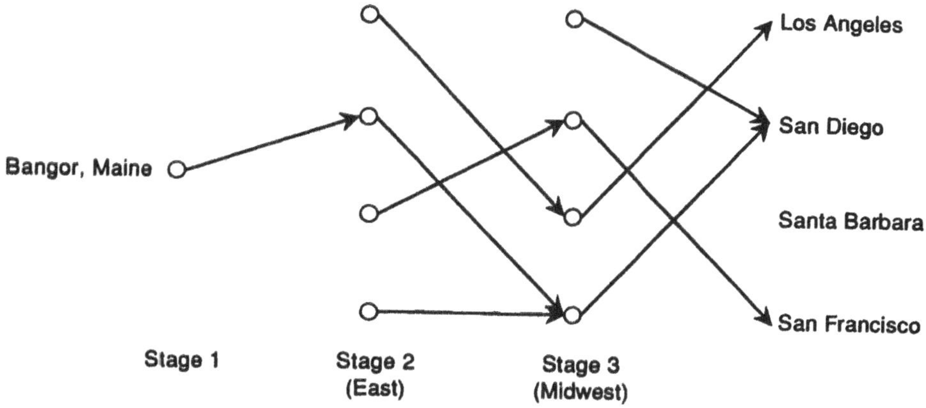

FIGURE 9.7. Dynamic program (first step of solution) (last step of solution).

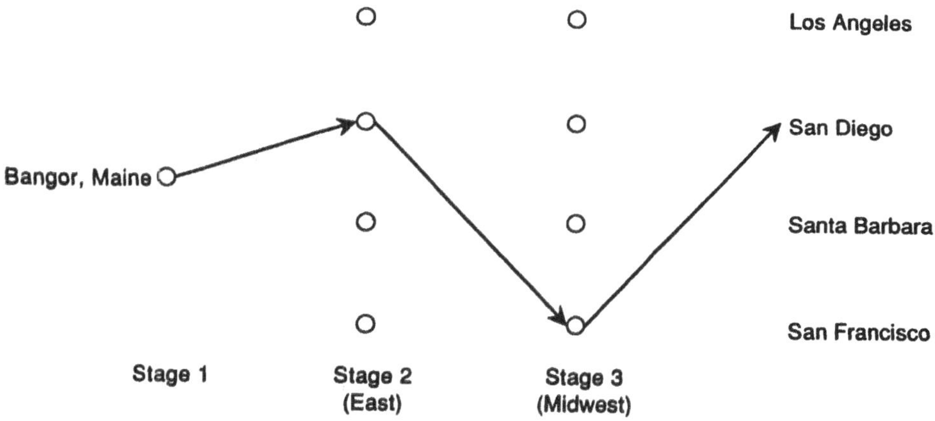

FIGURE 9.8. Optimal solution to dynamic program.

presented at each stage of the recursion. Unfortunately, dynamic programming does not address the issue of *how* to solve each of the one-stage problems. Dynamic programming refers to the *approach* of creating stages and states and defining recursions. Thus there are no general purpose algorithms that solve dynamic programs. Each application must be formulated anew and implemented on a computer from scratch.

Sensitivity analysis is generally not possible with dynamic programming. If problem parameters change, the dynamic program usually must be solved again. Dynamic programming, however, does provide, as its standard result, the optimal policy given every possible state in each stage. This information can be useful for understanding how the optimal policy depends on the current state.

One of the strengths of dynamic programming relative to other mathematical programming techniques is its flexibility; it can handle generic functional forms and can be adapted to problems in which uncertainty is a key factor. There can be uncertainty in moving from state to state. In the inventory example, the next month's inventory level will depend not on the forecasted demand, but upon the actual demand, which may be considered uncertain when the ordering decision is made. Thus any given purchase order level could land you at different inventory levels next month depending on which actual demand results. Uncertainty can also appear in the return from each decision. In the cross-country drive example, uncertain road conditions can make the driving time between cities uncertain. Dynamic programming can usually handle uncertainties such as these with minor changes to the modeling formulation. In this case it is called *stochastic* dynamic programming. For some problems with uncertainty, dynamic programming may be the only mathematical programming technique that is practical.

One weakness of dynamic programming is the lack of general software; each application must be coded separately. Another disadvantage is what is known as the *curse of dimensionality*. This refers to the unfortunate situation that as more information is put into the state description, the time required to find the optimal solution grows quickly. Suppose, for example, that the driving time, from city to city, depends heavily upon the weather along the highway route and the students want to know what routes to take under different weather conditions. Assuming we know how different types of weather affect the driving times, we can add weather information to the state. We can then solve the problem (i.e., determine the best alternative) at each stage for every *combination* of city and weather forecast. For example, the students might travel to a different city from Bangor if the forecast were rain than if it were sunny. Unfortunately, the curse of dimensionality implies that a slight change like this increases the computation time significantly. Adding two types of weather—rainy and sunny—to the state description doubles the total number of states in the dynamic program and will likely double the computation time. Dynamic programming also tends to require much ingenuity and experience to apply to realistic problems.

9.3.2.5 Goal Programming

The mathematical programming techniques we have discussed thus far address problems where the goal of the decision maker is expressed as a single objective that is to be maximized or minimized. Goal programming is a mathematical programming technique that recognizes that the concept of "optimizing" may be inappropriate for some types of decisions. There may be several concurrent, and possibly competing, objectives to be satisfied. A manufacturer, for example, may be interested in obtaining, simultaneously, a desired market share, a certain level of investment, and a reasonable profit.

When introducing the concepts of objective function and constraint, we stressed how they must be expressed explicitly, in mathematical terms. The same holds true for the goals considered by a goal program. Goals usually are defined as desired levels of some quantity of interest. A goal could be a desired level of revenue, a proposed utilization of resources, or a desired total job completion time. Returning to our earlier example, suppose the planks used by the carpenter to make chairs and ladders are unsuitable for any other products made by the carpenter. One goal may be to utilize as many of the planks as possible.

Goal programming is really a variant of linear programming. In an ordinary linear program, the objective function quantifies the goal of the single-objective decision maker. In goal programming, this objective function is replaced with a different function that quantifies the extent to which the decision-maker's several goals are achieved. A goal program is a linear program whose objective is to *maximize the extent to which the goals are achieved*.

But how does one compare achievement among several goals that are measured in different units? Some goals, like a desired profit, are easily measured in monetary units. Other goals may be measured in time, market share, pounds, or gallons. The usual method for handling this discrepancy is to apply weights to the goals that represent their relative importance. If the decision maker feels that missing the revenue goal by one dollar is twice as harmful as missing the total work-hours goal by one hour, then the former goal would be given a weight that is twice the weight of the latter. Alternatively, goals can be given identical weights, in which case progress toward each goal is considered to be of equal importance. When goals are, in this sense, comparable, the technique is referred to as *nonpreemptive* goal programming. This type of goal programming strives to achieve all goals simultaneously, subject to their importance as indicated by the relative weights.

The other major type of goal programming considers *preemptive goals*, which are goals that have a distinct ordering of importance. With preemptive goals, the goal program aims to satisfy the highest level goal as much as possible; then, holding the progress toward this goal as fixed, it searches for a solution that best achieves the second priority goal, etc., until as many of the goals are satisfied as possible. In preemptive goal programming each priority level is infinitely more important than the ones following it; the smallest progress toward the first-level goal is preferred to any progress toward the second-level goal, no matter how large. Goal programs can utilize both types of goals (both preemptive and nonpreemptive). These programs have multiple priority levels, each of which consists of one or more goals that are given relative weights within their priority level.

The priorities of the decision maker are very important in goal programming. The decision maker can express priorities using relative (or *differential*) weights or preemptive weights. Through the use of weights, goal programming often incorporates more subjective information than other forms of mathematical programming.

Many practitioners see goal programming as a more realistic model of how corporations and their managers make decisions, and consider this to be its chief advantage. The ability to simultaneously consider many predefined objectives rather than a single goal allows many different decisions to fit within a single mathematical programming formulation.

Solution techniques for goal programs use a modification of the standard simplex method. Unfortunately these procedures are more complicated than the standard simplex algorithm, so there are many fewer software codes available.

9.4 CASE STUDY: UNDERGROUND PIPELINE

An Application of Mathematical Programming to Chemical Process Safety

This application involves Case One, the underground pipeline, which was introduced in Section 5.2; this case is also used in both the Voting chapter (Section 6.4) and the Decision Analysis chapter (Section 11.4). In this section we illustrate the five steps described in Section 9.3.1.

9.4.1 Step 1: Formulate the Problem

Mathematical programming is an excellent method for addressing problems of resource allocation. To illustrate the application of mathematical programming adequately, we modify the underground pipeline case study as follows.

The failed pipeline came to the attention of the director of safety at the company's corporate office. Concerned that the pipeline risk may exist at the company's other six facilities, she decided to address the problem in a comprehensive fashion by securing a budget for the expressed purpose of reducing pipeline risk. A memo to the managers of the affected plants described the company's commitment to reducing the risks from the pipelines and sought recommendations for addressing the potential pipeline risk. Specifically, each manager was to study the risk at his or her facility and develop an "action plan" of corrective measures to address the risks adequately according to company guidelines. Each manager submitted a report that described the action plan, including its total costs and a description of the risk reduction benefit.

The six plant managers' action plans together had a total cost of slightly over $1.6 million. The director made a formal request to senior management for additional funds to implement the six plans. Senior management allocated $1 million to the director explaining that previous financial commitments to other important safety programs precluded them from securing the full $1.6 million. They asked the director to do the best that could be done with the available funds. The director then faced the problem of how to allocate the $1 million to obtain the best company-wide risk reduction benefit.

Clearly, the director could not fund all of the action plans submitted. One option available to her was to allocate the funds to a subset of the original action plans submitted. The result, however, would most likely leave some installations without any corrective actions at all. While funding no plan at a particular facility would not, in and of itself, be detrimental to the company, had the individual managers known the budget was limited, they might have submitted more economical plans that may have provided more cost-effective risk reduction. In a letter to each plant manager, the director explained the budget situation and asked that the manager develop an additional two, more economical, action plans. The director would decide which of the four alternatives—the original action plan, one of the two more economical alternatives, or no action plan—to fund at each plant.

How could the director use this information to allocate the budget to obtain the highest benefits across the company?

9.4.2 Step 2: Construct a Mathematical Model

To apply mathematical programming to this decision, the director needed to identify the three main components of a mathematical program: the decision variables, the objective function, and the constraints.

The immediate decision was which (if any) of the action plans to fund at each facility. For each action plan at each facility, the director was to make a yes/no decision whether to fund that plan. In mathematical programming, yes/no decisions are represented by binary decision variables whose values are restricted to 1 (yes) and 0 (no). To address the funding decision, the Director needed one binary decision variable for each action plan considered. If she denoted a facility by F and an action plan by A, then X_{FA} can be the decision variable that identifies whether Action Plan A is funded at Facility F. If X_{FA} equals 1, then the action plan is funded; if X_{FA} equals 0, the action plan is not funded. For example, a solution in which $X_{13} = 1$ corresponds to a decision in which Action Plan 3 at Facility 1 is funded.

The director's objective in making the funding decision was to reduce the total company-wide risk as much as possible. To use mathematical programming, it is necessary to express this objective in terms of the yes/no decision variables that are already defined. The director needed to explicitly characterize the relationship between *which* action plans are chosen and *how much* risk reduction would be accomplished. Risk, however, is usually expressed in terms of probabilities, and the standard mathematical programming approaches generally are less well-equipped to incorporate probabilistic inputs. The director instead used the expected fire losses and environmental cleanup costs averted over the subsequent 10 years to measure the benefits. This assigned to each action plan a benefit that represents the averted fire losses and environmental cleanup costs that the company would expect to save over the next 10 years if it funded that plan. The total benefit for funding a series of plans equals the sum of the averted costs for all plans funded. The director identified the net benefit for Action Plan A at Facility F as B_{FA}. The director calculated the total risk reduction benefit for any suggested allocation of the budget by summing up the B_{FA} for those action plans that are funded. For any plan, funded or not, the expression $B_{FA} \times X_{FA}$ gives the contribution of the plan FA toward the entire risk reduction effort. (Note that for any plan not funded, $X_{FA} = 0$ and $B_{FA} \times X_{FA}$ equals zero and contributes nothing.) Summing these expressions for all facilities and all action plans yields the total benefits:

Total benefits over all plans

$$= \sum_{\text{all F}} \sum_{\text{allA}} C_{FA} \times X_{FA} = B_{11} \times X_{11} + B_{12} \times X_{12} + \cdots + B_{63} \times X_{63}$$

Note that for any trial allocation of funds, the only benefits included in the sum are those for which $X_A = 1$ (i.e., those action plans that are funded). (We assume for this example that the director or the individual plant managers had a procedure for calculating the B_{FA} parameters. Although the quantities of these parameters are influential in the solution to the mathematical program, we will not discuss developing these inputs because it is not central to the understanding of how to formulate mathematical programs in general. The important issue is that some expression must be developed that relates the quantities of the decision variables to the goal of the decision maker.)

Finally, the director was concerned with the limits on, or requirements of, her selection of action plans. The most obvious constraint on her choice of action plans was that she must not spend more than her \$1 million budget. The sum of the costs associated with each action plan chosen could not exceed \$1 million. If we let C_{FA} represent the cost of each action plan, then this constraint can be expressed by:

$$\sum_{\text{all F}} \sum_{\text{allA}} C_{FA} \times X_{FA} \leq \$1{,}000{,}000.$$

Again, for any suggested allocation (i.e., any suggested set of values X_{FA} with each X_{FA} equal to either one or zero), this expression includes costs only for those action plans that are funded, i.e., for which X_{FA}= 1. The expression $C_{FA} \times X_{FA}$ equals 0 for any action plan not funded.

An additional set of constraints is necessary to force the selection of at most one action plan for each facility. There would be no point in funding more than one action plan at any site because the more expensive action plans at each facility are merely extensions of the more economical action plans. In terms of the decision variables, for any given facility (F), at most one of the X_{FA} decision variables can take on the value 1. Because each X_{FA} must equal either 0 or 1, we can express this constraint as

$$\sum_{\text{all F}} X_{FA} \leq 1$$

which, for the first plant, would be

$$X_{11} + X_{12} + X_{13} \leq 1.$$

Because the X_{FA} are restricted to 0 or 1, this constraint has the desired effect of forcing at most one of X_{FA} for each plant to be equal to 1. Note that there are six of these constraints—one for each of the six facilities. The total number of constraints is therefore seven: one budget constraint and six action plan constraints.

Now that these three items were defined, the director could express the mathematical programming formulation of this problem. The English and mathematical statements of the problem are shown in Table 9.2. Note that the mathematical statement involves all linear functions, yet the decision variables themselves are restricted to integer values. This mathematical program is an

TABLE 9.2
Two Statements of the Safety Director's Problem

	Statement of problem in words	Mathematical statement of problem
Decision	Decide Yes or No on each plan.	For each facility F and each plan A, $X_{FA} = 1$ if plan A is funded at Facility F $X_{FA} = 0$ if plan A is **not** funded at Facility F
Objective	Maximize the total risk reduction benefits.	Maximize $\sum_{\text{all F}} \sum_{\text{allA}} X_{FA} \times B_{FA}$ where B_{FA} is the benefit for plan A at Facility F
Constraints	Budget is limited to \$1 million. Each of the six facilities can be assigned, at most, one action plan.	subject to $\sum_{\text{all F}} \sum_{\text{allA}} C_{FA} \times X_{FA} \leq \$1{,}000{,}000$ (budget) $\sum_{\text{all F}} X_{FA} \leq 1$ (one for each Facility F) where C_{FA} is the cost for plan A at Facility F

TABLE 9.3
Parameters of the Safety Director's Problem

Facility		1	2	3
1	Cost Benefit	\$141,000 (C_{11}) 190 (B_{11})	\$210,000 (C_{12}) 300 (B_{12})	\$392,000 (C_{13}) 540 (B_{13})
2	Cost Benefit	\$ 40,000 (C_{21}) 50 (B_{21})	\$104,000 (C_{22}) 160 (B_{22})	\$164,000 (C_{23}) 240 (B_{23})
3	Cost Benefit	\$ 81,000 (C_{31}) 150 (B_{31})	\$152,000 (C_{32}) 280 (B_{32})	\$290,000 (C_{33}) 500 (B_{33})
4	Cost Benefit	\$165,000 (C_{41}) 220 (B_{41})	\$227,000 (C_{42}) 370 (B_{42})	\$312,000 (C_{43}) 450 (B_{43})
5	Cost Benefit	\$105,000 (C_{51}) 110 (B_{51})	\$142,000 (C_{52}) 160 (B_{52})	\$224,000 (C_{53}) 250 (B_{53})
6	Cost Benefit	\$141,000 (C_{61}) 420 (B_{61})	\$190,000 (C_{62}) 540 (B_{62})	\$250,000 (C_{63}) 680 (B_{63})

example of a pure integer linear program. Table 9.3 describes the cost and benefits of the action plans available at each facility based on information supplied by the plant managers. These numbers constitute the parameters of the mathematical program. To avoid confusion between dollars spent and dollars of benefit, we have expressed the benefits in thousands of dollars without dollar symbols attached. The mathematical statement of the problem in Table 9.2, combined with the parameters in Table 9.3, completely specify the mathematical program to be solved.

9.4.3 Step 3: Derive a Solution to the Model

Because the director decided to fund at most one action plan at each facility, there were four alternatives for each facility: fund Plan 1, fund Plan 2, fund Plan 3 or fund no plan. In total there are $4^6 = 4096$ possible allocations, only some of which satisfy the budget constraint. For this very small problem, one could explicitly enumerate and test all of the possible allocations by computer to see which one, among those that satisfy the budget, yields the most benefit. However, even for a small problem, it may be worthwhile to use a dedicated integer programming software package, especially if one wishes to expand the model in the future.

The optimal solution for the above problem is that X_{12}, X_{33}, X_{42}, X_{63} all equal 1 and the rest of the X_{FA} are equal to 0. The director interpreted this solution as recommending that she fund Action Plan 2 for Facilities 1 and 4; Action Plan 3 for Facilities 3 and 6 and no action plan for Facilities 2 and 5. The total cost of this allocation is $977,000 and the total benefit is 1850 (representing $1,850,000).

The integer programming approach finds the single solution that provides the maximum total benefit. It is interesting to compare its solution to other approaches that the director could have used to solve the problem.

A simple, but very naive, approach would have been simply to allocate an equal amount of funds to each facility. If the director followed this procedure, allowing each facility to get the best benefit out of its $166,000 share, the total benefit is only 1510, with Facilities 1, 4, and 6 funding Action Plan 1; Facilities 3 and 5 funding Action Plan 2; and Facility 2 funding Action Plan 3. The total cost would be $905,000. Those facilities that had additional funds left over could probably spend the remaining budget in a way that would provide additional benefits, but it is clear that this overall allocation method is much less efficient toward achieving company-wide risk reduction benefit (18% less benefit).

A more sophisticated procedure might entail calculating benefit–cost ratios for all of the plans and selecting plans with the highest benefit–cost ratio. Because at most one plan will be funded at each facility, once a plan is selected at a facility, the other plans at that facility are removed from consideration. This procedure can be repeated until either the budget is exhausted, or all of the facilities have plans funded. Applying this procedure to the director's problem results in the suggestion to fund Plan 1 at Facilities 3 and 6; Plan 2 at Facilities 1, 2, 4, and Plan 3 at Facility 5. This allocation costs $987,000 and results in 1680 in benefits, a 9% decrease in benefits over the optimal solution, at a slightly higher cost.

There are two unfortunate aspects to using these alternative procedures. First, they do not give the optimal solution. While the solutions they gave may seem close in terms of percentage, in decisions that involve large capital expenditures, a single percent may represent thousands or millions of dollars. Improvements that appear to be small in terms of percentage, may be huge in terms of absolute dollar amounts. Second, there is no way of knowing, without finding the optimal solution, exactly how far from optimal the solutions are. It is easy to create simple examples to show that both of these alternative procedures can produce rather poor approximations to the optimal solution.

9.4.4 Step 4: Test Sensitivity of the Solution

Suppose the director was dismayed that the optimal solution is to fund no plan at Facilities 2 and 5. One can simply change the constraint that requires *at most* one plan to be funded at each facility to a constraint that requires *exactly* one plan to be funded at each facility. This modification simply replaces the

"≤" symbol with a "=" symbol in each facility constraint. Such a change is trivial in most mathematical programming software. Solving the problem again with the modified constraint, the optimal solution is to fund Plan 1 at Facilities 1, 2, and 5; Plan 2 at Facility 3; and Plan 3 at Facilities 4 and 6. This allocation costs exactly $1,000,000 and provides 1760 in benefits. The director decided that this alternative solution was more appropriate than the original solution, because the intangible benefits from funding all of the plants that were worth the 5% decrease in benefits.

9.4.5 Step 5: Implement and Maintain the Solution

The director discussed the results with the both the senior management and the plant managers at the affected facilities. This helped verify that the tradeoff made by funding one plan at each facility was consistent with the goals and objectives of the senior management. The director then shared the results with the plant managers, notifying each of them about the approved action plan.

9.5 EXTENSIONS OF MATHEMATICAL PROGRAMMING

The techniques described in this chapter reflect the most basic mathematical programming techniques. There are, however, a wide array of mathematical programming formulations. All share the same structure, to optimize an objective under some constraints. Mathematical programming techniques differ in particular characteristics of either the decision variables or the functions used to describe the objective or constraints. Mathematical programming techniques not discussed in this chapter, such as geometric programming, separable programming, and convex programming, are variations of the general formulations described here.

Aside from dynamic programming, we have made little mention of uncertainty in this chapter. There is active research on the application of mathematical programming to problems characterized by significant uncertainty. One technique, called *chance constrained programming*, considers planning situations in which a constraint need not hold with certainty. A more difficult problem is that of *stochastic programming*, where there is uncertainty in the problem parameters. The difficulty with stochastic programming is that uncertainty in only a few parameters can cause the mathematical program to increase in size very rapidly. Unfortunately, even small problems can become intractable when uncertainty is considered. This is due to the fact that the usual approach to solve mathematical programs with uncertainty is to reformulate them as equivalent mathematical programs without uncertainty and solve them with the standard Simplex Method. To adequately consider alternative possible futures, however, the equivalent formulations have many more constraints and decision variables than the original problem. Usually, to solve such problems, solution procedures must exploit a problem's special structure. Much research has gone into developing efficient procedures to handle uncertainty, but as yet there is no procedure that will easily solve a general stochastic program.

9.6 IMPLEMENTATION NEEDS

Mathematical programming requires detailed information about the problem being modeled. Objectives and goals and all significant constraints or requirements on the decision must be recognized and stated explicitly.

The successful use of mathematical programming techniques require a combination of training, experience, and artistry. This is true for all aspects of mathematical programming, from formulating the problem to solving it using a computer and interpreting sensitivity analyses.

Other significant resources required are time and computers (with appropriate software). Realistic applications of mathematical programming require time to formulate, solve and test the solution, invariably with the aid of a computer. In some cases this process might require coding a specialized algorithm for the application at hand, which adds to the resource requirements. Depending on the technique being used and the size of the problem being handled, computer needs may vary. For typical linear programs, today's computers and mathematical programming software can solve very large problems, with thousands of constraints and variables. For other types of mathematical programming the problem size must be kept much smaller.

9.7 SUMMARY

9.7.1 Strengths and Weaknesses of Mathematical Programming

Mathematical programming's acceptance in industry stems from its several strengths. Its greatest strength is that it provides optimal solutions to complex problems. In doing so, mathematical programming techniques find the single best alternative out of millions, billions or even an infinite number of possible alternatives for the decision maker. Not only do the algorithms provide this benefit, but they exist in computer codes (for many mathematical programming techniques) that are extremely efficient, allowing such problems to be solved easily. This technology has provided companies with real, quantifiable benefits.

Even when an algorithm has difficulty finding a single best solution, it may nonetheless provide a feasible alternative that the decision maker may not have considered. Resource allocation problems that are highly constrained may be difficult to solve. The capability of finding a single feasible solution may be a significant benefit in and of itself.

Also, mathematical programming is a decision aid that can be applied to planning problems that exist in the context of other decisions. Mathematical programming can be embedded in a decision process that also utilizes other decision aids. For example, a company might wish to open a new factory in a new city. Other decision aids may be more appropriate for determining which city is the best location, perhaps voting or multiattribute utility analysis. Given a particular location decision, mathematical programming can be utilized to help decision makers plan the operation of that facility, from needed manufacturing capacities to job assignments to capital outlays.

To use mathematical programming, decisions and the constraints on the decisions must be stated in very precise mathematical language. This is both a benefit and a weakness. It is a benefit because it clarifies the decision-maker's alternatives and goals while providing the opportunity to exploit sophisticated quantitative techniques. It is a disadvantage because, with the possible exception of goal programming, there is little flexibility to handle more abstract problems that may be less suitable to mathematical modeling.

Mathematical programming consists of sophisticated mathematical techniques and solution procedures. This means that it requires specialized knowledge, experience and a certain degree of artistry to formulate problems, understand and interpret solutions, and effectively communicate results. This can provide difficulties; there can be resistance to mathematical programming solutions in an organization where there is limited understanding or appreciation for these methods.

With the exception of dynamic programming, the stochastic side of mathematical programming, the side that concerns uncertainty, is less well-developed than the deterministic side, the side that assumes problem parameters are known with certainty.

9.7.2 Suggestions for Further Information/Study

In-depth discussions of mathematical programming can be found in introductory management science and operations research textbooks. Cook and Russell (1985) provide a particularly readable introduction containing many real-life case studies and applications. Hillier and Lieberman (1990) contribute additional information on algorithms. Excellent introductions can also be found in Wagner (1975) and Gould, Eppen, and Schmidt (1991).

9.8 REFERENCES

Ackoff, R. L. (1956). The Development of Operations Research as a Science. *Operations Research* 4:256f.

Charnes, A. A., and W. W. Cooper (1961). *Management Models and Industrial Applications of Linear Programming*. 2 vols. New York: John Wiley and Sons.

Cook, T. M., and R. A. Russell (1985). *Introduction to Management Science*. Englewood Cliffs, NJ: Prentice-Hall.

Daellenbach, H. G., J. A. George, and D. C. McNickle (1983). *Introduction to Operations Research Techniques*. 2nd ed. Boston: Allyn and Bacon.

Dantzig, G.B. (1963). *Linear Programming and Extensions*. Princeton, NJ: Princeton University Press.

Gould, F.J., G.D. Eppen, and C.P. Schmidt (1991). *Introductory Management Science*. Englewood Cliffs, NJ: Prentice-Hall.

Hillier, F., and G. J. Lieberman (1990). *Introduction to Mathematical Programming*. New York: McGraw-Hill.

Wagner, H. M. (1975). *Principles of Operations Research*. 2nd ed. Englewood Cliffs, NJ: Prentice-Hall.

10

PAYOFF MATRIX ANALYSIS

*Most major business decisions are taken as a result of animal spirits—
of a spontaneous urge to action rather than inaction, and not
as the outcome of a weighted average of quantitative
benefits multiplied by quantitative probabilities.*
—John Maynard Keynes

10.1 PURPOSE OF CHAPTER

Uncertainty is an unavoidable and important aspect of most decisions on chemical process safety. It is hard to predict perfectly how equipment and workers will perform, what trends the business will experience in the future, what actions will be taken by regulators and the public, and many other factors that influence decisions. Many decision-making approaches address uncertainty in an informal fashion. For example, conservative values are sometimes used to increase the margin of safety, or decisions that decrease the variation in future outcomes or that minimize the chance of poor outcomes are arbitrarily favored. Decision makers are often uncomfortable with these informal approaches. The resulting decision has a less than rigorous foundation, and the basis for this decision is not entirely explicit. An informal approach does not make clear what the chances are of a good or bad result. Finally, the result may be to reject alternatives that a more careful analysis would show to be desirable.

Payoff matrix analysis is one of three decision aids discussed in this book that deals explicitly with uncertainty. The others are decision analysis and multiattribute utility analysis. All three have a rigorous, logical foundation based on probability and utility theory.

Payoff matrix analysis has several advantages. It is the simplest of the three approaches that explicitly deal with uncertainty. It can be used for quick decision making and is effective in group situations. It does not require software or extensive training. Finally, payoff matrix analysis provides a useful introduction to the concepts employed in decision analysis and multiattribute utility analysis.

This chapter explains the use of payoff matrix analysis and gives two simple examples. The chapter also provides an introduction to skills such as probability and value assessment that will be useful in decision analysis and multiattribute utility analysis. The chapter provides enough information on payoff matrix analysis to evaluate its usefulness for a particular problem, and the chapter will provide the reader with the skills needed to apply this technique to simple problems. Additional training and the readings provided at the end of the chapter will help one apply payoff matrix analysis to complex problems or in group situations. This chapter begins with an overview and history of payoff matrix analysis. We describe the basic approach through a simple example, and illustrate extensions of the approach through a more realistic example. Finally, we discuss the strengths and weaknesses of the approach and provide references for further study.

10.2 OVERVIEW OF PAYOFF MATRIX ANALYSIS

10.2.1 Introduction to Payoff Matrix Analysis

Payoff matrix analysis is the simplest method that explicitly recognizes the distinction between possible actions, uncertainties, and values. There are eight steps in a payoff matrix analysis (summarized in Figure 10.1):

1. Develop a clear problem statement.
2. Generate alternatives by identifying possible actions that may be taken to resolve the problem.
3. Determine uncertainties and their likelihood of occurrence.
4. Specify possible outcomes caused by each combination of alternatives and uncertainties.
5. Assign a value to each outcome.
6. Calculate the expected value of each alternative by weighting the outcome values by the probability that the outcome occurs, and summing over the possible outcomes for that alternative.
7. Review the analysis and results for completeness and credibility; if the analysis is not satisfactory, repeat all or parts of the analysis.
8. Make a recommendation. The recommendation may include choosing one of the alternatives, or doing a more thorough analysis with a different decision aid.

Payoff matrix analysis is so named because the alternatives, uncertainties, probabilities, and values are recorded in a matrix. The columns of the matrix refer to the alternatives, and the rows of the matrix refer to the uncertainties. These steps will be discussed in greater detail later in this chapter.

Decision analysis and multiattribute utility analysis employ the same concepts as payoff matrix analysis and a similar series of steps. The differences

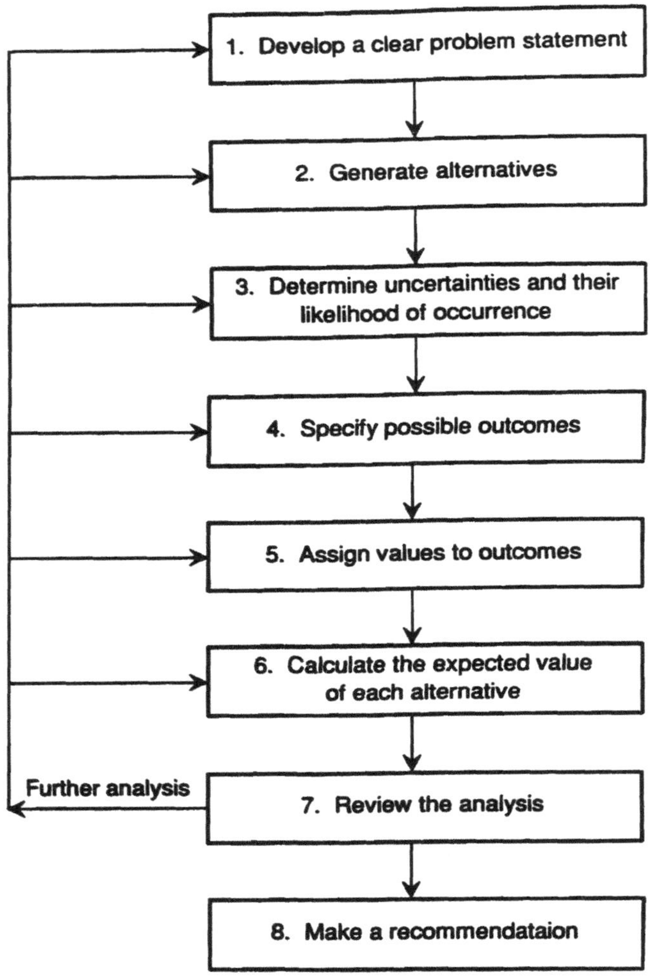

FIGURE 10.1. Steps in the payoff matrix analysis.

are in the level of detail and the complexity that can be accommodated. Decision analysis and multiattribute utility analysis emphasize decomposing the problem into its fundamental elements, thinking about the individual elements, and constructing the complete analysis from these elements. The payoff matrix approach deals with fewer and, for complex problems, more aggregated pieces. For simple problems, the payoff matrix approach provides a full analysis. For more complex problems, the payoff matrix approach provides a quick overview and a rough solution. In such problems, it is necessary to make simplifications in order to apply the technique. For complex problems, decision analysis and multiattribute utility analysis are likely to provide more insights and more thorough, robust solutions.

10.2.2 History of Payoff Matrix Analysis

The history of quantitative decision analysis spans several centuries. Mathematicians such as Pascal, Fermat, and Huygens established the concepts of probability and expectation in the mid-1600s. Daniel Bernoulli, in the early 1700s, made use of the concept of utility (the value of an outcome to a decision maker) to explain behavior in risky situations, such as buying insurance. In the 1940s, von Neumann and Morgenstern formalized the principles of utility and of decision making based on expected utility.

Because they provide a very simple and clear means of recording value and probability information and an easy format for calculation, payoff matrices are popular teaching tools. For example, in his book on decision making, Lindley (1971) uses payoff matrix analysis extensively, referring to a payoff matrix as a decision table. Lindley uses payoff matrices to illustrate simple decision-making principles, but does not present them as a practical tool for making significant decisions.

For practical problems, payoff matrix analyses, using criteria other than values weighted by probabilities, have been used frequently. The earliest example that we have found is by Belyaev (1976). He discusses an application to a Soviet power system expansion decision. Martin (1979) describes a simple, but complete and practical method for decision making using payoff matrices that consider probability. Hanna (1979) describes a chemical process safety application. More recently the Electric Power Research Institute (EPRI) has advanced the concept of payoff matrix analysis through the CATALYST process for group decision making (EPRI 1992).

10.2.3 Decisions/Situations Most Appropriate for Payoff Matrix Analysis

Because of its simplicity, payoff matrix analysis is suitable for situations in which there are relatively few alternatives and uncertainties, the time for analysis is constrained, and/or group decision making is desired. In group decision-making situations, the simplicity of the method allows decision makers to concentrate on communicating ideas, concerns, and background information rather than the analytic details. A wide variety of issues can be addressed when assigning values and probabilities, but the method does not include a rigorous process to determine values and probabilities or to record the details of an analysis. A payoff matrix analysis can be a good first step of a more detailed analysis, or a way to decide whether further analysis is warranted.

10.3 EXPLANATION OF PAYOFF MATRIX ANALYSIS

10.3.1 Key Concepts

Several terms that are used in this chapter may be unfamiliar to readers. We recommend that the reader review the discussion of "Uncertainty" in Section 2.6.

Key concepts are associated with several steps of the process illustrated in Figure 10.1. These concepts are also important for decision analysis and multiattribute utility analysis.

A key concept associated with generating alternatives (Step 2) is that of a *decision*. As used in payoff matrix analysis, a decision is an action that irrevocably commits resources or forecloses an opportunity. *Alternatives* are the possible actions that can be taken. Decisions are defined by two or more alternatives. Decisions are differentiated from concerns, worries, and regrets, which may bother us, but which do not in themselves commit resources or foreclose an opportunity. The payoff matrix approach is designed to result in action and is therefore not applicable when dealing with concerns, worries, and regrets. More significantly, determining whether or not a problem involves an actual decision can help focus a decision maker's energy, increase effectiveness, and reduce anxiety.

Estimating uncertainties (Step 3) introduces the *subjective* view of probability as distinguished from the *objective* view. In the subjective view of probability, probabilities are personal beliefs about the world; in the objective view of probability, probabilities are physical characteristics of the world. The objective view defines probability as the long-run frequency with which an event occurs in an infinite number of trials. Under this view, probability is obtained from historical frequency data and is only relevant for repetitive events. However, the subjective view of probability recognizes that the world often forces us to act in unique situations. To predict future events in these situations, we must consider not only historical data but also more general knowledge about how the world works. Probabilities are an unambiguous and very useful form in which to communicate our predictions or beliefs. From the subjective view, probabilities do not refer to the past. When we say that outcomes have a 50% probability, it means that we expect half of these outcomes will occur in the future.

Assigning subjective probabilities is not a natural activity for most people. For this reason, accomplishing Step 3 involves mental exercises that may be unfamiliar. These are discussed in Section 10.3.2.4.

The concept of *value* is important in estimating the value of possible outcomes (Step 5). Value is the desirability of an outcome to a decision maker. In many cases, the concept of value is simple. For instance, we frequently make decisions about how much we are willing to pay for an item. The higher our value for something, the more we are willing to pay for it. However, it may be hard to assign values to complex future outcomes, particularly outcomes with multiple characteristics or attributes. As with probabilities, the discussion of this step will suggest ways to assign values to outcomes so that they are useful to the analysis.

Finally, the concept of *expected value*, is introduced by Step 6. The expected value of an alternative can be thought of as a weighted average of the values of all possible outcomes, where the weight for each outcome is the probability of occurrence of that outcome. The payoff matrix approach chooses the alternative that maximizes expected value. Expected value is discussed in more detail in Section 2.6.2 and is illustrated in the examples in this chapter.

10.3.2 Description of Payoff Matrix Analysis

10.3.2.1 An Illustrative Example

A simple example will be used to explain payoff matrix analysis.

> Bob is planning a chemical process safety course for his company. There are two dimensions to his decision: what size of hotel meeting room to rent, and how many meals and binders to order. Since this is the last course that will use these binders, any excess binders will be worthless. A hotel meeting room for 30 costs $500, and a hotel meeting room for 50 costs $750. Binders cost $20 each, and meals cost $60 per attendee.

The following paragraphs describe each step of a payoff matrix analysis of Bob's decision in more detail.

10.3.2.2 Step 1: Develop a Clear Problem Statement

A clear problem statement will answer the question, "What is the problem that will be addressed through the analysis?" The problem statement should address a single problem. Because it is easier to solve several smaller independent problems than one large, complex problem, it is a good practice to focus on the simplest problem that is independent of other concerns. By independent, we mean that the value of solving this problem will not be significantly affected by other activities. For Bob, the problem statement might be:

What preparations should I make for the upcoming safety course.

10.3.2.3 Step 2: Generate Alternatives

The decision maker next determines the decision that must be made, and specifies the decision by describing the available alternatives. The decision maker must control or be able to implement each alternative considered. As far as possible, the alternatives should be specific with respect to cost, other important resources, and timing. Bob states the decision for this simple example as, "For how many participants should preparations be made?" Bob specifies the following alternatives:

1. Prepare for low attendance (20)
2. Prepare for medium attendance (30)
3. Prepare for high attendance (40)
4. Prepare for very high attendance (50).

10.3.2.4 Step 3: Determine Uncertainties and Their Likelihood of Occurrence An uncertainty is something that is unknown and uncontrollable, or not perfectly known or controllable, such as a quantity or the outcome of a future event. In Bob's problem, the number of participants who enroll in the course is an uncertainty. It is important for the decision maker to consider the range of uncertainties, but to also concentrate on those that are most important to the decision.

States describe the distinct possible occurrences or results of each uncertainty. A payoff matrix analysis is difficult unless the conditions in the future can be summarized by 10 or fewer possible states. The states must be mutually exclusive (i.e., nonoverlapping) and collectively exhaustive (i.e., cover the range of all possible future occurrences). Definitions of states must be precise in the sense that the definition must be understood by everyone participating in the analysis. Terms such as "soon," "small," "large," and "few" have widely differing interpretations and can lead to misunderstanding if not specifically defined. Quantitative definitions are more appropriate. For example, "a small accident in the near future" is a poorly defined state. "One nonpermanent injury in the next year" is a better definition.

For the simplest form of payoff matrix analysis, the uncertainties considered in the analysis must be probabilistically independent of decisions. That is, the decision should not change the probability of a state occurring. With some modifications to the simplest approach, this restriction can be relaxed. In the next example in this chapter the restriction is relaxed.

In our example, Bob determines that the most important uncertainty is the number of people who enroll for the course. This is independent of the number of participants for which Bob prepares. Bob specifies the following states for this uncertainty:

1. Low interest (16 to 25 sign-ups)
2. Medium interest (26 to 35 sign-ups)
3. High interest (36 to 45 sign-ups)
4. Very high interest (46 to 55 sign-ups).

Now Bob needs to estimate the probability of each state. A probability is a number that expresses the likelihood of occurrence of a possible state of an uncertainty. By definition, a probability must range from 0 to 1, and the sum of probabilities for all possible states of an uncertainty must be 1. When estimating probabilities, the decision maker should use all available past information, but should not hesitate to use his or her own best judgment. In most cases, statistics, logic, and experience all provide some guidance for estimating probabilities. A good way to begin a probability assessment is to list the states in order from the most to the least likely one, and then quickly make an initial assignment of probabilities. If there is little information on which to base this estimate, the decision maker may initially assign the same probability to each state.

Next, it is useful to establish the relative probabilities of the most likely, least likely and a middle state. After revising these reference probabilities, the probabilities assigned to the other states can be adjusted. As before, the probabilities must sum to one. A useful mental exercise for thinking about probabilities is to compare the state probabilities to familiar and simple bets. For example, suppose we set up a bet in which Bob wins $100 if 46 to 55 participants sign up for the course. Likewise, we can set up another bet in which Bob receives $100 if a coin toss results in "heads." We assume the chance of "heads" is 50%. If Bob prefers to take the bet of 46 to 55 sign-ups over the coin toss resulting in "heads," we know that Bob's probability of 46 to 55 sign-ups is greater than 50%. If Bob assigns a probability of 50% to a state, Bob should be indifferent between receiving a hundred dollars if that state occurs or receiving $100 if a coin toss results in "heads." Other bets that are familiar or easy to visualize are: drawing a spade from a pack of cards (25%), rolling a six with a single die (17%), drawing a white ball from a bag containing one white and nine black balls (10%). Thinking about and assigning probabilities is a skill that is essential to payoff matrix analysis, decision analysis, and multiattribute decision analysis. Good references on probability assessment are Merkhofer (1987), von Winterfeldt and Edwards (1986) and Behn and Vaupel (1982).

In our example problem, Bob knows that the average attendance at the last 4 classes has been 25, and class sizes have been growing. After considera-

TABLE 10.1
Payoff Matrix with Alternatives, States, and Probabilities

		Alternatives			
States	*Probabilities*	*Prepare for low attendance (20)*	*Prepare for medium attendance (30)*	*Prepare for high attendance (40)*	*Prepare for very high attendance (50)*
Low interest (16 to 25)	20%				
Medium interest (26 to 35)	35%				
High interest (36 to 45)	40%				
Very high interest (46 to 55)	5%				

tion of these data, Bob assigns the probabilities shown in the second column of the payoff matrix for Bob's problem, Table 10.1.

The first column of the table lists the possible states of the uncertainty (Number of sign-ups). In the second column, the probability that each of these states occurs is recorded. Bob's alternatives are listed across the top. In the next step, outcomes will be entered for each of the alternatives in each of the possible states.

10.3.2.5 Step 4: Specify Possible Outcomes

An outcome is the specific result of a decision and the resolution of an uncertainty or series of uncertainties. The next step in the process is to specify the outcomes of different alternatives and states. Outcomes may have multiple criteria, or attributes. In our example, important criteria might include: the number of attendees, the number of people turned down, and the cost. These criteria are listed in the third column of the matrix of Table 10.2. The number of attendees and the number of people turned down depend on how many sign up and how many Bob is prepared to accept. For simplicity in the calculations, Bob chooses to represent the number of people who sign up with four discrete points instead of ranges:

TABLE 10.2
Payoff Matrix with Important Criteria

			Alternatives			
States	*Prob-abilities*	*Criteria*	*Prepare for low attendance (20)*	*Prepare for medium attendance (30)*	*Prepare for high attendance (40)*	*Prepare for very high attendance (50)*
Low interest (20)	20%	# Attend # Turned down Cost	20 0 $2100	20 0 $2900	20 0 $3950	20 0 $4750
Medium interest (30)	35%	# Attend # Turned down Cost	20 10 $2100	30 0 $2900	30 0 $3950	30 0 $4750
High interest (40)	40%	# Attend # Turned down Cost	20 20 $2100	30 10 $2900	40 0 $3950	40 0 $4750
Very high interest (50)	5%	# Attend # Turned down Cost	20 30 $2100	30 20 $2900	40 10 $3950	50 0 $4750

1. Low interest (20 sign-ups)
2. Medium interest (30 sign-ups)
3. High interest (40 sign-ups)
4. Very high interest (50 sign-ups).

Some criteria, such as the number of participants who attend or are turned down, are naturally quantifiable. Often a mathematical model, such as a spreadsheet or an equation, is needed to determine outcomes. For example, the cost criteria of the outcome in Bob's decision can be calculated with the following equation:

$$\text{Cost} = \text{Cost of Room} + \text{\# of Planned Participants} \times (\text{Cost per Binder} + \text{Cost per Meal Plan}),$$

or

$$\text{Cost} = \$500 + \text{\# Planned Participants} \times (\$20 + \$60)$$
for 30 or fewer planned participants

$$\text{Cost} = \$750 + \text{\# Planned Participants} \times (\$20 + \$60)$$
for more than 30 planned participants.

For the situation in which Bob prepares for 20 participants, the cost calculation is:

$$\text{Cost} = \$500 + 20 \times (\$20 + \$60) = \$2100.$$

This is placed in the fourth column of Table 10.2. The cost is the same regardless of the number of participants who actually attend.

10.3.2.6 Step 5: Assign Values to Outcomes

The next step in payoff matrix analysis is to assign a single numerical value to each outcome. A value measure is a single quantity (e.g., dollars) that represents the desirability of an outcome that may have multiple attributes. Several methods are available for assigning values to the outcomes. We will describe two methods. The first method is very simple and flexible but requires careful consideration by the decision maker. The second method is more mechanical and often quicker; however, the results are not as reliable.

Method 1: Direct Assessment

A good reference on this method is von Winterfeldt and Edwards (1986). The following steps are involved:

1. Rank all the outcomes from the best to the worst
2. Assign the worst outcome 0 value
3. Consider how much you would pay to assure the best outcome rather than the worst, and assign this value to the best outcome

4. Pick a middle ranked outcome and make a similar value assignment by comparing the relative desirability of the outcome to the best and worst outcomes
5. Similarly assign values to all other outcomes.

Bob uses the direct assessment method to assign the values shown in Table 10.3. He does this by comparing the outcomes of Table 10.2. Bob feels that the best outcome will occur if he prepares for 50 participants and has 50 participants. The worst outcome will occur if he prepares for 50 participants and only 20 attend. Further, Bob feels it would be worth $5000 to assure the best outcome rather than the worst. Therefore, he assigns the worst outcome a 0 and the best $5000 as shown in Table 10.3. Bob then assigns the other outcomes by making relative comparisons to the best and worst outcomes.

To check the assigned values, Bob performs a mental exercise involving hypothetical gambles. Initially, Bob picks an outcome that is ranked intermediate in value between the best and worst extremes—for example, preparing for high attendance and having only medium interest. Next, Bob considers a situation in which he can receive this mid-outcome or a 50–50 gamble on the best vs. worst outcomes. This initial gamble is illustrated at the top of Figure 10.2. Bob decides that he prefers the gamble to the mid-outcome. (This preference for the gamble is designated by ">" in Figure 10.2.) Bob then repeats the hypothetical exercise but with a *lower* probability on the best outcome. (If Bob preferred the mid-outcome, he would place a *higher* probability on winning the best outcome.) Bob continues adjusting the hypothetical gambles until he is indifferent between the gamble and the mid-outcome. As illustrated at the bottom of Figure 10.2, Bob feels that he would be indifferent between a 60–40 gamble on the worst and best outcomes and the outcome associated with preparing for high attendance and having medium interest. (This indifference

TABLE 10.3
Analysis of the Class Sizing Problem

		Alternatives			
States	*Probabilities*	*Prepare for low attendance (20)*	*Prepare for medium attendance (30)*	*Prepare for high attendance (40)*	*Prepare for very high attendance (50)*
Low interest (20)	20%	2500	1500	1000	0
Medium interest (30)	35%	2000	2500	2000	1500
High interest (40)	40%	1000	2000	3500	3000
Very high interest (50)	5%	500	1000	2500	5000

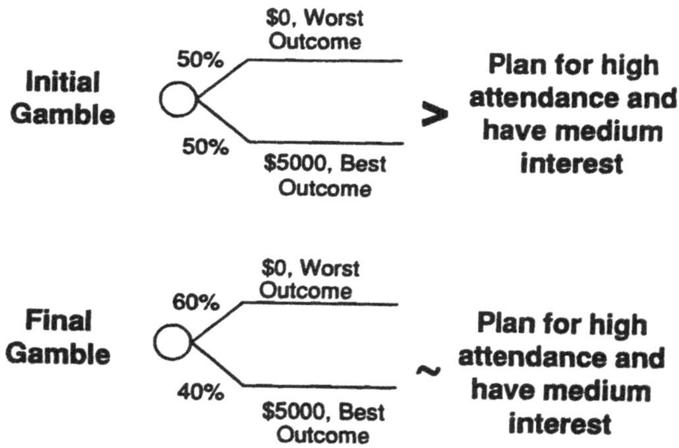

FIGURE 10.2. A mental exercise to check value assignments using hypothetical gambles.

is indicated by "~" in Figure 10.2.) This final gamble gives the value for the mid-outcome.

The value assigned to the mid-outcome should be approximately the value of the best outcome times the probability of winning the best outcome in the final gamble. In equation form:

$$\text{Value of Mid-Outcome} = \text{Probability of Winning Best Outcome} \times \text{Value of Best Outcome.}$$

For Bob, the calculation is:

$$\text{Value of "prepare for high interest and have medium attendance"} = 40\% \times \$5000 = \$2000.$$

As shown in Table 10.3, this is the value that Bob enters in the cell corresponding to high interest, medium attendance.

It is useful to perform the mental exercise with several middle ranked outcomes, and then review all of the assigned values based on the results.

In some situations, particularly those involving risks, the decision maker may feel uncomfortable making the initial assignment of dollar values to the best versus the worst outcome. In this case, simply assign the best outcome a value of 1, and the worst outcome a value of 0. Then one can use the mental gambles to assign a few intermediate values; and, finally, fill in other intermediate values.

Method 2: A Mechanistic Approach

The following method for assigning values is suggested by Martin (1979). This method of evaluating outcomes is a mechanistic approach. We recommend that when this approach is used, the final payoffs be carefully reviewed. When reviewing the payoffs, one should first make sure that outcomes with nearly equal payoffs are nearly equally desirable. Payoffs that are the average of two other payoffs provide another good test of the relevance of the payoff values. One should be indifferent between having the average payoff with certainty or taking a gamble on a flip-of-the-coin between the two other payoffs.

This method begins by ranking the criteria associated with outcomes by importance, and then weighting the criteria using the formula:

$$\text{Weight} = (n - \text{Rank} + 1)/k$$

where n is the number of criteria and k is the sum of the ranks.

Table 10.4 illustrates this criteria for the class-size problem. Since there are three criteria, the sum of the ranks is 6, $1 + 2 + 3 = 6$. By the formula, the second-ranked criterion will receive a weight of:

$$(3 - 2 + 1)/6 = 0.33.$$

Next, each alternative is ranked on each criterion for each possible state. In this round, the better alternative is given the higher rank. Table 10.5 illustrates for the class-size example and the state that 40 participants attend.

TABLE 10.4
Ranks and Weights of Criteria

Criteria	*Rank*	*Weight*
Turning away participants	1	0.50
Number of participants	2	0.33
Cost	3	0.17

TABLE 10.5
Ranks of Each Action under State-of-the-World 40 Potential Participants

	Number of Participants Planned For			
Criteria	*20*	*30*	*40*	*50*
Turning away participants	1	2	3	3
Number of participants	1	2	3	3
Cost	4	3	2	1

In the first line of the table, the most participants are turned away when the plan is for 20 participants. This is the worst case so it receives a score of 1. The next most participants are turned away when the plan is for 30 participants. If 40 or 50 participants are planned for and the 40 participants attend, no one is turned away; therefore, these are equally good outcomes and each receives a 3. Other criteria are treated similarly.

In the next step, the ranks are multiplied by the criteria weights. Then the weighted rankings are added to determine the value of the action under each possible state. Table 10.6 continues the example and contains the final payoff values for one state: 40 participants attend.

In Table 10.7, all the final payoff values are presented and the expected value of each action is calculated. The calculation of expected value is the same as under the method presented in Section 10.3.2.7.

The mechanistic method has several problems, as illustrated by this example. First, there is no clear rule for determining the absolute rankings. For

TABLE 10.6
Ranks of Each Action under State That 40 Participants Attend

	Number of Participants Planned For			
Criteria	*20*	*30*	*40*	*50*
Turning away participants	0.50	1.00	1.50	1.50
Number of participants	0.33	0.66	1.00	1.00
Cost	0.68	0.51	0.34	0.17
TOTAL	1.51	2.17	2.84	2.67

TABLE 10.7
Outcome Values and Expected Values

		Number of Participants Planned For			
Potential Participants	*Probability*	*20*	*30*	*40*	*50*
20	20%	1.51	1.34	1.17	1.00
30	35%	1.51	2.17	2.00	1.83
40	40%	1.51	2.17	2.84	2.67
50	5%	1.51	2.17	2.84	3.49
EXPECTED VALUE		1.51	2.00	2.21	2.08

example, assume there are three alternatives; and on Criteria A the alternatives rank 1, 2, 3; and on Criteria B the alternatives are tied. The final answer can be determined by whether the ranks on the tied criteria are set at (1, 1, 1) or (3, 3, 3). We have adopted the rule that ties are ranked by the lowest rank number. Next, because the value under each possible state is determined separately, illogical values across states can occur. This can be seen in the example by examining the column corresponding to the action "Plan for 20" and the states that 20, 30, 40, or 50 participants attend. The alternative "Plan for 20" has equal weights in each of these states. Yet these are successively worse situations.

10.3.2.7 Step 6: Calculate the Expected Value of Each Alternative

This step calculates the expected value of each alternative using the values and probabilities estimated in earlier steps. An expected value is the mathematical expectation of the numerical values associated with the outcomes of an uncertainty or series of uncertainties, calculated by summing the possible numerical values times their probabilities, over all possible numerical values. (See the discussion of "Uncertainty" in Section 2.6 for further detail.)

Table 10.8 shows the expected values of Bob's alternatives. As an illustration of the calculations Bob did, the expected value of the "Prepare for low attendance (20)" was calculated as follows:

Expected Value

$$= (20\% \times 2500) + (35\% \times 2000) + (40\% \times 1000) + (5\% \times 500) = 1625.$$

The action with the highest expected value is, tentatively, the preferred alternative. In the example, the preferred alternative is to prepare for high attendance (40).

TABLE 10.8
Analysis of the Class Sizing Problem Showing Expected Value

		Alternatives			
States	*Probabilities*	*Prepare for low attendance (20)*	*Prepare for medium attendance (30)*	*Prepare for high attendance (40)*	*Prepare for very high attendance (50)*
Low interest (20)	20%	2500	1500	1000	0
Medium interest (30)	35%	2000	2500	2000	1500
High interest (40)	40%	1000	2000	3500	3000
Very high interest (50)	5%	500	1000	2500	5000
EXPECTED VALUE		1625	2025	2425	1975

10.3.2.8 Step 7: Review the Analysis

The next step is to review the analysis and return to earlier stages if necessary. As part of this review, the decision maker should vary the values and probabilities to see if the recommended alternative changes. In Table 10.9, Bob reviews his analysis with higher probabilities placed on lower levels of interest.

Even with the altered probabilities shown in column two of Table 10.9, the highest expected value alternative is still to prepare for high attendance (40). This gives Bob more confidence in this alternative. If sensitivity analysis resulted in a change to the highest expected value alternative, Bob would need to review the inputs to which the solution is sensitive.

10.3.2.9 Step 8: Make a Recommendation

The final step in the analysis is to make a recommendation. This may include recommending that the highest expected value alternative be chosen, or that further analysis be done. In some cases, the recommendation may be to do a more thorough analysis with a different decision aid. In reporting and presenting the results of a payoff matrix analysis, it is helpful to have a record of the process of creating the matrix. Often, insights gained during this process are the most valuable parts of the analysis. In a group analysis, it is valuable to have participants provide a "story" about actions and effects of each strategy and scenario combination.

10.3.3 CATALYST, Payoff Matrix Analysis for Groups

10.3.3.1 Introduction

CATALYST is a form of payoff matrix analysis that is intended for group decision making. The method was developed by the Electric Power Research

TABLE 10.9
Analysis of the Class Sizing Problem Showing Expected Value under Altered Probabilities

		Alternatives			
States	*Probabilities*	*Prepare for low attendance (20)*	*Prepare for medium attendance (30)*	*Prepare for high attendance (40)*	*Prepare for very high attendance (50)*
Low interest (20)	25%	2500	1500	1000	0
Medium interest (30)	40%	2000	2500	2000	1500
High interest (40)	30%	1000	2000	3500	3000
Very high interest (50)	5%	500	1000	2500	5000
EXPECTED VALUE		1750	2025	2225	1750

Institute (EPRI) and has proven to be a useful approach for many problems. In the chemical process industry, CATALYST is most useful for group decision making and problems with many impacts in addition to safety. Nonsafety impacts might include impacts on corporate finances, management, and image. CATALYST is similar to other methods of group analysis that use scenarios of future events to help decision making, but is more firmly tied to expected utility theory. The method is explained in detail in the "CATALYST Facilitator's Guidebook" (EPRI 1992). In this section we describe CATALYST and identify some key differences from simple payoff matrix analysis. In Section 10.4, the use of a CATALYST-type process in a group context will be explained in more detail through a case study.

The primary purpose of a CATALYST session is to promote communication and understanding among stake holders and experts interested in a problem. If the time for analysis is limited, CATALYST can help a group reach a timely and logical decision. However, CATALYST is more often used to gain an initial understanding of a problem and to develop a logical plan for further analysis.

There are typically five to fifteen participants in a CATALYST session; it is difficult to have an effective session with more than fifteen participants. The participants should include both decision makers and technical experts. The group should represent all parts of the organization that have information about the problem or that will participate in solving the problem.

CATALYST sessions are usually led by a trained facilitator. The facilitator is very important to the success of the session. The facilitator:

- Must be seen as neutral with respect to the problem
- Must be skilled in leading groups
- Must not analyze, form opinions, or make decisions.

The CATALYST process is often conducted in several passes. First, there is a preliminary pass over one or two hours. This is a quick "run through" to familiarize participants with the process and to uncover the most important issues. The next pass generally involves two or three days with the group. This pass is most successful when conducted over a continuous period and at a location where participants are not disturbed. Additional passes are usually conducted by a smaller analysis team, using a more sophisticated decision aid.

The CATALYST process follows the same steps as the simple payoff matrix analysis.[1] However, it builds upon the simple process in several ways. The next subsection will discuss strategies, scenarios, and a technique for

1 In the CATALYST guidebook, Step 1 and Step 2 are discussed as part of CATALYST's Stage 1: Problem Definition. The CATALYST guidebook divides Step 3 into two parts: Stage 2: Uncertainty Identification and Stage 3: Scenario Development. Finally, the CATALYST guidebook divides Step 4 into Stage 4: Strategy Development, Stage 5: Incremental Risk Analysis, and part of Stage 6: Evaluation. Also, the method for dealing with strategy dependent uncertainties shown here is somewhat different than that described in the CATALYST guidebook.

incorporating uncertainties that are dependent on the decision. These are the key extensions to the simple payoff matrix analysis.

10.3.3.2 The Use of Strategies and Scenarios

For more complex problems, alternatives are usually composed of a number of separate actions that work together to solve a problem. Such complex alternatives are often referred to as strategies. To be more precise, a strategy is defined as a set of alternatives across several decisions. For example, on a cloudy morning you may need to decide "how will I go to work today," and "should I take my umbrella." The alternatives for "how will I go to work today" might be "drive my car," "take the bus," and "ride with my neighbor." The alternatives for "should I take my umbrella" might be "take my umbrella" and "do not take my umbrella." "Take the bus" and "do not take my umbrella" represents one possible strategy. As a safety example, consider dealing with a fire hazard. One strategy for controlling the risk would be to reduce ignition sources, maintain the availability of fire control equipment at current levels, and increase the level of fire prevention training. A second strategy for controlling the risk would be to take no action to reduce ignition sources, increase the availability of fire control equipment, and increase the level of fire prevention training.

For complex problems, just as actions must be combined to form strategies, uncertainties must be combined to form scenarios. A scenario is a set of states that cover all the uncertainties important to the decision. For example, two simple uncertainties are "when will the bus arrive," and "will it rain today." The states of "when will the bus arrive" might be "10 minutes" and "20 minutes." The states of "will it rain today" might be "rain" and "no rain." The combination of "the bus arrives in 20 minutes" and "rain" comprises one possible scenario. As a second example, consider a fire hazard. Two uncertainties associated with the hazard might be the chance of a flammable liquid leak and the presence of an igniting flame or spark. One scenario is that the flammable liquid leaks and that there is no igniting flame or spark. An alternative scenario is that the flammable liquid leaks and that it is ignited by a flame or spark.

Scenarios describe a logical future course of events. Because scenarios must be carefully created, this step is more extensive than for simple problems. For complex problems, there are several distinct substeps. These are to:

1. Identify and select key uncertainties
2. Define states
3. Develop scenarios and assign scenario probabilities.
4. Assign probabilities to decision-dependent uncertainties such as risk.

A brainstorming exercise begins the process to *identify and select key uncertainties*. The group lists uncertainties that represent significant threats or opportunities and typically posts the lists around the meeting room for easy reference. This brainstorming session is usually fast-paced, and surfaces many ideas. Group members learn a great deal about each others' concerns and expectations during these sessions.

Next, the group selects the key uncertainties. A useful mental exercise is to imagine an oracle with perfect knowledge of the future. However, the oracle charges $1 million per prediction, and the group has a $4- to $7-million research budget. The most important uncertainties are those to which the group allocates budget for prediction. After allocating the budget, the group will review the key uncertainties to ensure that each represents a fundamental concern and that each is important enough to retain in the analysis.

After key uncertainties are identified, the next substep is to *define states* for each uncertainty. The process for defining states is the same as in simple payoff matrix analysis.

The third substep is to *develop scenarios and assign scenario probabilities*. This substep begins with the group identifying dependencies that exist among the uncertainties and decisions. Probabilistic dependence exists between two uncertainties if knowing the resolution of one uncertainty would change your assessment about the resolution of the other. For example, suppose that the probability of rain is 20% when there is no information on the cloud cover, and that the probability of rain is 50% when the day is cloudy. Then the likelihood of rain is dependent on the cloud cover, and probabilistic dependence exists between rain and cloudiness. The identification of such dependencies will be important for determining the probabilities of scenarios.

Probabilistic dependence may also exist between a decision and an uncertainty. If a decision changes your probabilities of an uncertainty's states, the uncertainty is dependent on the decision. For example, decisions to implement safety measures reduce the probability of accidents; therefore, an accident uncertainty is dependent on decisions to implement safety measures. The payoff matrix approach requires that uncertainties that are probabilistically dependent on decisions be treated differently from uncertainties that are independent of decisions. Decision-dependent uncertainties are handled in the next substep.

There are a number of ways to develop scenarios and assign scenario probabilities. Usually the process begins with a single uncertainty and builds scenarios by adding other uncertainties one at a time. As the states of an added uncertainty are combined with the prior scenarios, the number of scenarios grows. For example, if we begin with the uncertainty on whether it will rain, our initial scenarios correspond to the two states "rain" and "no rain." When the uncertainty on bus arrival is added, the number of scenarios grows to four:

1. rain, bus arrives in 10 minutes
2. rain, bus arrives in 20 minutes
3. no rain, bus arrives in 10 minutes
4. no rain, bus arrives in 20 minutes

To limit the number of scenarios, less important scenarios are dropped and scenarios with similar impacts are grouped together. The scenario probabilities can be assessed after the addition of each uncertainty or following the creation of final scenarios. Probabilities can be assigned to scenarios using the techniques described in the introductory example (see Section 10.3.2.4).

The CATALYST process includes more sophisticated techniques for developing scenarios and assessing scenario probabilities. These techniques have been proven in practical applications and may result in greater accuracy.

However they are created, the final set of scenarios and their probabilities should meet the following criteria:

- The number of scenarios should be limited to four to eight.
- The scenarios should illuminate key threats and opportunities for the organization.
- Scenarios should span the range of possible futures.
- Any risk that significantly affects an individual strategy should be included.
- The group must be able to assign a probability to each scenario.
- The relative probabilities must make sense. If Scenario A has a probability of 10% and Scenario B has a probability of 20%, the group should believe that Scenario B is twice as likely as Scenario A.
- It must be impossible for any two scenarios to occur simultaneously.
- The sum of the scenario probabilities must equal 1.

The last substep is to *assign probabilities to decision-dependent uncertainties such as risk*. Analysis of such uncertainties is very important in most safety studies. To address these uncertainties, rows for the decision-dependent uncertainties are added to the bottom of the table. As illustrated in Table 10.3, "Analysis of the Class Sizing Problem," in the simple payoff matrix each row has one probability, but the values in a row differ according to the alternative. The key difference in the rows that will be added to deal with decision-dependent uncertainties is that each row has *one value but the probabilities across a row differ according to the alternative.*

TABLE 10.10
Probabilities of Decision-Dependent Uncertainties

States	Values	*Alternatives*		
		Take bus, as normal	*Take bus, leave 10 minutes early*	*Drive, leave 10 minutes early*
Arrive on time		85%	10%	5%
Arrive 0 to 10 minutes late		95%	3%	2%
Arrive more than 10 minutes late		98%	1%	1%

This is illustrated in Table 10.10 for a decision on travel to work involving a risk of arriving late. The value column is left blank because these are assessed in a later step. The probabilities are assigned using the techniques illustrated with the introductory example in Section 10.3.2.4. Each column contains probabilities assuming that the action at the top of the column is taken.

10.4 CASE STUDY: CHLORINE LOADING FACILITY

An Application of Payoff Matrix Analysis to Chemical Process Safety

This section describes a hypothetical application of the payoff matrix approach to a decision regarding a chlorine loading facility near a residential area. The basics of this case are introduced in Section 5.3. The case is also discussed in Section 8.4, where cost–benefit analysis is applied to the problem.

We modify this case to assume that only a qualitative risk analysis had been completed prior to the decision. The risks associated with the facility had become a major concern. A plant foreman asked the safety manager whether chlorine production would be shut down because of the problem. The operations manager at the plant believed that there was no problem and that the risk analysis has overstated the dangers from the loading. The plant manager was concerned that there could be a sudden restriction placed on chlorine shipments due to the risk.

The safety manager realized that safe and reliable loading of chlorine was important to both this plant's economics and to operations elsewhere in the firm. Any solution that did not have widespread support in the organization seemed likely to encounter severe opposition and would be hard to carry out. The best solution might depend on factors such as the demand for chlorine, public actions in the future, and the plant's future physical arrangement; and these factors were highly uncertain and beyond the safety manager's expertise. The safety manager wanted to pull together a cross-section of plant staff, have them share their knowledge, and jointly arrive at a plan for action. These aspects of the problem made it suitable for the application of a payoff matrix approach in a group setting.

Through close personal involvement in the prior risk analysis, and with responsibility to promote and implement safety measures, the safety manager realized the possibility of being seen as a biased facilitator. The safety manager arranged for the meeting to be moderated by an experienced facilitator from the company's human resources department.

In their first telephone conversation the facilitator, plant manager, and safety manager discussed who should participate in the problem session. They agreed to include the:

- Plant manager
- Engineering manager
- Operations manager

- Safety manager
- Regional logistics coordinator.

Each of these individuals had special knowledge to contribute to the analysis. The plant manager was aware of future plans for the facility and the strategic role of the facility in the company. The plant manager was also viewed as the person who best understood the public's attitudes towards the facility. The plant manager could contribute to the discussion of the demand for chlorine, and had considerable knowledge about storage and the flexibility this gives to both supply and demand. The engineering manager would be responsible for carrying out any engineered solution and was an expert on costs. The operations manager was also an engineer and an expert on all aspects of producing, storing, loading, and transporting chlorine. The safety manager had led the earlier risk analysis, was most familiar with the safety aspects of design, and was also most knowledgeable about the accepted standards for safety within the firm. The regional logistics manager could supply additional information on the demand for the plant's chlorine and on shipping and storage options.

The facilitator and the participants agreed to gather relevant information and then hold a two-day session off-site to discuss problems associated with the chlorine loading facility.

10.4.1 Step 1: Develop a Clear Problem Statement

The first step at this meeting was to develop a problem definition. To help scope the problem, the group determined the time horizon and specified the range of future alternatives to be included in the scope of the analysis. A brief discussion resulted in the following problem statement:

> What actions, if any, should we take to reduce the risks from the chlorine loading operation, while minimizing disruption to chlorine deliveries and to the production of products that depend on chlorine?

This statement limited the problem to one issue while addressing key concerns about the situation.

10.4.2 Step 2: Generate Alternatives

The development of strategies begins by brainstorming a list of alternatives. During the brainstorming phase alternatives are not judged and are recorded with very little comment. After brainstorming, the possible alternatives are reviewed and strategies are composed. Typically, from a long list of possible strategies, the group must choose from two to six of the strategies for detailed analysis.

This step ends by choosing the time frame for the analysis and the decision criteria. For a safety problem, criteria might include injuries, direct costs, and corporate impacts. The group should concentrate on a few criteria that cover the most important impacts of the decision.

The alternatives generated for the chlorine loading problem were to:

1. Shut Down the plant
2. Do Nothing
3. Do More Study
4. Switch to Hypochlorite
5. Reduce Pipe Size
6. Install Isolation Valves
7. Enclose the Chlorine Loading system
8. Use an Alternate Transportation Method for chlorine
9. Move the Loading Operation elsewhere within the plant boundary
10. Move Flammable Cars away from the chlorine loading site.

The group decided that ten alternatives were far too many to cover in a two-day session. The group selected a subset that they could discuss fully during the session, and that would provide distinctly different approaches to the problem. Any alternative that the group tentatively accepted would be further refined in an engineering analysis. The group restricted its discussion to the following alternatives:

1. Do Nothing
2. Move the Flammable Cars
3. Move the Loading Operation
4. Implement Low-Cost Measures including reducing pipe size and installing isolation valves.

The group considered it unrealistic to pursue the alternatives of Shut Down, Use an Alternate Transportation Method, Switch to Hypochlorite, and Enclose the Chlorine Loading. The Do More Study alternative would not be included in the detailed analysis; however, it would be examined during the review at the end of the session. The alternatives, though not detailed, all involved a definite commitment of resources, were under the control of the decision makers, and would significantly affect the problem under consideration. Thus they were all valid alternatives.

Finally in this step, the group considered the time frame of the problem and criteria. They decided to examine the impacts that the solutions would cause over a 5-year period. This horizon was driven by the plant manager's understanding that the plant would undergo a significant reorganization and rebuilding about 5 years after the time of the analysis. The group chose costs and injuries as the important decision criteria.

10.4.3 Step 3: Determine Uncertainties and Their Likelihood of Occurrence

To apply the payoff matrix approach to a complex, real-world problem such as this one, the group must carefully construct scenarios of the future and assign probabilities to these scenarios. This process can be described in four key substeps:

3.1: Identify and select key uncertainties
3.2: Define states
3.3: Develop scenarios and assign scenario probabilities.
3.4: Assign probabilities to decision-dependent uncertainties such as risk.

These substeps of the case study are described below.

10.4.3.1 Identify and Select Key Uncertainties

The members of the group brainstormed for uncertainties. In doing so, they thought about the uncertain economic and operational factors in the future that they would like to know today in order to make the decision. They developed the following list of uncertainties:

1. Value of chlorine production
2. Public concern about safety
3. Need for space on site for new facilities
4. Accident severity
5. Construction of new plant at another location
6. Price of nearby land.

The group realized that not all of the uncertainties could be given detailed consideration in this analysis, nor were they equally important. The next task was to choose the key uncertainties to focus on when creating scenarios. All others would be assumed to be in a nominal or base-case state. The group chose the following items from the initial list:

1. Value of chlorine production
2. Need for space on site for new facilities
3. Accident severity
4. Construction of a new plant at another location.

The other uncertainties were less important to include in scenarios. Because the public does not understand the technologies of chlorine transfer, the group assumed that changes in public attitude would affect all of the alternatives similarly. Also, because the price of nearby land had been rising in a stable pattern, land price was considered known.

10.4.3.2 Define States

Next the group specified the possible states of the uncertainties. States describe the distinct possible occurrences or results of each uncertainty. For the key uncertainties, the group created a set of states describing the possible future occurrences. For the other uncertainties, one nominal or base-case state is needed; when alternatives are assigned values later in the analysis, these uncertainties are assumed to be in the nominal state in all scenarios. The states used to represent each uncertainty are listed in Table 10.11.

10.4.3.3 Develop Scenarios and Assign Scenario Probabilities

The facilitator next asked the group to point out any of the key uncertainties that would be strongly affected by strategies. As noted previously, decision-dependent and decision-independent uncertainties are treated somewhat differently. The group noted that each strategy greatly affects the types and severity of accidents that could occur. Thus, *accident severity* is a decision-dependent uncertainty and was analyzed separately from scenarios.

The group must combine the individual decision-independent uncertainties into the scenarios that appear on the left-hand side of the payoff matrix. Creation of scenarios begins with the identification of probabilistic depend-

TABLE 10.11
Example Problem Uncertainty States

Uncertainty	*States*
1. Value of chlorine production	• Remains at present level • Grows at 10% per year over the 5-year period
2. Need for space on site for new facilities	• No need for increased space • Need for 3 additional acres in three years
3. Accident severity	• No accident • 1 to 3 injuries • 4 to 15 injuries • 16 to 50 injuries
4. Construction of new plant at another location	• No new plant • Company builds new plant producing a similar array of products
5. Public concern	• There will be a moderate increase in the public's interest in the plant. News articles will discuss environmental and safety issues at the plant, but there will not be public protests. Chlorine loading will not be singled out as an important issue.
6. Price of nearby land	• Prices will grow at inflation plus 2% a year.

encies among the uncertainties. As noted earlier, recognition of probabilistic dependencies is important in constructing and setting the probability of scenarios. When constructing scenarios, it is simplest to begin with an uncertainty that is considered to cause or precede other uncertainties; for example, we think of clouds preceding rain. Then we consider uncertainties that are influenced. A useful graphical tool to help think about dependency is an influence diagram. An influence diagram graphically represents the relationships between uncertainties. Ovals are used in an influence diagram to indicate uncertainties, and arrows are used to indicate dependency.[2]

Figure 10.3 is the influence diagram that the group developed for the key uncertainties in this analysis. At the top of Figure 10.3, is the uncertainty regarding the construction of a new plant producing a similar range of products. If the new plant were built, it would be less likely that the value of chlorine would grow or that additional space would be needed at this site. Arrows indicate that *new plant* influences *need for space* and *value of chlorine*. If the value of chlorine does increase, this may indicate a general increase in demand for the company's products and would increase the probability that additional space will be needed. This influence is indicated by the arrow from *value of chlorine* to *need for space.*

It is usually not possible to have a scenario for each combination of the states of the uncertainties, since the remaining analysis would be unmanageable. The goal in scenario construction is to cover the distinctly different futures that can occur, while eliminating scenarios that will result in similar impacts. In our example, after discussing the uncertainties and the influence of these uncertainties on each other, the group decided that it would examine three scenarios. The scenarios and the probability that the team assigned to each one were:

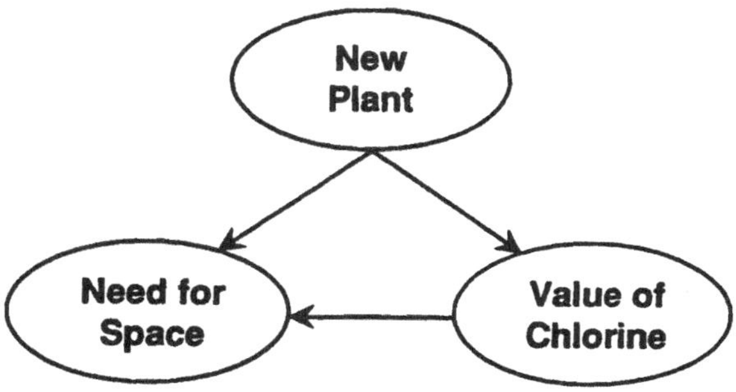

FIGURE 10.3. An influence diagram for the chlorine decision.

2 In this context, influence and dependency are synonymous. Hence, the name influence diagram. Since influence diagrams are a core tool of Decision Analysis, they are explained in detail in Chapter 11, Decision Analysis (Section 11.3.2.3).

Scenario 1: No Significant Changes (probability 50%):
- New plant
- No need for additional space on site
- No increase in the value of chlorine.

Scenario 2: Value Growth (probability 20%):
- No new plant
- No need for additional space on site
- Value of chlorine grows at 10% per year over the 5-year period.

Scenario 3: Maximum Impacts (probability 30%):
- No new plant
- A need for a minimum of 3 acres for expanded facilities at the end of three years
- Value of chlorine grows at 10% per year over the 5-year period.

The group wrote a brief "story" to further describe and define each scenario. For example, the description of the "Maximum Impacts" scenario was:

> Strong growth occurs in the demand for the company's products; however, the company is reluctant to make the investment in a new plant. This results in chlorine production and loading becoming increasingly important to the company's operations. In addition, there is an acute demand for land at the site to accommodate expansion of several operations. These expansions are desired approximately three years from today.

10.4.3.4 Assign Probabilities to Decision-Dependent Uncertainties

The group also assessed probabilities for the decision-dependent uncertainty *accident severity*. These probabilities are shown in Table 10.12.[3] Because of their decision-dependence, the probabilities differ across the strategies. Probabilities were assessed by the methods described in Section 10.3.2.4.

10.4.4 Step 4: Specify Possible Outcomes

After constructing the scenarios, the payoff matrix is drawn and the outcomes are specified. The purpose of this step is to create insights, maximize the value of strategies, and allow the evaluation and comparison of the strategies.

In many cases, a planned strategy based on one assumed scenario will exist prior to the analysis. If such a strategy–scenario combination exists, the analysis begins with this combination. The group discusses the strategy–sce-

3 The probabilities shown in Table 10.2 differ from the probabilities presented in Chapter 8 for this case. Only a qualitative risk analysis was performed prior to the payoff matrix analysis. The probabilities here and in Chapter 8 should not be compared.

TABLE 10.12
Accident Probabilities

Accident Severity	*Value*	*Do Nothing*	*Move the Flammable Cars*	*Move the Loading Operation*	*Implement Low-Cost Measures*
No accident		0.889	0.89	0.989	0.989
1 to 3 injuries		0.1	0.1	0.01	0.01
4 to 15 injuries		0.01	0.01	0.001	0.001
16 to 50 injuries		0.001	0.001	0.0	0.0

nario combination in detail and records a brief outline of the outcomes over the study horizon. The group carefully notes what actions in a strategy occur prior to, and what actions occur after, significant events. The group carefully considers how the strategy can adapt to future events. The group then moves on to other strategy–scenario combinations in the matrix. After the group has reviewed all strategy–scenario combinations, the group should discuss any new strategies that the analysis has suggested.

The matrix for the case is shown in Table 10.13. The scenarios are listed on the far left, the scenario probabilities in the second column at the left, and the strategies at the top. Important future events are listed within the appropriate matrix boxes in italics.

10.4.5 Step 5: Assign Values to Outcomes

In this step, the group assigns a value to each cell of the matrix. In determining the values, the group keeps in mind that this is a preliminary evaluation and that excessive accuracy would overstate their knowledge. Values are assigned by consensus or by voting if necessary. Dissenting opinions are often recorded as part of the analysis documentation. For the strategy–scenario combinations, values can be determined in the manner described earlier for the basic payoff matrix approach, Section 10.3.2.6.

In the case, the group considered the Do Nothing–Value Growth combination to be the best outcome and the Move the Loading Operation–Maximum Impacts combination to be the worst outcome. The other outcomes were ranked between these. The group assigned the worst strategy–scenario combination a value of $0 and the best combination a value of $8 million. Intermediate values were then assigned. The values the group assigned are shown in Table 10.14.

Next, values were assigned to the accidents described at the bottom of the matrix. When the accidents involve injuries or fatalities, this is a particularly difficult task; however, quantification is necessary to effectively use resources

TABLE 10.13
Alternative-Scenario Descriptions for Chlorine Loading Facility Problem

		Alternatives			
Scenarios	*Probability*	*Do Nothing*	*Move the Flammable Cars*	*Move the Loading Operation*	*Implement Low-Cost Measures*
No changes	0.5	No current cost, production not disrupted, no long-run problems	Moderate present cost, very brief disruption of production, no long-run problems	High cost, large disruption as production is moved, no long-run problems	Low present cost, some disruption of production, no long-run problems
Value Growth	0.2	No current cost, production not disrupted, no long-run problems	Moderate present cost, very brief disruption of production, no long-run problems	High cost, large disruption as production is moved, somewhat higher disruption cost, no long-run problems	Low present cost, some disruption of production, slightly higher disruption cost, no long-run problems
Maximum Impacts	0.3	No current cost, production not disrupted, *new space required*, no long-run problems	Moderate present cost, very brief disruption of production, *new space required*, space problem exists but is manageable	High cost, large disruption as production is moved, disruption cost heightened by the increased value, *new space required*, major space problems, costly future disruption during expansion	Low present cost, some disruption of production, slightly higher disruption cost, *new space required*, no long-run problems

and achieve the greatest benefit per dollar spent. The evaluation of injuries and fatalities is discussed in some detail in Section 2.5.

A very simple approach is to assign a value per injury or fatality and calculate the impact of all accidents based on this figure. In this approach, if one severe injury is valued at $4 million, two would be valued at $8 million, and so on. In some organizations, such values may be standardized. Many organizations feel that the impact of accidents grows more quickly than a linear pattern. Thus, if one severe injury is valued at $4 million, 10 severe injuries might be valued at $60 million. Finally, some individuals and groups may find it easier to think in terms of lotteries on injuries or fatalities. For example, if the group is willing to spend $40,000 to avoid a 1 in 1000 chance

TABLE 10.14
Alternative-Scenario Evaluations

		Alternatives			
Scenarios	*Probability*	*Do Nothing*	*Move the Flammable Cars*	*Move the Loading Operation*	*Implement Low-Cost Measures*
No changes	0.50	$7.5MM	$5.5MM	$4.0MM	$7.0MM
Value Growth	0.20	$8.0MM	$6.0MM	$3.5MM	$7.5MM
Maximum Impacts	0.30	$7.5MM	$3.0MM	$0.0 MM	$7.0MM
Accident Severity	*Value*				
No accident	0	0.889	0.89	0.989	0.989
1 to 3 injuries	–8	0.1	0.1	0.01	0.01
4 to 15 injuries	–38	0.01	0.01	0.001	0.001
16 to 50 injuries	–132	0.001	0.0	0.0	0.0
EXPECTED VALUE		$6.4 MM	$3.7 MM	$2.6 MM	$7.0 MM

of 10 severe injuries, a value of $40,000 times 1000 or $40 million would be placed on 10 severe injuries. (The values used in this discussion are illustrative only, not recommendations.)

In the case, the group agreed on a uniform value of $4 million per injury. The median injury level in each range was used to calculate values. For example, an accident resulting in four to fifteen injuries was evaluated at $(4 + 15)/2 \times \$4$ million = $38 million. Values are entered as negative numbers because these are "costs" rather than benefits. The values assigned to injuries appear in the second column of the lower portion of Table 10.14.

10.4.6 Step 6: Calculate the Expected Value of Each Alternative

The expected value of each alternative was calculated; these appear in the bottom row of Table 10.14. The alternative with the highest expected value is to Implement Low-Cost Measures.

The specifics of the expected value calculation can be illustrated using the values for the strategy Implement Low-Cost Measures. The calculation is shown in Table 10.15. The entries in each row under this strategy are multiplied by entries from column two. In each row, this is the multiplication of a value and a probability. The resulting weighted values are summed to get the expected value.

TABLE 10.15
Expected-Value Calculation for "Implement Low-Cost Measures" Alternative ($MM)

Uncertainties	*Probability*		*Value*		*Weighted Value*
No Changes	0.50	×	7	=	3.5
Value Growth	0.20	×	7.5	=	1.5
Maximum Impacts	0.30	×	7	=	2.1
No accident	0.989	×	0	=	0
1 to 3 injuries	0.01	×	–8	=	–.080
4 to 15 injuries	0.001	×	–38	=	–.038
16 to 50 injuries	0.0	×	–132	=	0
			EXPECTED VALUE		7.0

10.4.7 Step 7: Review the Analysis

The group conducted sensitivity analysis, reviewed their work, and made plans for future steps.

The group considered changes in the probability of the Maximum Impacts scenario, in the probability of accidents, and in the value assigned to accidents. The high expected value for the Implement Low-Cost Measures strategy was stable.

The facilitator recorded insights and concerns from the analysis as follows:

- An examination of the outcomes and the values indicates that the decision is driven by the cost of relocating the chlorine loading and the concern over future needs for space.
- The Implement Low-Cost Measures strategy does not differ significantly from the Move Chlorine Loading strategy in its ability to reduce accidents. A quantitative risk analysis would help to confirm this conclusion.
- Because Do Nothing and Implement Low-Cost Measures are close in expected value, it may be reasonable to delay action until more is known about the possibility of a new facility.

In planning for further analysis the following questions are considered:

- What is the cost of gathering more information?
- How likely is it that additional information will change the decision?
- How much might an improved decision increase the value to the decision maker?

10.4.8 Step 8: Make a Recommendation

The final step is to recommend that the organization either pursue further analysis of the decision, or undertake a specific alternative.

The recommended actions developed by the group were:

- Determine land requirements for an expanded facility more accurately
- Gather more data on the likelihood of constructing a new facility
- Do a more detailed risk analysis.

The plant manager decided that the group would pursue the recommended actions before making a final decision.

10.5 EXTENSIONS

To improve group decision-making sessions, it is helpful to incorporate other group facilitation techniques. An excellent book on group facilitation is *How to Make Meetings Work* (Doyle and Straus 1976).

10.6 IMPLEMENTATION NEEDS

10.6.1 Data/Information Requirements

Because payoff matrix analysis employs subjective inputs, the formal data and information needs are modest. The decision maker and experts asked to participate must be willing to make the probability and value estimates required by the analysis. Experts familiar with each of the alternatives and the uncertainties identified can contribute to assigning probabilities and values. Their contributions will only be relevant if the decision maker has faith that their subjective judgments are accurate and unbiased. Calculated risks, costs, and other quantities can appear as criteria in the cells of the matrix, and extensive data and sophisticated models may be used to calculate these outcome values. However, payoff matrix analysis is generally considered a simple "rough-cut" technique, and such tools are seldom used.

10.6.2 Personnel/Training Required

Very little training is needed to perform payoff matrix analyses. However, an understanding of the definitions of uncertainty and decisions, of probability and probabilistic independence, and of judgmental biases is useful.

If the analysis is performed by a group, the facilitators should have strong skills in working with groups. It is useful to have an experienced facilitator who is viewed as a neutral manager of the process by all participants. Training in group facilitation is useful. The "CATALYST Facilitator's Guidebook" can provide additional training that facilitators may require.

10.6.3 Other Resources Needed

No special resources are needed to use the basic payoff matrix. Computations can generally be performed with a hand calculator or, for larger problems, with a computer spreadsheet.

Group sessions, such as CATALYST, work best when the group can meet off-site and devote two to three days exclusively to the analysis. Easels with large writing pads, overhead projectors, and chalk or white boards are useful during group sessions. It is also valuable to be able to pin or tape notes taken during the sessions on the walls for easy viewing.

10.7 SUMMARY

10.7.1 Strengths and Weaknesses of Payoff Matrix Analysis

Key strengths of payoff matrix methods are their simplicity and usefulness for performing quick analyses. These methods allow a "back-of-the-envelope" analysis that suffices for some problems and generates important insights for most problems. The methods employ expected utility theory, which has a logically sound basis. Other strengths of approaches like CATALYST are that they work well in group settings, help decision makers think about downstream decisions and future flexibility, and provide an excellent starting point for more detailed analyses.

A significant weakness of payoff matrix analysis is the use of different procedures for incorporating strategy-independent and strategy-dependent probabilities. Further, the method for handling strategy-dependent uncertainties is somewhat cumbersome. Payoff matrix analysis works best when the strategies change the outcomes or consequences, but not the risks.

Methods like CATALYST share several problems with other methods based on scenarios. Many problems are difficult to simplify to the degree necessary for the method. Also, there is no theoretical basis for the process of creating scenarios.

10.7.2 Suggestions for Further Information/Study

The references listed in Section 10.8 can provide more detail and training on payoff matrix analysis. Of particular value here are the "CATALYST Facilitator's Guidebook" (EPRI 1992) and *Applied Decision Analysis*, by D. W. Bunn (1984).

10.8 REFERENCES

Behn, R., and J. Vaupel (1982). *Quick Analysis for Busy Decision Makers*. New York: Basic Books.

Belyaev (1976). Payoff Matrix Technique. *Energy (Oxford)* 15(7–8):631–643.

Bunn, D. W. (1984). *Applied Decision Analysis*. New York: McGraw-Hill.

Doyle, M., and D. Straus (1976). *How to Make Meetings Work*. New York: Jove Books.

EPRI (1992). "CATALYST — A Group Process for Strategic Decision Making: Facilitator's Guidebook." Prepared by Applied Decision Analysis and The Brattle Group. TR-100583, Research Project 2631, Final Report. Palo Alto, CA: Electric Power Research Institute (September).

Hanna, R. C. (1979). Apply Decision Theory to Hazard Evaluation. *Hydrocarbon Processing* December):133-150.

Lindley, D. V. (1971). *Making Decisions*. London: Wiley.

Martin, M. P. (1979). Decision Making: The Payoff Matrix. *Journal of Systems Management* 30(1):14–18.

Merkhofer, M. W. (1987). Quantifying Judgmental Uncertainty: Methodology, Experiences, and Insights. *IEEE Transactions on Systems, Man, and Cybernetics* SMC-17(5):741–752.

von Winterfeldt, D., and W. Edwards (1986). *Decision Analysis and Behavioral Research*. Cambridge: Cambridge University Press.

11

DECISION ANALYSIS

In the rare instances in which I choose to be rational, decision analysis is most useful.
—Howard Raiffa

11.1 PURPOSE OF CHAPTER

Process safety decisions often involve numerous alternatives, each with complex uncertainties about what might occur if each alternative were chosen. Decision Analysis (DA) can be used to analyze these decisions using a logical framework that rigorously accounts for the risks and rewards of each alternative, and for the desire of the decision maker to avoid risk. This feeling toward risk, or "risk attitude," can be represented explicitly by a mathematical function. This mathematical function can be applied to many decisions, thereby allowing the organization to deal with risk consistently from one decision to another.

The primary goals of DA are to:

- Identify the alternative that maximizes the decision-maker's expected rewards given personal preferences and uncertainties
- Provide the decision maker with insights into the decision problem
- Use the insights to provide a direction for action.

Through a decision analysis we try to gain an understanding of the problem and the relationships between the various elements of the problem, not simply find the best alternative. The insight provided by the analysis helps direct the decision maker toward an action to take. The best action may not be to adopt one alternative immediately, but may involve further analysis, refinement of alternatives, or gathering more information to reduce the uncertainty with which the decision maker is faced. An attractive feature of DA is that it provides a way to calculate the monetary value of obtaining more information before making a decision. This enables the decision maker to determine how much it is worth to pay for more research or information gathering.

This chapter introduces the reader to the principles and techniques of decision analysis. A demonstration of the decision aid is given using the underground pipeline example.

The chapter assumes that the reader is familiar with the concepts presented under "Uncertainty" in Chapter 2 (Section 2.6). In addition, the reader may find it beneficial to read Chapter 10 (Payoff Matrix Analysis) before proceeding.

11.2 OVERVIEW OF DECISION ANALYSIS

11.2.1 Introduction to Decision Analysis

Decision analysis provides a formal process for structuring, analyzing, and communicating decisions in a complex, uncertain environment. This process is called the *decision analysis cycle* and is described in greater detail in Section 11.3.2. The decision analysis cycle breaks the decision into components and structures them into a *decision model*. The critical issues can then be identified through analyzing the model. To create a decision model, three basic questions are asked as shown in Figure 11.1. A decision analysis logically integrates what one can do, what one knows, and what one wants into the decision model. Once created, the decision model is used to analyze the decision and provide the decision maker with insights and a direction for action. Among the key features of the analysis, the decision maker is provided with:

- The optimal decision policy, which indicates the course of action that the decision maker should take if the decision has to be made without further information or analysis
- A sensitivity analysis indicating the extent to which the various elements of the problem affect the optimal decision strategy
- The value of gathering more information to reduce uncertainty about the possible results under each alternative
- The value of being able to control events or aspects that introduce uncertainty into the decision
- A means of communicating the problem and the decision clearly.

Each of these features will be explored in detail in this chapter.

Decision analysis is a useful, practical technique, with a strong theoretical foundation. It is founded on the principles of probability theory and a set of axioms based on utility theory. The key result from the theory is that alternatives, or possible actions, are evaluated based on the likelihood of the possible outcomes of the actions and the decision maker's relative preference for these outcomes. Although the theory is extensive, this chapter focuses on the practical side of DA.

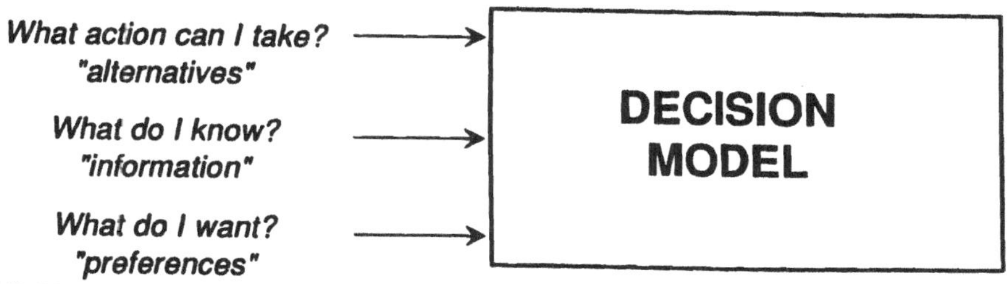

FIGURE 11.1. Basic questions one must ask before creating a decision model.

11.2.2 History of Decision Analysis

Decision analysis is a well-established mathematical technique. Its intellectual roots go back to the early 1700s. At that time, Daniel Bernoulli introduced the idea that decisions should incorporate the likelihood of the possible consequences of the decision and the relative desirability of those consequences to the decision maker. In the 1940s, John von Neumann and Oskar Morgenstern described the desirability of uncertain consequences with a mathematical expression called a utility function. The work of von Neumann and Morgenstern laid the foundation for a decision theory indicating how rational individuals should make decisions under uncertainty. The theory includes a set of axioms for rational decision making which are generally accepted as being logically sound. These axioms have been stated by various individuals in different ways, but each with the same implications for decision making. If one accepts these axioms, then it logically follows that every possible consequence of a decision can be assigned a utility and the decision maker should prefer the alternative which yields the highest expected utility. The expected utility is the mathematical expectation, or expected value, of the possible utilities (i.e. the sum of the utilities of the possible consequences weighted by the likelihoods or probabilities of these consequences). The concept of mathematical expectation is explained in Chapter 2. This chapter will focus on the practical application of DA, not the theory. However, since the axioms are so crucial to the theoretical rigor of DA, we provide a brief description of these axioms in the boxed-off section that follows. Alternative descriptions of the axioms and further readings on decision theory can be found in Howard (1968); Keeney and Raiffa (1975); Luce and Raiffa (1957); Merkhofer (1987a); North (1968); Pratt, Raiffa, and Schlaifer (1965); and Savage (1954).

The axioms of rational decision making provided the theoretical foundation for decision analysis and paved the way for applying decision analysis to real problems. Decision analysis as a formal discipline grew out of systems analysis and operations research in the 1960s. Decision analysis practice within private industry and government expanded greatly in the 1970s and still continues to grow. Decision analysis is used in a variety of industries, including chemical, transportation, oil, electric utilities, pharmaceuticals, health care, law, education and many others. The literature cites numerous applications including those in the areas of

- Capital investment
- Health and safety
- Management of environmental waste
- Marketing
- Product introduction
- R&D planning
- Risk management
- Siting of facilities

- Social policy
- Strategic planning.

AXIOMS FOR RATIONAL DECISION MAKING

The existence of a utility function follows from a set of axioms for rational decision making which are generally accepted as logically sound. These axioms are normative, not descriptive: they express what is required for rational decision making. This may differ from the decision making behavior that is frequently observed in real life. These axioms have been stated by various individuals in many different ways, but each has the same implications for decision making. If you accept Axioms 1–5 below, then it follows that:

- There exists a utility function, $u(x)$, that describes the strength of your preference for each possible outcome, x, of a decision.
- The higher your utility, the more you prefer the outcome.
- In situations involving uncertainty, the expected utility (i.e., the utilities of the outcomes weighted by their likelihoods or probabilities of occurrence) is the appropriate index for comparing the desirability of alternatives.

The axioms can be examined in terms of "risky opportunities." A risky opportunity is a situation in which there is a set of outcomes, or "payoffs," each having an associated probability. A risky opportunity can be represented as shown in Figure 11.2. Each branch of the opportunity has an outcome, O_i, placed at the end, and a probability of occurrence, p_i, placed under the branch. Examples of risky opportunities in life are betting on a football team, tossing a coin to see who buys lunch, and purchasing auto insurance. For a toss of a six-sided die, the opportunity may look as shown in Figure 11.3.

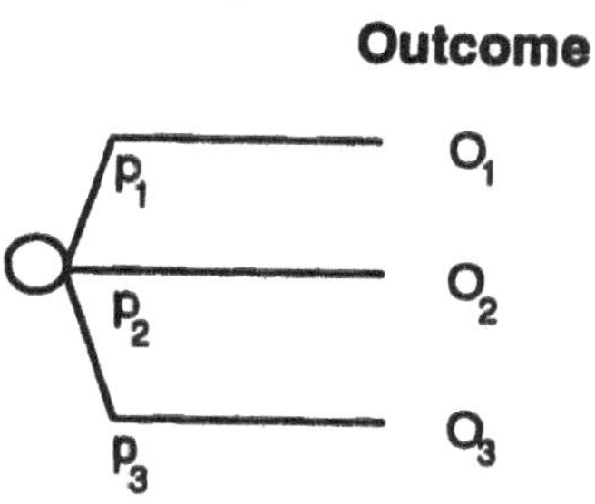

FIGURE 11.2. A risky opportunity with three outcomes.

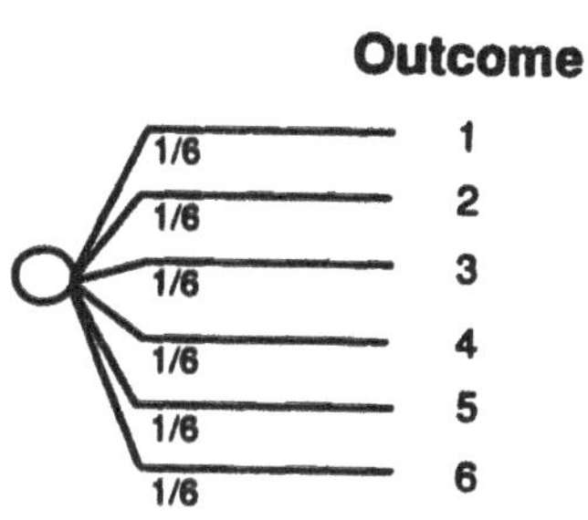

FIGURE 11.3. A risky opportunity with six outcomes.

The axioms can be stated in terms of risky opportunities, outcomes, and probabilities as follows:

Axiom 1: ***Outcomes can be ranked by their desirability, and their rankings are transitive; if outcome A is preferred to B and B to C, then A is preferred to C.*** For example, suppose there is a year-end lottery at the office, and there are three possible prizes: a new computer, a trip to Hawaii, or a turkey. If you prefer the computer to the trip to Hawaii and the trip to Hawaii to the turkey, then Axiom 1 says that you must prefer the computer to the turkey.

Axiom 2: ***If A is preferred to B and B to C, where A, B, and C are all possible outcomes, then there is some probability p such that you would be indifferent between receiving B for sure and having a risky opportunity with the probability p of receiving C and the probability 1 − p of receiving A.*** This is shown graphically in Figure 11.4 below:

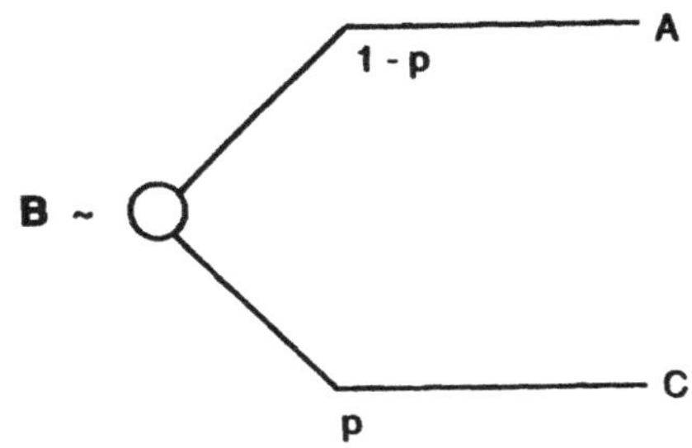

FIGURE 11.4. Indifference between outcome B and a risky opportunity involving A and C.

B is referred to as the certain equivalent of the risky opportunity. The "~" represents indifference between B and the risky opportunity with possible outcomes A and C.

Axiom 3: ***You are willing to substitute the certain equivalent and the risky opportunity for each other.*** Using the example of Axiom 2, this means you would be willing to give up B for the risky opportunity with possible outcomes A and C as shown.

Axiom 4: ***When there are two possible outcomes, a higher probability of the best outcome is preferred to a lower probability of the best outcome.*** As an example, suppose you prefer sunny weather to rainy weather. In one situation you face a 30% chance of sunshine and a 70% chance of rain as shown in Figure 11.5. In another, you face a 60% chance

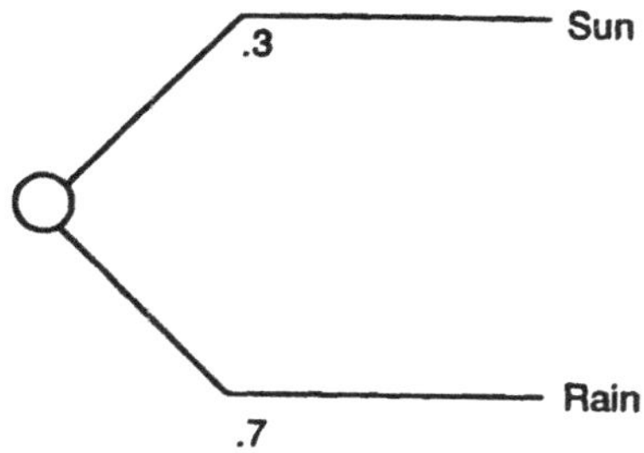

FIGURE 11.5. A risky opportunity involving sun and rain.

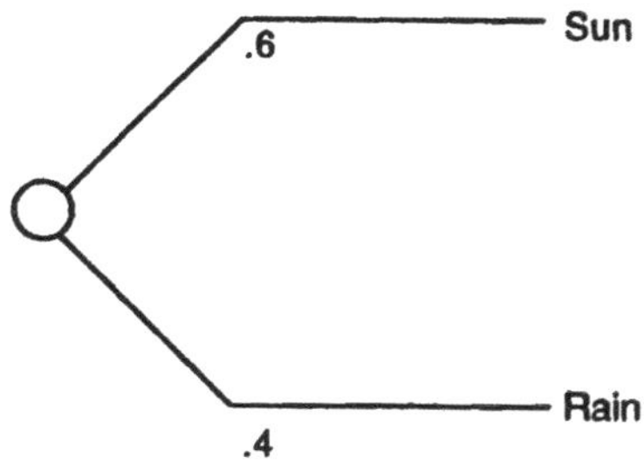

FIGURE 11.6. A second risky opportunity involving sun and rain.

of sunshine and a 40% chance of rain as shown in Figure 11.6. Axiom 4 says that you must prefer the situation in Figure 11.6 to that of Figure 11.5 because there is a higher probability of the best outcome (i.e., sun).

Axiom 5: *The laws of probability can be used to reduce a sequence of risky opportunities to a single risky opportunity.* This axiom can be illustrated by the two situations shown in Figures 11.7 and 11.8. In Figure 11.7, there are two risky opportunities and four possible outcomes. In Figure 11.8, there is just one risky opportunity with the same four outcomes. Using the laws of probability, we can calculate the probability of the outcomes resulting from multiple events by multiplying the probabilities of the individual events. If we do this for the risky opportunity in Figure 11.7, we see that it reduces to the single risky opportunity shown in Figure 11.8.

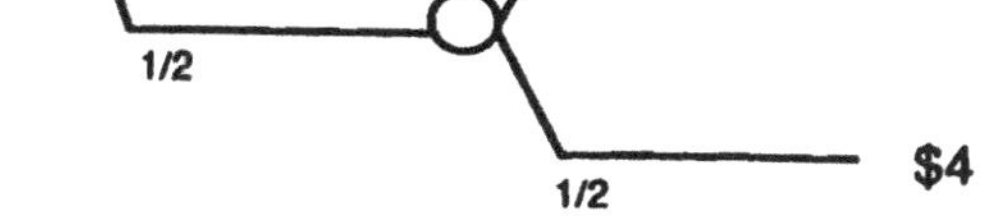

FIGURE 11.7. Two risky opportunities involving four outcomes.

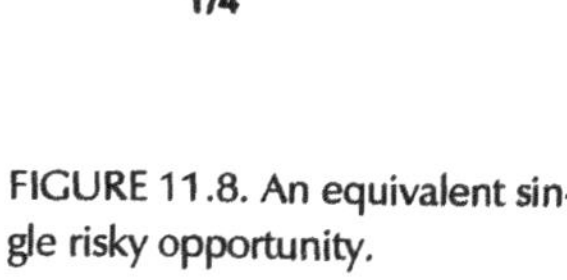

FIGURE 11.8. An equivalent single risky opportunity.

Specific applications of DA that may be of particular interest to the reader include:

- Design of hazardous products (Owen 1981)
- Fire risks in oil refineries (Paté-Cornell 1985)
- Safety backfits on nuclear reactors (Brown and Ulvila 1988)
- Smoke detection laws and their impact on fire loss (Jensen, Tome, and Darby 1989)
- PCB risks from transformer fires (Boyd et al. 1985)
- Petroleum exploration (Newendorp 1975)
- Petrochemical expansion (Braunstein 1983)
- Hazard evaluation (Hanna 1979)
- Hazardous waste site assessments in real estate acquisition (Noonan and Vidich 1992)
- Nuclear plant safety modifications (Bohn 1992)
- Component inspection and replacement in a power system (Mauney 1993).

In addition to continued growth in application, DA research and theory have sustained lasting attention. The set of tools available to facilitate DA has been expanded over the years to include various graphical aids for problem formulation and structuring, as well as software packages to aid in analysis. In addition, DA has become an area of academic study at universities, indicating that further growth in research and application is likely.

11.2.3 Decisions/Situations Most Appropriate for Decision Analysis

Decision analysis is especially suited to problems that involve considerable uncertainty, complex alternatives, or the need to justify the recommended decision. Decisions are often difficult because they contain a great deal of uncertainty. Decision analysis provides a method for addressing uncertainty by incorporating it explicitly and quantitatively into the evaluation of alternatives. It allows the decision maker to work with limited information and knowledge rather than requiring or assuming that uncertainty will be eliminated at the time of the decision.

The process and mathematics of DA help the decision maker address complexities that can result from numerous interrelated factors, numerous alternatives, or complex value tradeoffs. It is also ideal for decisions that are complex as a result of dependencies among aspects of the problem. Another complexity addressed by DA is the presence of "downstream decisions," that is, eventual decisions that will be made after earlier decisions are made and partial outcomes observed. Decision analysis allows the analyst to address multiple related decisions and to investigate the impact of decisions and uncertain future events on downstream decisions.

Decision analysis also performs well as a decision aid when a decision must be justified either within an organization or externally. The rigorous, logical methodology of DA results in well-documented, defensible decisions. This makes it an especially useful decision aid when the problem involves a large, "once-in-a-lifetime" allocation of resources. However, it is also useful in making smaller-scale, repetitive decisions in which a consistent approach toward handling risk is desired.

11.3 EXPLANATION OF DECISION ANALYSIS

11.3.1 Key Concepts of Decision Analysis

In order to introduce the methodology of decision analysis, one must first understand some basic definitions and concepts. A good place to start is with the building blocks of decision analysis—decisions, uncertainties, and values.

- A *decision* is a selection among alternatives. Alternatives are possible actions. For instance, if you face the decision of how to commute to work tomorrow, your alternatives may include driving and riding the train.
- An *uncertainty* is something that is unknown and uncontrollable, or not perfectly known or controllable, such as a quantity or the outcome of a future event. Uncertainties can have different conditions, or *states*, that exist or could occur. The actual state that exists or will occur is uncertain at the time of the decision. For example, in deciding how to commute to work, the severity of freeway congestion may be one of your important uncertainties. You might think of this uncertainty as having three possible states—light, medium, or heavy congestion. Each state of an uncertainty has an associated probability that represents the decision-maker's belief about the likelihood that each state occurs. For instance, you may believe there is a 0.3 chance that freeway congestion is light, a 0.2 chance that it is medium, and a 0.5 chance that it is heavy. Concepts of uncertainty are described further in Section 2.6.
- *Values* are what the decision maker views as important in evaluating the desirability of the outcomes of a decision. An *outcome* can be thought of as a particular occurrence resulting from a decision and an uncertainty or series of uncertainties. Outcomes depend on which alternative is chosen and the resulting states of the uncertainties. As an example, if you value both your time and money, the outcome of the commute decision will have two aspects—commute time and commute cost. To get a single measure of value for the outcome, you might assign a dollar value to the commute time, and add it to the commute cost to obtain the value for that outcome expressed as a total cost. Value measures are frequently, but not always, expressed in monetary terms.

Figure 11.9 shows the components of the commuter's decision. The decision maker has two alternatives (Drive, Ride Train) and one uncertainty (freeway congestion) with three possible states (light, medium, heavy). The outcome is measured in terms of commute time and commute cost. Commute time and commute cost are combined in a single value measure—the total cost of the trip in dollars. Each dollar value will bring the decision maker a certain amount of satisfaction, or utility. Ultimately, the decision maker will choose the alternative which yields the highest expected utility. Recall from Figure 11.1 that DA incorporates what one can do (in this example, the alternatives of Drive or Ride Train), what one knows (in this example, uncertainty about the severity of freeway congestion), and what one wants (in this example, values for commute time and cost). The rest of this section will describe how alternatives, uncertainties, and values fit together to form the decision model that is used to analyze the decision.

11.3.1.1 Probability Fundamentals

Probability is an important part of decision analysis. Decision analysis uses probability as a quantitative representation of the decision-maker's uncertainty. It is a *subjective* assessment of the likelihood of occurrence of an event. By subjective, we mean that it represents personal belief, experiences, and knowledge, not simply objective data or statistics. For instance, in the commuter's decision, the decision maker must assess the likelihood of the various degrees of freeway congestion. At 7 A.M. you might assign a 0.7 probability to the chance that freeway congestion is going to be heavy at 8 A.M. However, your neighbor may turn on the news, hear a traffic report indicating that there is little freeway congestion right now, and assign a 0.2 probability that congestion will be heavy at 8 A.M. Your probability assessment can be different

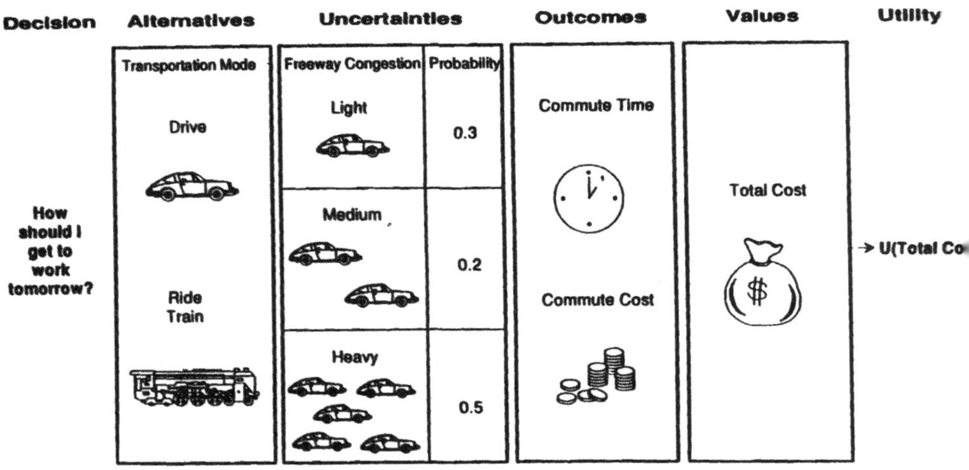

FIGURE 11.9. The commuter's decision.

from your neighbor's because probability is not a fixed state of the world, but a reflection of one's current knowledge of it. The rest of this chapter will assume that the reader is familiar with basic probability concepts including the use of cumulative distributions and expected value. For a review of these concepts see the discussion of "Uncertainty" in Chapter 2 (Section 2.6). To review specific terminology, refer to the Glossary.

Probabilities can be presented graphically using decision trees. Figure 11.10 shows a tree for the commute decision. A rectangle is used to represent a decision, and uncertainties are represented by circles. In this example the commute decision has one branch for each alternative, Drive and Ride Train. If the commuter drives, he faces the freeway congestion uncertainty represented by the circle. The three possible states of congestion are placed on the branches with their probability of occurrence. There are no delays if the commuter rides the train.

11.3.1.2 Outcomes

As defined earlier, an outcome is a particular occurrence resulting from a decision and an uncertainty or series of uncertainties. In other words, an outcome is the consequence incurred at the end of a decision tree path. For example, Figure 11.11 shows the outcomes for the commute decision. If the commuter chooses to Drive and freeway congestion is light, the outcome he faces consists of a 5-minute driving time and a $2 commute cost. Similarly, the outcome of Ride Train is a 15-minute commute time and $1.50 commute cost. More complex decisions might require detailed mathematical models to calculate outcomes. However, in our example we can simply assign an outcome to each decision tree path.

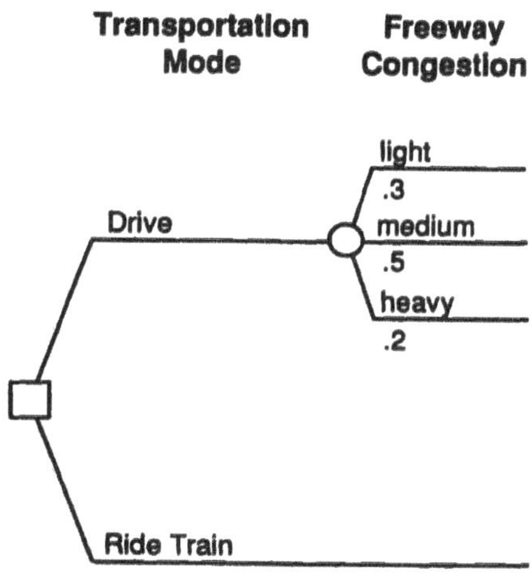

FIGURE 11.10. Decision tree for commuter's decision.

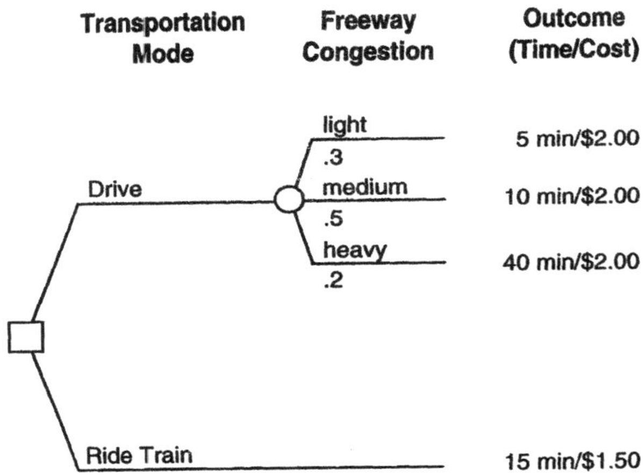

FIGURE 11.11. Decision tree showing outcomes for commuter's decision.

11.3.1.3 Value Function

To evaluate alternatives, the decision maker must assign a single numerical value or *value measure* to each possible outcome. This is done through a value function. There are situations in which the outcome has multiple attributes, such as in the commute decision where the outcomes are in terms of time and cost. Decision analysis forces the decision maker to combine the multiple attributes of the outcome into a single measure of value through a value function. This is illustrated for the commute decision in Figure 11.12. The outcome is composed of both commute time and commute cost. If we determine that the decision maker values his time at $15 per hour, the value function can be expressed by the following equation:

$$\text{Total Cost} = \text{Commute Cost} + (\$15/\text{hour} \times \text{Commute Time})$$

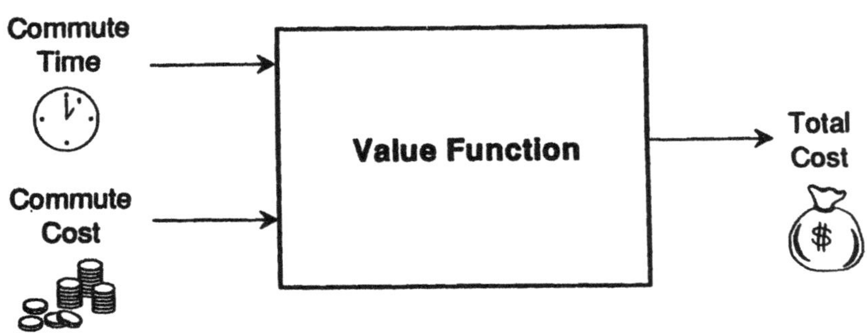

FIGURE 11.12. Model of value function for commuter's decision.

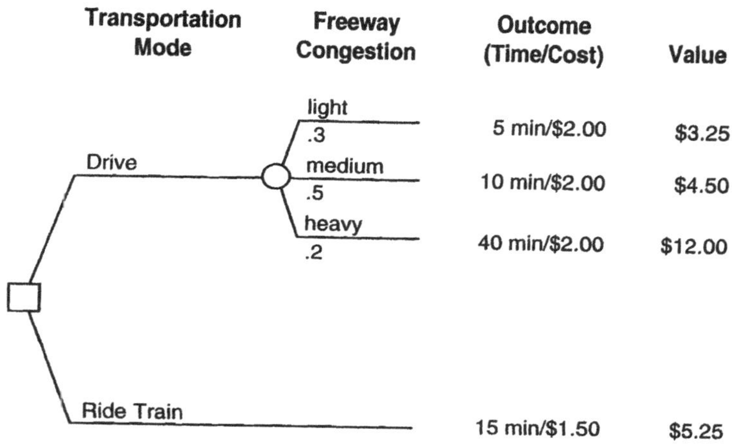

FIGURE 11.13. Decision tree showing values for outcomes of commuter's decision.

Using the function, the values for each outcome are now added to the tree in Figure 11.13.

In general, the value measure is in terms of money, but this is not required to perform an analysis. The value measure can use a different unit of measure that the decision maker finds convenient. But in decision analysis, all attributes of the outcome must be combined into one value measure. An alternative method of dealing with multiple attribute outcomes is through multiattribute utility analysis, which is described in Chapter 12.

11.3.1.4 Utility

Once outcomes have been mapped into a value measure, it is necessary to consider the decision-maker's preference for various amounts of the value measure. Utility is a measure of the decision-maker's strength of preference for a given quantity. It captures the relative worth of something to the decision maker. The higher the decision-maker's preference for a value, the higher the utility associated with that value. Even though an individual may prefer more of something, there may be less additional satisfaction with each additional unit received. This is indicated by a utility function with a decreasing slope as shown in Figure 11.14. Utility curves typically have an exponential shape such as that in Figure 11.14. Utility is a relative measure that is usually calibrated to a scale from 0 to 1. The least preferred value is assigned a utility of 0 and the most preferred is assigned a utility of 1. All values in between are relative to the end points. Utility functions enable the decision maker to incorporate personal or organizational attitudes toward risk-taking into the decision. Section 11.3.2.6 includes a discussion of how attitudes toward risk-taking are integrated into a decision analysis through a utility function. A utility function

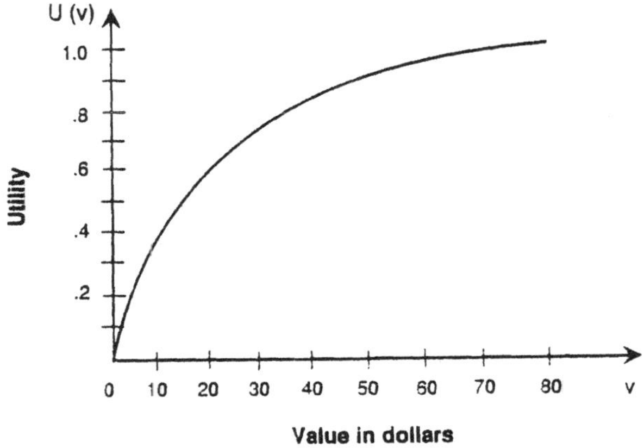

FIGURE 11.14. A utility function for dollars.

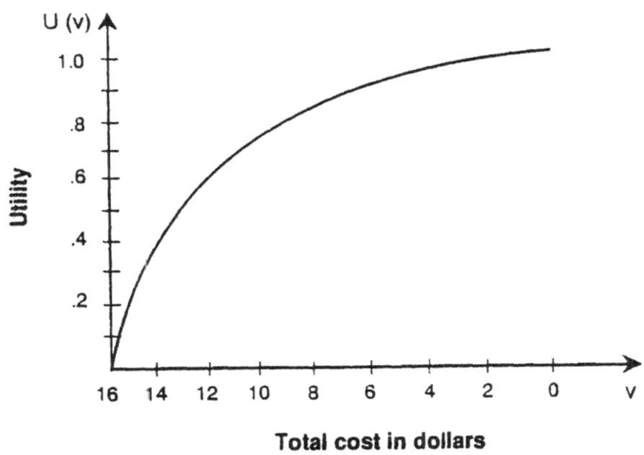

FIGURE 11.15. Utility function for commute cost.

for a commuter is shown in Figure 11.15. The value measure is in terms of total cost in dollars. Each outcome of total cost has a corresponding utility. The utilities can now be incorporated into our tree as shown in Figure 11.16.

11.3.1.5 Expected Utility

If utility is the measure of worth to a decision maker, then we would like to choose the alternative that will yield the outcome with the highest utility. Since future outcomes are uncertain, decision analysis bases the evaluation of alternatives on the mathematical expectation of utility, or *expected utility*. Expected utility is calculated by summing the products of the probabilities of

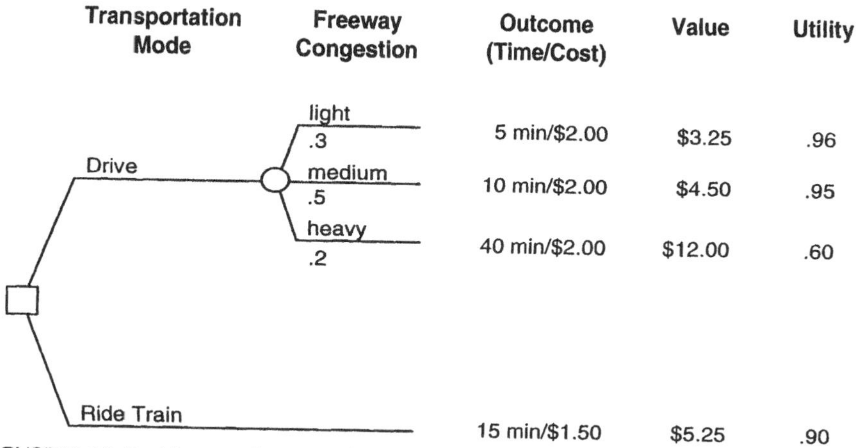

FIGURE 11.16. Decision tree showing utility for outcomes of commuter's decision.

each outcome times the corresponding utilities. To illustrate the calculation of expected utility, consider the decision tree depicted in Figure 11.16. The tree shows the commuter's two alternatives—either to Drive or to Ride the Train. For the Drive alternative there is a .3 chance of receiving a utility of .96, a .5 chance of receiving a utility of .95, and a .2 chance of receiving a utility of .6. Multiplying the utilities by the respective probabilities, we get:

For the Drive alternative:

$$\text{Expected utility} = (.3 \times .96) + (.5 \times .95) + (.2 \times .60) = .88$$

The expected utility for the Ride the Train alternative is simply the single utility value of .9, since the single outcome under this alternative has a probability of 1 and utility of .9. This expected utility of .9 is greater than the .88 expected utility for driving. In order to maximize utility, the best alternative for the commuter is to Ride the Train.

The example illustrates a unique feature of DA: alternatives are evaluated by decomposing them into possible outcomes, the likelihood of the occurrence of each of the outcomes, and the value that the decision maker associates with each of the outcomes. Unlike many other decision aids, DA separates the probabilities and values for outcomes, instead of evaluating alternatives based on a single measure that does not explicitly consider each of these elements. Had we asked the decision maker to simply tell us how much he prefers to take the car and the train, he would have internally had to integrate his uncertainties and values. Doing this can be extremely difficult, and there would be no documentation of the basis for his choice.

11.3.2 Description of Decision Analysis

The preceding section provided an overview of the process and key concepts of decision analysis. This section describes the DA process in more detail.

11.3.2.1 Overview of the Decision Analysis Cycle

The decision analysis cycle guides the decision maker through the formal steps involving modeling, analysis, information gathering, and communication. The input to the cycle is an unstructured decision problem. The output is ultimately a recommended course of action, though not always a choice from among the original alternatives. For example, one may decide to gather more information, analyze a new alternative, or do more analysis. The insights gained from performing the DA will help guide us on how to proceed.

A diagram of the cycle is shown in Figure 11.17. There are five main phases of the process:

- Formulation phase
- Deterministic phase
- Probabilistic phase
- Evaluation phase
- Communication phase.

The first phase, formulation, is used to define and scope the decision problem. In the deterministic phase, a decision model is constructed to represent the relationships between the alternatives, outcomes, and the value measure. Using the decision model, the analyst examines the sensitivity of the variables in the problem to determine which ones have the greatest impact on the decision model. In the probabilistic phase, the possible states of the uncertainties and their probabilities are assessed and incorporated into the decision model. The model is then analyzed in the evaluation phase to determine the optimal decision policy, the value of gathering more information, the value of being able to control one of the factors in the decision, and the recommended course of future action. If necessary, more information is gathered and the model is refined in another pass through the cycle. In the final phase of the cycle, the results and insights are communicated to those involved in making the decision and/or those who have a stake in the results.

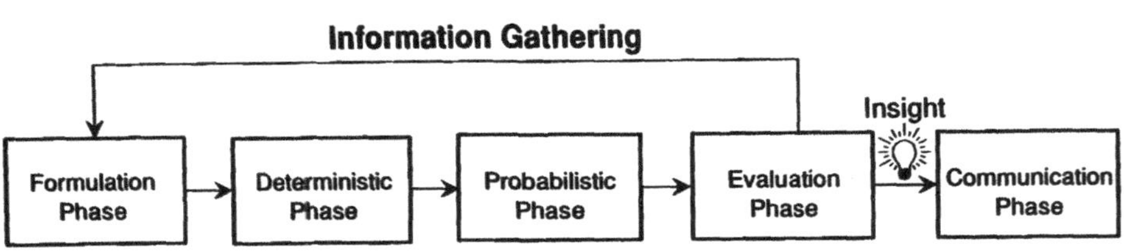

FIGURE 11.17. The decision analysis cycle.

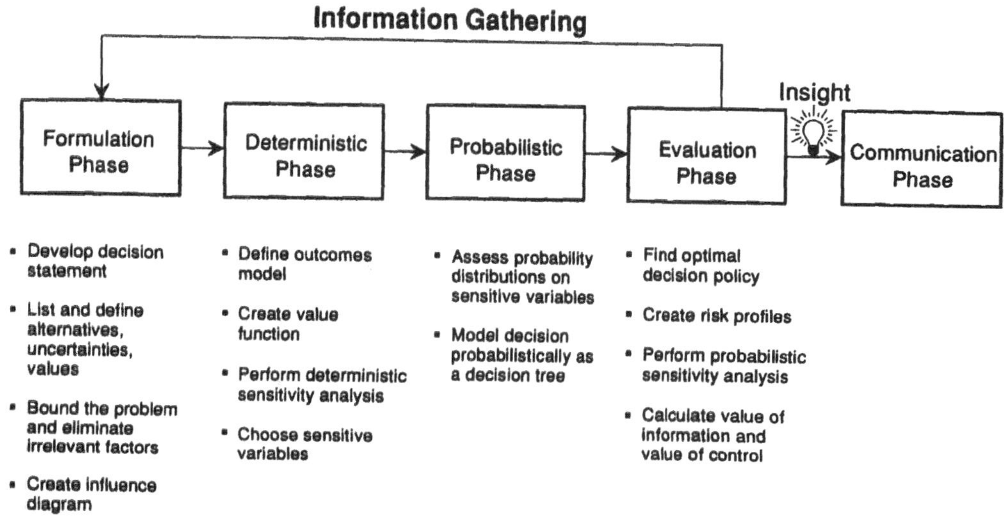

FIGURE 11.18. Tasks included in the DA cycle.

The remainder of this section will describe the roles of the various participants in the DA process and detail the steps of each of the phases. Figure 11.18 presents the key tasks in each of the phases. It is intended to serve as an overview and reference for the detailed descriptions of the phases.

11.3.2.2 Participants in the Decision Analysis Process

The persons involved in the phases of the DA cycle can be classified as decision makers, analysts, and experts. The decision makers are responsible for setting the ground rules and guiding the analysis. They select the experts and analysts and review all of the critical elements of the analysis. The decision makers also determine the value measures and express their preferences for outcomes and risk. In addition, the decision maker actually makes the decision. Analysts are responsible for building the decision model, obtaining unbiased assessments of the inputs to the model, performing the analysis of the model, ensuring the results are accurate, and helping support effective communication of the results. The decision maker and analysts are often collectively referred to as the decision team. Experts help the decision team model the relationships between variables, provide probability assessments for uncertainties about which they are knowledgeable, and provide background information and other inputs to the model. On small problems, the decision maker will often serve as the analyst and as an expert. Although the roles can overlap somewhat, if the same person plays more than one role, the potential for bias and conflict should be recognized.

11.3.2.3 Formulation Phase

The purpose of the formulation phase is to define the decision problem and to develop a more precise understanding of the alternatives, uncertainties, and values. These tasks are extremely important since they lay the groundwork for the rest of the analysis. Many believe that this phase is the most important and illuminating phase of the entire analysis. Even if the entire analysis is not conducted, the formulation phase alone can add structure and insight to the decision-making process.

The steps of this phase are shown under the first box of Figure 11.18. In the first step, the analyst or analysis team develops a clear statement of the decision. Next, the alternatives, uncertainties, and values are identified and given precise definitions. In the third step, the problem is bounded and irrelevant factors eliminated.

Once the definition and scope of the decision have been determined, it is helpful to represent the decision problem graphically. To do this, decision analysts have developed a useful tool, the *influence diagram*. The influence diagram depicts the relationships between decisions, uncertainties, and values. The elements of such a diagram are shown in Figure 11.19. Rectangular nodes represent *decisions* in which the decision maker will choose among alternatives. Oval nodes represent *uncertainties* which are unknown quantities or future events that have more than one possible state. Rounded rectangles are deterministic nodes representing quantities that are known or can be calculated at the time of the decision. *Value* nodes are a subset of deterministic nodes, since the value measure is assigned or calculated using the value

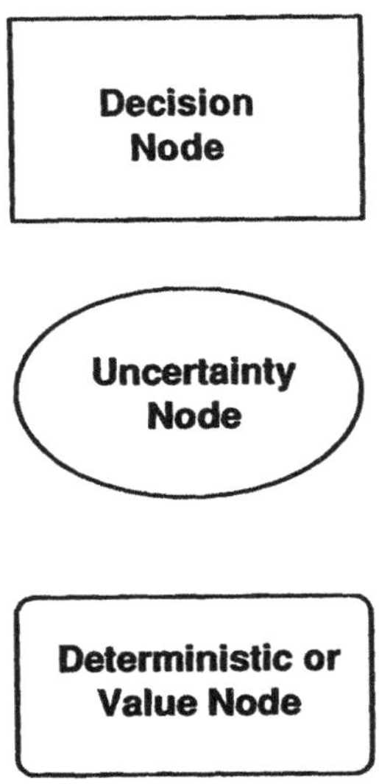

FIGURE 11.19. Components of an influence diagram.

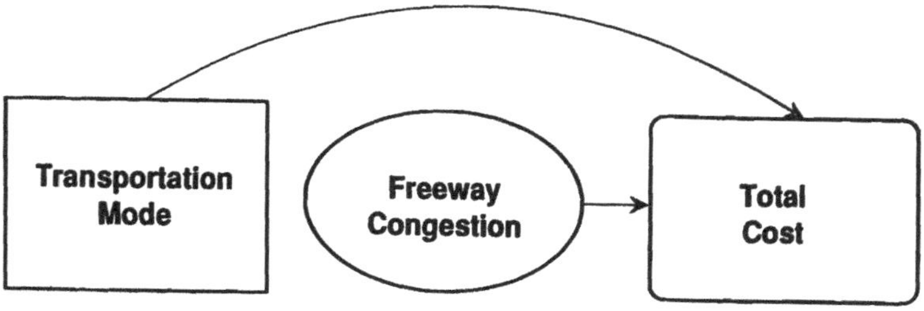

FIGURE 11.20. Influence diagram for commuter's decision example (Section 11.3.1).

function. Arrows between the nodes indicate either conditionality (i.e., one element is dependent upon another or is calculated from another) or sequentiality (e.g., one decision or event occurs after another). The influence diagram for the commuter's decision is shown in Figure 11.20. The decision is what mode of transportation to take, and the uncertainty is the freeway congestion. The arcs leading from transportation mode and freeway congestion to total cost indicate that total cost *depends on* or is *influenced by* the mode of transportation and the state of freeway congestion.

To illustrate a more complex influence diagram, consider the decision of a fictitious company we will call Chemcorp. We will return to this example several times in the following sections to illustrate the entire DA process. Chemcorp primarily participates in long-term research on pesticides. Scientists at Chemcorp have recently developed a new pesticide for protecting vegetables from damaging pests. While the final work is being completed in the R&D lab, management is considering how best to proceed with production of the new product. Their existing production facilities have a capacity to produce 4 million units of the pesticide annually, but this might not be sufficient if demand is high, so they are considering an alliance with National Pesticides. In addition, they have the option of building a new production facility. However, an investment in a new facility might not pay off if demand is low.

National Pesticides has made two offers to Chemcorp. The first alliance is a joint venture in which production is done by Chemcorp and marketing is done by National Pesticides in return for a 50% share in net profits. The second alliance is a royalty partnership in which National Pesticides takes over both the production and marketing, and provides royalty payments of 2% of the sales revenue to Chemcorp. Chemcorp expects the demand for the product to be higher if National Pesticides markets the product using their larger, more experienced sales staff.

The analysis team will consider only a 2-year time frame since innovation in this field is very rapid. In the formulation phase, they have come up with a list of decisions they face:

How do we produce the product?

How do we market the product?
Should we form an alliance with National Pesticides?
Should we build a new plant?

After further scoping of the problem, Chemcorp has developed a single decision statement for the analysis: "How should we produce and market the new pesticide?" The alternatives they are considering are as follows:

A. Produce the pesticide in the existing facility and market the product ourselves
B. Build a new plant to produce the pesticide and market the product ourselves
C. Accept National Pesticides' offer to pay the entire cost of a new plant and market the pesticide in return for a 50% share in net profits
D. Accept National Pesticides' offer to undertake both the production and marketing of the product and receive a royalty payment of 2% of the sales revenue.

After formulating the decision statement and alternatives, the analysis team brainstormed to identify uncertainties and values. The uncertainties include:

- First-year demand
- Growth in demand over time
- Cost of manufacturing new pesticide
- Cost to build new plant
- Market share
- Market size
- Time to construct new plant
- R&D opportunity cost.

Values that the company considers in its activities include:

- Profits
- Worker satisfaction
- Environmental impacts
- Worker safety
- Long-term economic viability.

After bounding the problem and eliminating irrelevant factors from the brainstorm lists, Chemcorp developed the influence diagram shown in Figure 11.21 to represent their problem. For this particular decision, Chemcorp has chosen profits over a 2-year period as their single value measure, and this is represented by the *Profits* node. The rectangle, *New Pesticide Decision*, represents the decision among the four alternatives outlined previously. An arrow goes from the decision node to the *First-year Demand* node since the first-year demand depends on which company markets the product. The growth in

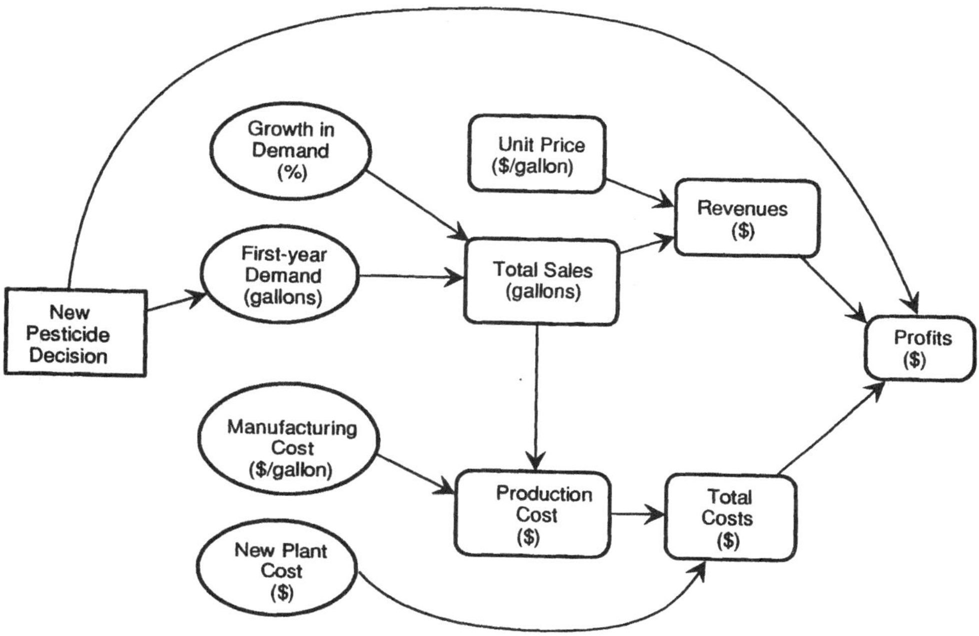

FIGURE 11.21. Chemcorp influence diagram.

demand from the first year to the second (*Growth in Demand*), the per gallon cost of manufacturing the pesticide (*Manufacturing Cost*), and the cost of building a new plant (*New Plant Cost*) are uncertainties that are framed as independent of the alternative chosen and therefore, no arrow is drawn to them from *New Pesticide Decision*. For instance, Chemcorp believes the per gallon cost of manufacturing will be the same regardless of who manufactures the pesticide. Chemcorp's decision does not influence the manufacturing cost, although whether or not Chemcorp manufactures the pesticide itself does impact the profits. This relationship is captured in the arrow between the decision and the *Profits* node.

Total Sales represents the number of gallons sold over the 2-year period and can be calculated from the first-year demand and the growth in demand from the first year to the second. Chemcorp is using a fixed, known price for the decision. Therefore, unit price is a rounded rectangle. *Revenues* is a rounded rectangle because it can be calculated by multiplying *Unit Price* by *Total Sales*. The arrows from *Unit Price* and *Total Sales* into *Revenues* indicate that *Unit Price* and *Total Sales* are used in the calculation of *Revenues*. Likewise, *Production Costs* can be calculated by multiplying *Total Sales* by *Unit Manufacturing Cost*, assuming that all units produced are sold. *Total Costs* are then found by summing the production cost and the cost of building a new plant. Finally, as indicated by the arrows into the *Profits* node, profits depend on total costs and revenues.

The influence diagram can serve as a guide throughout the DA process. It provides a visual representation of the elements of a decision and the important dependencies. The process of developing the diagram is perhaps more important than the diagram itself. It helps the analysis team understand and organize the problem prior to developing a quantitative model.

11.3.2.4 Deterministic Phase

At the conclusion of the formulation phase, the decision model consists of a graphical representation of the decision problem. In the deterministic phase, the influence diagram is used to guide creation of a mathematical form of the decision model. The model is referred to as "deterministic" at this stage because it does not account for uncertainty and risk, but assumes a single future state for each of the uncertain variables. After the mathematical model has been constructed, sensitivity analysis is performed on the value measure by changing each of the model variables. The results determine which variables of the model will be treated as uncertainties. These variables will be assigned states and probabilities in the next phase. The rest of this section will explain the deterministic phase in greater detail.

The Value Function

At this phase of the analysis, the mathematical decision model will consist of a value function that maps outcomes into a single value measure. It is an equation or set of equations used to calculate the value measure associated with an outcome. In order to determine what the outcome is, one can ask "If I could see into the future, on what basis would I judge the success of this decision?" The answer may be profit, market share, worker safety, environmental impact, or a combination of various factors. Sometimes outcomes will have multiple attributes with different dimensions on which value is measured. In these cases, the value function must combine these multiple attributes into a single measure of value. For example, in the commute decision, there were two attributes to the outcome—time and money. Commute time and commute cost were combined into total cost, using a value function expressed by a single equation. By combining multiple attributes into a single measure of value, the decision maker is making an explicit statement about the relative worth of the attributes. As mentioned previously, Chemcorp has chosen profits as their single value measure. Therefore the outcome can be measured solely in monetary terms, resulting in a single-attribute problem.

If an influence diagram has been created, the value function can be thought of as the equation or set of equations corresponding to the influence diagram. The value function for the Chemcorp decision consists of a set of equations used to calculate the profits achieved under each of the four alternatives. For example, if Chemcorp chooses Alternative B, "Build a new plant to produce the pesticide and market it ourselves," the equations used to calculate profits are as follows:

Profits = Revenues – Total Costs

Revenues = Unit Price × Total Sales

Total Sales = First-year Demand
+ [(1+ Growth in Demand) × First-year Demand]

Total Costs = Production Cost + New Plant Cost

Production Cost = Manufacturing Cost × Total Sales

In a real application, the terminal value of the plants would usually be taken into account in the value model. The terminal value is the worth of an asset at the end of the time period. For example, Chemcorp may need to spend $15 million on a new plant, but they may have an asset worth more than the existing building at the end of the time period. This additional worth would be accounted for in a terminal value. Terminal value can be negative. For instance, there may be environmental requirements that make the plant and the land unusable without a sizable investment.

Spreadsheets are often a good tool for creating a value function. Spreadsheet models may have already been developed in a company prior to a decision analysis. If so, they can be incorporated into the decision model. Sometimes more complex models such as simulation or optimization models are needed.

Time Value of Money

If it is appropriate, the decision maker may want to incorporate the time value of money into the decision model at this point. The "time value of money" indicates how a series of benefits received or costs paid over time are valued by the decision maker. Generally, a decision maker will prefer benefits received sooner over the same benefits received later. Similarly, costs incurred now are less desirable than those incurred later. Because of the time value of money, a decision maker may, for example, prefer an alternative that is less profitable than a second alternative, but for which the profits are obtained sooner.

The effect of the time value of money is incorporated into the decision model through discounting, as described under "Economic Evaluation Principles" in Section 2.2. Generally, an analysis uses a constant discount rate, but this is not a requirement. A different discount rate may be used for each period under consideration. Typically, a corporate decision maker will prefer to use a corporate discount rate that has been adopted for all corporate activities, or to adopt one if it does not currently exist. The appropriate discount rate should reflect the opportunity cost of the required resource allocation. Usually, this will be the rate of return on investments of similar magnitude, and with similar risks and time horizons to the alternatives under consideration. For simplicity, we will not use a discount rate in the Chemcorp model (i.e., profits received in the second year will not discounted). For additional discussion on the time value of money, see Section 2.2.2.

Inputs to the Decision Model

There are two types of inputs to the decision model, and a single output, or value measure. The two types of inputs are inputs that are known for certain and inputs that are uncertain.

Inputs that are known for certain may come from published data and/or accepted or standard methods of calculation. For example, in a decision of what type of material to use to build something, the price of the materials may be found in a supplier's catalog. Another example is a known parameter of a recognized engineering equation or model (e.g., the freezing point of water, the speed of light, the weight of aluminum). Uncertain inputs are based on the best estimates of experts after reviewing available information. Their true values will be determined by the resolution of an uncertainty in the future, or they may never be known.

In decision analysis, rather than assigning a single numerical estimate to an uncertain variable, the uncertainty is specified through a probability distribution. In the deterministic phase, a probability distribution over an uncertain variable is estimated by assessing low, nominal, and high settings for the variable. Usually, the nominal setting is defined as the value at which there is a 50% chance that the true value of the variable is lower. Likewise, there is a 10% chance that the true value will be below the low setting and a 90% chance that the true value will be below the high setting. These probabilities are *cumulative* probabilities (see Chapter 2 for an explanation of probability concepts). The objective in estimating a probability distribution is to get a range of possible numerical quantities that a variable can have. Then the impact of the quantities in the range on the decision outcome can be assessed in the next phase.

The uncertain inputs in the Chemcorp decision correspond to the uncertainties shown in the influence diagram of Figure 11.21. For the analysis, Chemcorp will use a fixed price of $20 per gallon which was obtained from a similar pesticide on the market. *First-year Demand* refers to sales if Chemcorp does the marketing. We will assume that sales would be 50% greater if National Pesticides completes the marketing. Table 11.1 shows the initial ranges chosen for the uncertain variables. Chemcorp experts in different areas provided the input. For instance, the marketing manager assessed the *First-year Demand*. The first row of the table expresses the marketing manager's range of uncertainty for this variable. The low, nominal, and high settings are those that the marketing manager believes have a 10%, 50%, and 90% cumulative probability, respectively. For example, the marketing manager believes

TABLE 11.1
Uncertain Inputs for the Chemcorp Decision Model

Uncertain Variable	*Low (10%)*	*Nominal (50%)*	*High (90%)*
First-year Demand (MM gallons)	1	3	6
Growth in Demand (%)	–70	20	100
Manufacturing Cost ($/gallon)	8	9	11
New Plant Cost ($MM)	10	15	25

there is a 10% chance that *First-year Demand* will be less than or equal to 1 million gallons.

Deterministic Sensitivity Analysis
Since uncertain inputs are expressed by subjective assessments (i.e., low–nominal–high estimates), it is helpful to know how important the range of these estimates is to the results of the analysis. This can be examined through sensitivity analysis. Sensitivity analysis is performed on uncertain variables to examine how changes in the variables affect the value measure associated with the outcomes of each alternative. Variables which have little or no effect can be input with their nominal setting for the probabilistic phase. Variables that are found to be sensitive will be assessed formally by obtaining a complete probability distribution in the probabilistic phase and incorporating this complete description of the uncertainty into the analysis. In addition, sensitivity analysis may identify dominated alternatives. Dominated alternatives are alternatives that show inferior results to another single alternative for all settings of the uncertain inputs. These alternatives need not be included in subsequent analysis, since another alternative is always preferred. In summary, sensitivity analysis helps simplify the analysis, reduce the number of alternatives, and provide insights that guide subsequent work.

One type of sensitivity analysis commonly performed in the deterministic phase is called deterministic sensitivity analysis. The first step in performing deterministic sensitivity is to use all uncertain variables at their nominal setting and determine the nominal value of each alternative, as given by the value function. The alternative with the highest value will be considered the preferred "nominal" alternative. The alternatives for the Chemcorp example and their nominal values are shown in Table 11.2. With all variables at their nominal settings, Alternative A has the highest profit, $72.6 million. Next, one uncertain variable is selected. The value measures for each alternative are recomputed twice, once with the selected variable at its low setting, and once with the selected variable at its high setting. In each computation, all other

TABLE 11.2
Nominal Values for the Chemcorp Alternatives

Alternative	*Nominal Profit ($MM)*
A	72.6
B	57.6
C	54.5
D	4.0

variables are kept at their nominal settings. This recomputation is done for each variable, setting each low and then high, one variable at a time. The results of a deterministic sensitivity analysis on the Chemcorp *First-year demand* variable are shown in Table 11.3. The chart shows that if demand were 1.0 million gallons and all other variables were at their nominal, then Alternative A would yield a $24.2 million profit. If demand were 6.0 million gallons, Alternative A would yield a $88 million profit. Similarly, the profit under the other alternatives is calculated for sales at both a high and a low setting. The values in **boldface** are the highest profits under each setting of the variables. For example, if sales were low, the preferred alternative is A with a $24.2 million profit. However, if sales were high, Alternative B would yield the highest profit ($130 million). This change in the preferred alternative as the variable changes within the specified range is crucial. It indicates that the decision is sensitive to the uncertainty in the variable. Therefore, *First-year demand* is referred to as a *sensitive variable.*

Interpreting Sensitivity Results

The results of the sensitivity analysis on remaining Chemcorp variables are shown in Table 11.4. These results can be represented visually with a *tornado diagram,* as shown in Figure 11.22. The vertical center line of the tornado is

TABLE 11.3
Results of Sensitivity Analysis on First-year Demand

	Profit ($MM)			
	Alternative A	*Alternative B*	*Alternative C*	*Alternative D*
First-year demand (MM gallons)				
Low = 1.0	**24.2**	9.2	18.2	1.3
High = 6.0	88.0	**130.2**	108.9	7.9

TABLE 11.4
Sensitivity Table for Chemcorp Decision

	Profit ($MM)			
	Alternative A	*Alternative B*	*Alternative C*	*Alternative D*
First-year Demand (MM Gallons)				
Low = 1.0	**24.2**	9.2	18.2	1.3
High = 6.0	88.0	**130.2**	108.9	7.9
Growth in Demand (%)				
Low = −70	**42.9**	27.9	32.2	2.3
High = 100	77.0	**84.0**	74.3	5.4
Manufacturing Cost ($/gallon)				
Low = 8	**79.2**	64.2	59.4	4.0
High = 11	**59.4**	44.4	44.6	4.0
New Plant Cost ($MM)				
Low = 10	**72.6**	62.6	54.5	4.0
High = 25	**72.6**	47.6	54.5	4.0

placed at the setting of the preferred alternative in the nominal scenario. For Chemcorp, the preferred nominal alternative was A with a profit of $72.6 million. Each bar represents the change in the profits yielded by the preferred alternative as the corresponding variable changes, with all other variables at their nominal settings. The color change in a bar indicates that as the value of the variable changes from nominal to one of the extremes, the best alternative changes. For example, profit varies from $24.2 million to $130.2 million as *First-year demand* varies from 1.0 million gallons to 6.0 million gallons. The left side of the top bar is placed at $24.2 million profit, and the right side is placed at $130.2 million. As indicated at the ends of the bar, A is the preferred alternative when *First-year demand* is low, whereas B is preferred when demand is high. The diagram is called a tornado diagram because it resembles a tornado when the variables are ordered with those that cause the greatest swing in value at the top and those that cause the least swing at the bottom.

Once each variable has been varied from low to high, the *sensitive variables* must be selected. There are two different types of variables that are generally considered to be sensitive variables:

- Variables that cause the preferred alternative to change as the setting of that variable changes from low to high
- Variables that have a large impact on the resulting value under one or more of the alternatives.

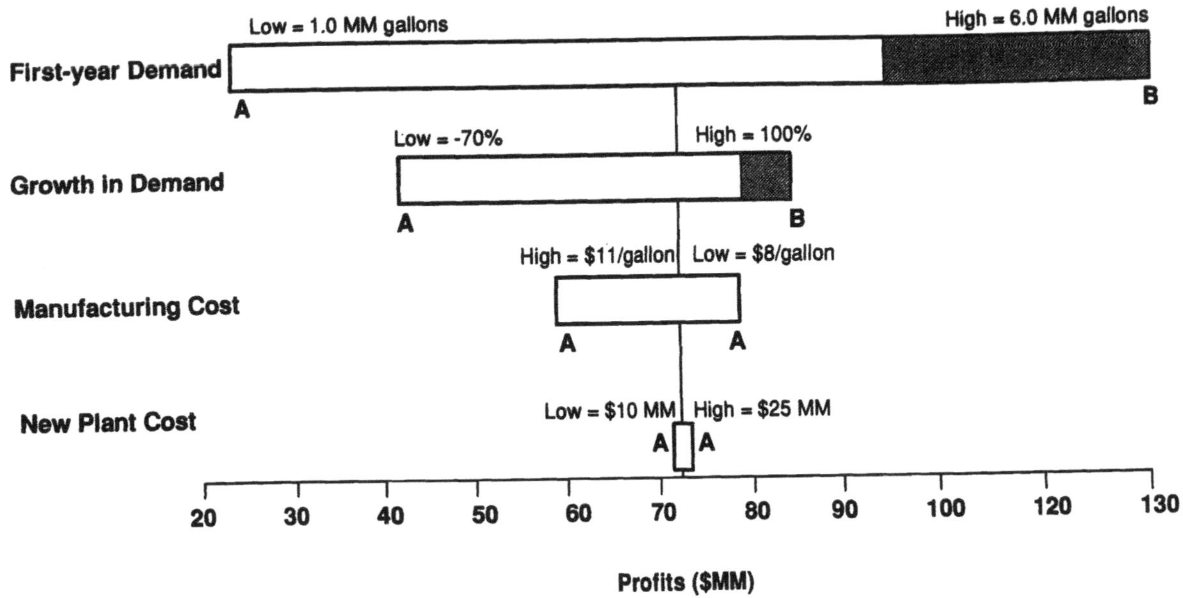

FIGURE 11.22. Tornado diagram for Chemcorp decision.

Under the first criterion on page 317, if the preferred decision at either extreme setting for a variable is different from the nominal decision, the decision is sensitive to that variable. For example, we see in Figure 11.22 that when *First-year demand* is at its low setting, Alternative A has the highest value measure, whereas Alternative B has the highest value measure when *First-year demand* is at its high setting. *First-year demand* is chosen as a variable to which the decision is sensitive. Likewise, *Growth in demand* is a sensitive variable. However, for *Manufacturing cost* the alternative does not change. Alternative A is the best alternative under each setting of *Manufacturing cost*.

Under the second criterion above, a variable may be considered to be sensitive if varying it from its low to high setting causes a significant change in the resulting value measure of an alternative, even though the preferred alternative may not change. There is no strict rule for choosing sensitive variables in this case. This type of sensitivity can be examined using the length of the bars of the tornado diagram. If the length of the bar for a variable is large relative to the other variables, it will often be considered sensitive. In effect, the length of the bars represents the contribution of the uncertainty in a variable to the value measure. The objective of sensitivity analysis is to simplify the model as much as possible while still capturing a sufficient amount of the decision-maker's uncertainty. If varying a variable over its range causes no change, or a relatively small change, compared to other variables, and it does not cause the preferred alternative to change, then that variable can be fixed at its nominal setting for the rest of the analysis. For example, in Figure 11.22, we see that varying *New plant cost* from $10 million to $25 million does not cause the preferred alternative to change and does not impact the profits. It will be kept in the analysis at its nominal setting.

By using a variable at its nominal setting, a formal probability distribution will not have to be assessed for the variable, saving analysis time and resources. Depending on the number of variables and the complexity of the model, the analyst may use judgment about how many variables to use at their nominal setting. These steps are also driven by the magnitude of the decision and the time and resources available. For the Chemcorp decision we have chosen *First-year demand* and *Growth in demand* as sensitive variables. As seen in the tornado diagram, they capture a significant portion of the uncertainty. *Manufacturing cost* and *New plant cost* will be set to their nominal values. This will eliminate some of the uncertainty in the analysis, but is unlikely to significantly impact the results.

If two or more variables are closely related or may vary together, joint sensitivity analysis should be performed. For example, if the decision maker believes that the chance that the *New Plant Cost* is high (or low) is affected by whether the *Manufacturing cost* is high (or low), joint sensitivity analysis on *New plant cost* and *Manufacturing cost* should be performed. Joint sensitivity analysis involves varying more than one variable at a time. With limited time and resources, the analyst must use judgment about which variables to analyze.

Sensitivity analysis can eliminate some alternatives from further analysis. The results in Table 11.4 indicate that for **all** settings of the uncertain variables, Alternative D has a lower profit than the other alternatives and therefore is dominated by all other alternatives. Alternative C is not optimal in any one setting of the variables. However, it is not inferior to another single alternative in **all** settings of the variables and is therefore not dominated. If an alternative is dominated, it can be dropped from further consideration. On the other hand, if an alternative clearly dominates all others, it may be appropriate to select it as the best alternative at this point, without continuing the analysis.

11.3.2.5 Probabilistic Phase

The output of the deterministic phase is a decision model that specifies the value under each alternative for any given setting of inputs. In the probabilistic phase, the analyst incorporates a full range of possible outcomes, including their probabilities of occurrence, into the decision model. This is done by obtaining assessments of the probability distributions on the sensitive variables and then representing the distributions with a discrete number of points for evaluation purposes. The end product is a discrete set of possible outcomes, each with a probability of occurring and a value as determined by the value function. The outcomes, probabilities, and values, are then represented graphically in a decision tree. The decision tree representation will facilitate later evaluation and communication of the decision.

Expert Assessments

The first step of the probabilistic phase is to identify the experts who will provide assessments of the probability distributions on the sensitive variables identified in the deterministic phase. If reliable long-run frequency data are available, the expert may use it to form a distribution. However, while some information may be available, rarely is it complete, accurate, or relevant enough to be used directly as the probability distribution. Instead, the distribution will most often need to be assessed using the judgment of one or more experts. Sometimes the distributions used to specify variables are of a standard functional form in which the expert would need only to identify the mean and the variance. However, if there is no standard form, a distribution must be created.

The formal process of assessing probability distributions is termed *encoding*. Once the appropriate expert is identified, the expert's knowledge of this variable is captured by assessing a cumulative probability distribution. For example, the distribution on *First-year demand* may be encoded by an expert with marketing expertise from inside or outside the organization. The expert is encouraged to use all available data in addition to personal knowledge and experience. Since probability is not intuitive to most individuals, encoding a probability distribution is not a trivial task. Decision analysts have developed interviewing techniques that help motivate the expert, adjust for biases and judgmental errors that can lead to an inaccurate assessment, and provide an accurate representation of the expert's knowledge. For best results, the encoding process must be guided by an independent, unbiased analyst. See Merkhofer (1987b) for a detailed discussion of probability assessment.

Typically, several points on a distribution will be assessed, and a curve will be fitted through the points to create a continuous cumulative distribution. An example distribution on *First-year demand* is shown in Figure 11.23. The points marked with asterisks were elicited by an analyst through assessment techniques. Then a continuous curve was fitted through the points.

The next step is often to *discretize* the assessed probability distribution. Unless the curve can be described mathematically and software is to be used in the evaluation phase, probability distributions must be specified by a discrete number of settings. The discrete points are chosen so as to best approximate the original continuous distribution. In Figure 11.24, the sales distribution has been discretized and represented by three settings, 1.2, 3.0, and 5.5, with the probabilities 0.25, 0.5, and 0.25, respectively. These points were chosen by using the Gaussian quadrature method explained by Miller and Rice (1983). In another common method called vertical integration, the analyst divides a distribution into intervals and picks a representative point in each interval. The details of these methods will not be discussed here.

Decision Tree Representation

Once all sensitive variables have been discretized, the decision model may be represented graphically as a decision tree. Decision trees represent the simplified decision problem to be solved at this stage. In addition to showing the

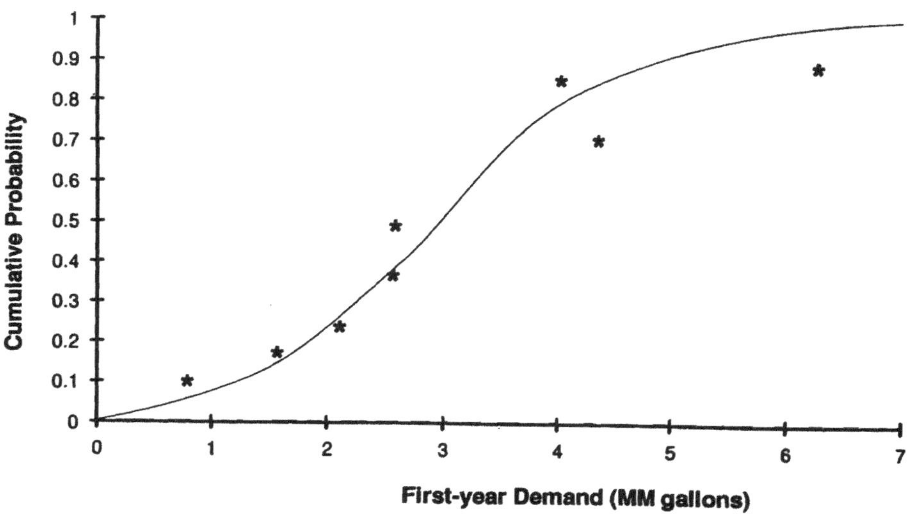

FIGURE 11.23. Cumulative probability assessed for first-year demand.

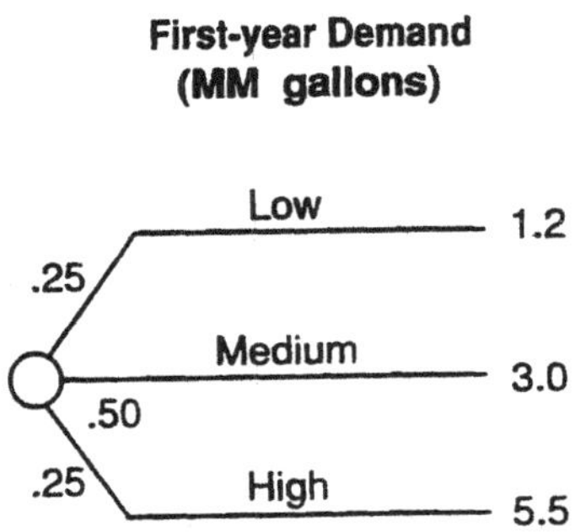

FIGURE 11.24. Discrete representation of the probability distribution on first-year demand.

relationships between decisions, uncertainties, and values as in an influence diagram, the decision tree:

- Shows the chronology of decisions and uncertain events
- Shows all the possible scenarios, or *paths*, for which an outcome is calculated
- Shows the probabilities and expected values associated with each possible path.

The decision tree elements are similar to the influence diagram elements. Rectangles represent decisions, and ovals represent uncertainties. Outcomes and/or values are placed at the end of each path of the tree. Decision trees are arranged chronologically from left to right, starting with a decision. Returning to the Chemcorp example, Figure 11.25 shows the decision tree. Note that there are three alternatives and two uncertainties and profit is the value measure. The decision node depicts the decision, and the branches from this node represent alternatives. Uncertainties in the tree are represented by branches with the discrete probability estimates. Each point at the end of a path represents a single possible outcome. One can calculate a value for each outcome by using the value function (in this case the profit equations) and the settings of the variables along a given path. By multiplying the probabilities of the events along a path leading to an endpoint, one can also obtain the probability for each outcome. This is called a *joint probability* and is the probability that **all** events along the path occur. The joint probabilities are shown in the last column of Figure 11.25.

11.3.2.6 Evaluation Phase

Once outcomes have been assigned values and probabilities, the optimal decision policy can be determined and evaluated. In addition, a decision-maker's risk attitude can be taken into account if it is important in the decision. This section describes these steps of the evaluation phase.

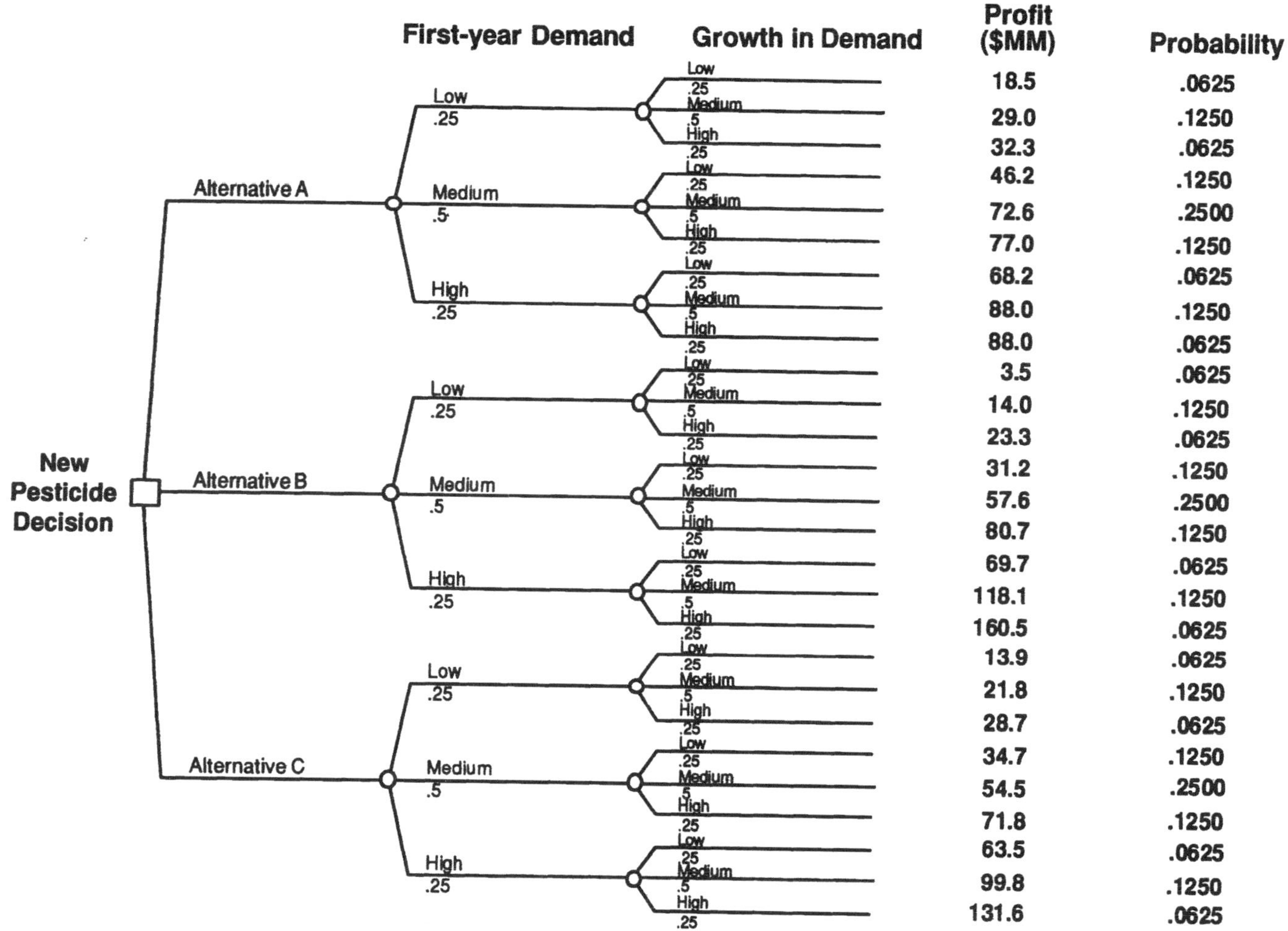

FIGURE 11.25. Decision tree for Chemcorp example.

Risk Profiles

The first step of the evaluation phase is to create a cumulative probability distribution for each alternative based on its discrete outcomes and associated probabilities. This is often referred to as a risk profile. (See Chapter 2 for a description of cumulative probability distributions.) Figure 11.26 shows a risk profile for Alternative A of the Chemcorp example. The risk profiles can be used to examine the range of possible outcomes. A point on the curve represents the probability that the value will be less than or equal to the value at that point. For example, in Figure 11.26, the distribution for Alternative A indicates that there is a 25% chance that the profit will be less than $40 million. The vertical line is placed at the expected value of the alternative. The calculation of expected value is illustrated by the tree rollback method described later. Risk profiles help answer questions about an alternative such as:

- What is the highest profit possible if we choose this alternative?
- How much can be lost?
- What is the probability of losing money (profit less than zero)?

The risk profiles for more than one alternative are often placed on the same graph for comparison. For example, Figure 11.27 shows Alternatives A and B of the Chemcorp example. The expected values are extremely close. However, Alternative B has a higher risk, indicated by the larger range of possible outcomes. If the risk is not of concern, often the decision will be made at this point by choosing the alternative with the highest expected value. Otherwise, risk concepts will be incorporated into the analysis by constructing a utility function as explained in the next section. Before decisions are made based on

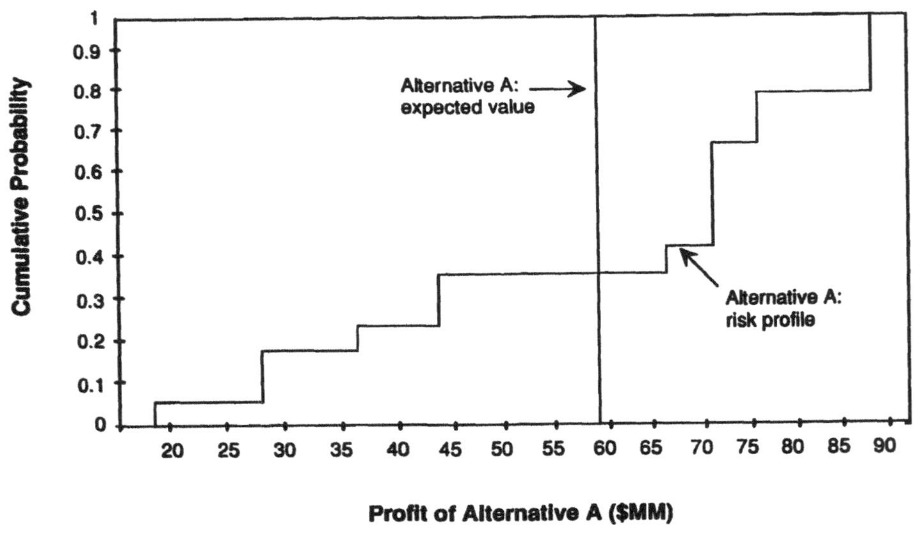

FIGURE 11.26. Risk profile for Alternative A of Chemcorp example.

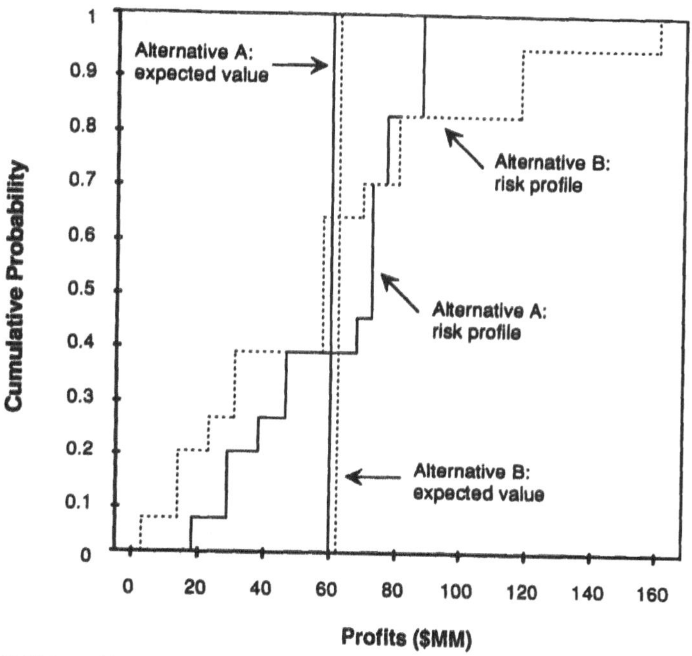

FIGURE 11.27. Risk profiles for Chemcorp alternatives.

expected value, the risk profiles should always be constructed to examine the risk of extreme outcomes under each alternative.

The examination of the risk profiles may lead us to discard one or more alternatives that are dominated. An alternative is dominated by a second alternative if the second alternative has a risk profile that lies entirely to the right of the first alternative's profile. This is often referred to as *stochastic dominance.* An example is shown in Figure 11.28. For every value, the cumulative probability of the dominated alternative is higher than the cumulative probability of the dominating alternative. That is, for every value, there is a smaller chance that the dominated alternative will actually have a higher final value. If one alternative dominates all others, then that alternative is optimal. There is no need to calculate expected values or construct a utility function.

Risk Concepts

If risk is an important consideration in the decision, it should be incorporated into the analysis before performing the tree rollback. Therefore, we will discuss it prior to explaining the tree rollback. This section explains how risk is accounted for in a decision analysis through a utility function. We present a hypothetical example of a woman named Sally who is faced with a decision involving gambling, and with money at stake. The gambling analogy is often

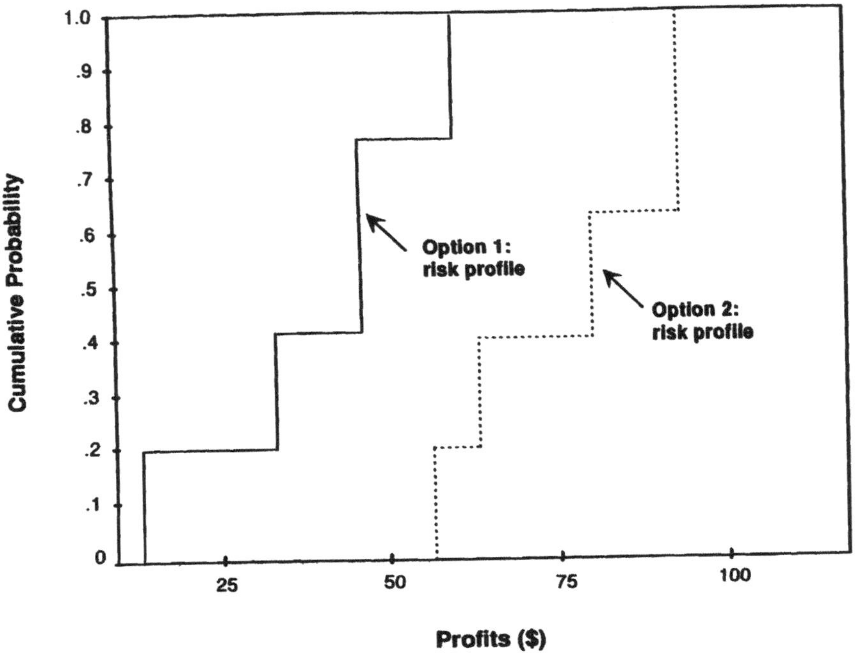

FIGURE 11.28. Risk profiles indicating that Option 1 is dominated by Option 2.

used to present risk because it is easy to relate the uncertainty in the gamble to risk taking. However, these concepts can be extended to any situation that involves uncertainty.

A decision-maker's attitude toward risk can be incorporated into the decision analysis by creating a utility function that maps the values determined by given outcomes into utilities for those values. Utility is a measure of the decision-maker's strength of preference for a given quantity. The higher the decision-maker's preference for a value, the higher the utility associated with that value. The graph in Figure 11.29 illustrates Sally's utility function with dollars as the value measure. To relate the utility curve to the concept of risk, suppose that we offer Sally a gamble involving the flip of a coin: if the coin lands heads she wins $1000; if it lands tails she wins nothing. Sally believes each outcome has a 50% chance of occurring, or a 0.5 probability, as illustrated in Figure 11.30. To determine what this gambling opportunity is worth to Sally, we ask the question "What is the maximum amount that you would be willing to pay for the option to take the gamble?" Sally is willing to pay $375 to take the gamble. The expected value of this gamble can be calculated by multiplying the possible outcomes by their likelihood of occurrence:

$$\text{Expected value} = p_1 \times v_1 + p_2 \times v_2 = (0.5 \times \$1000) + (0.5 \times \$0) = \$500$$

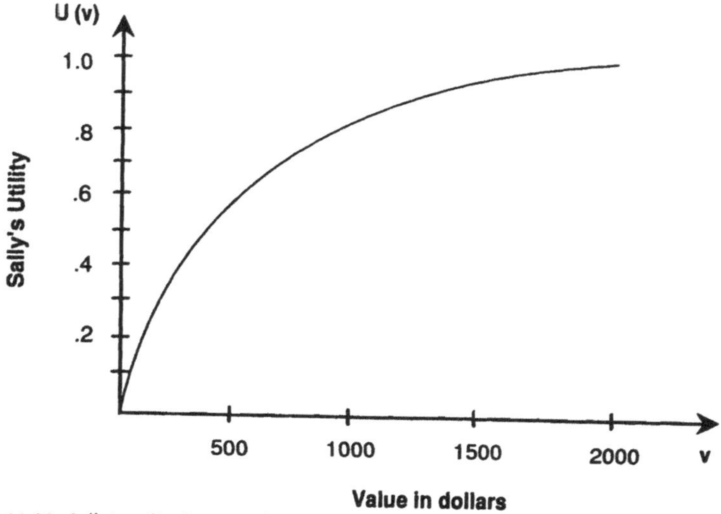

FIGURE 11.29. Sally's utility function for dollars.

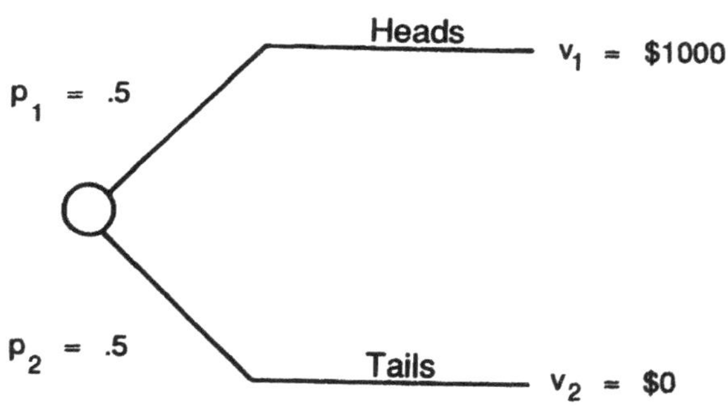

FIGURE 11.30. Coin toss gamble.

The expected value of this gamble is $500, but Sally is only willing to pay $375 for it. Like Sally, most decision makers would not be willing to pay as much as $500 for this gamble. The prospect of winning $1000 and therefore being $500 richer after the $500 cost is deducted does not outweigh the risk of a $500 net loss. Most people are risk averse and would offer significantly less than $500. Risk aversion is one way to express a person's risk attitude, or feeling toward risk. The difference between the expected value and Sally's willingness to pay for the gamble is captured in the utility curve. Using the utility curve in Figure 11.29 we can obtain the utilities for the dollar values in the gamble as shown in Figure 11.31. To get the worth of this gamble to Sally, we calculate her expected utility:

$$\text{Expected utility} = p_1 \times u_1 + p_2 \times u_2 = (0.5 \times 0.8) + (0.5 \times 0) = 0.4$$

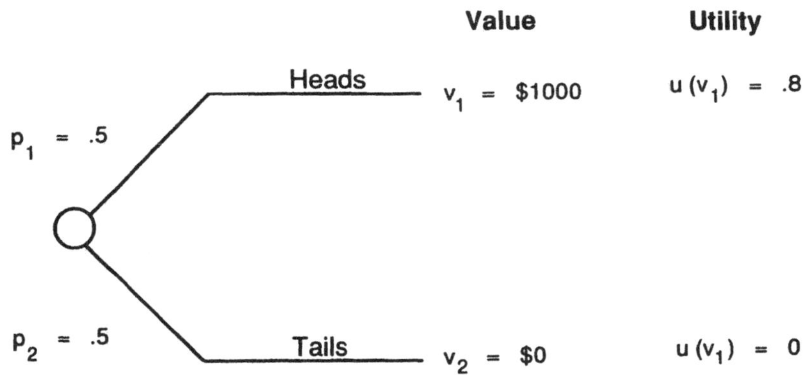

FIGURE 11.31. Coin toss utility assessment.

Using the utility function (Figure 11.29), we find the monetary value associated with a utility of 0.4 to be \$375. This value is referred to as the *certain equivalent*. It is the value at which an individual is indifferent between having the monetary value for certain, or taking the gamble. An individual is *risk-averse* if the certain equivalent is less than the expected value for a gamble. An individual is *risk-preferring* if the certain equivalent is greater than the expected value. A risk-preferring utility curve is shown in Figure 11.32a. An individual whose certain equivalent is always equal to the expected value has a linear utility curve as shown in Figure 11.32b and is called *risk-neutral*. Individuals or organizations may be risk-neutral for relatively small gambles, but are generally risk-averse when substantial stakes are involved. Risk-preferring behavior is uncommon.

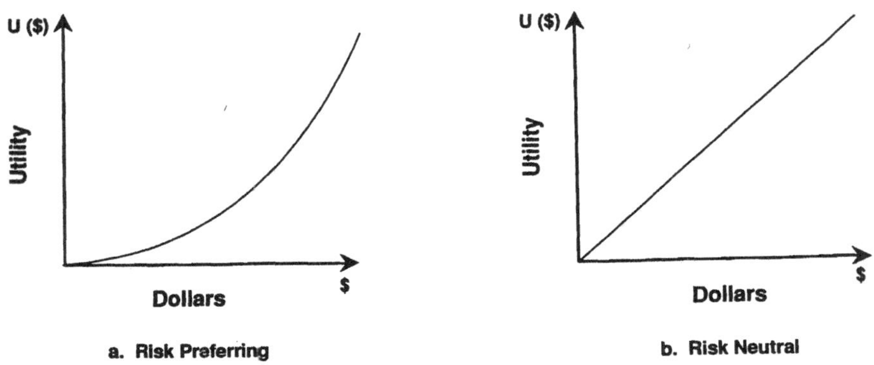

FIGURE 11.32. Utility functions for risk-preferring and risk-neutral decision makers.

Decision analysis incorporates risk attitude into decision making by treating uncertain events as gambles with outcomes and probabilities. By expressing the likelihood of the outcomes through probabilities, and mapping values for outcomes into utilities, a decision maker can calculate the expected utility of each alternative of a decision. It follows from the axioms of rational decision making that the preferred alternative is the alternative that yields the highest expected utility. If the decision maker is risk-neutral, using expected utility will yield the same preferred alternative as using expected value.

Assessing Risk Attitude

If the decision maker is risk-averse and risk attitude is important to the decision, a utility function must be constructed to account for the decision-maker's risk aversion. A utility function is constructed by a trained analyst who asks the decision maker a series of questions that elicit preferences for different gambling situations. Techniques for developing a utility curve are available, but the details are beyond the scope of this chapter. See von Winterfeldt and Edwards (1986) for more details on assessment techniques. Experience has shown that most individuals and companies demonstrate risk aversion in their decisions, and that a simple form of the utility function (exponential) can be used to express their utility for dollars. However, for ordinary decisions that involve only a small fraction of a company's total resources, the effect of risk aversion is probably slight and need not be taken into account. If the decision involves possible values that are small relative to the net worth of the company, the utility curve for the problem can be approximated by a linear function, and therefore, the decision is usually evaluated on an expected value basis. However, if the decision involves a significant portion of the company's resources, incorporating risk attitude into the decision is important.

Rolling Back the Tree

The *tree rollback* method is a way to visually represent the calculation of mathematical expectation. Specifically, in DA we determine the expected value or expected utility of the alternatives depicted on our decision tree. Each endpoint of the tree represents an outcome under a different setting of the variables, depending on the branches that lead to that endpoint. A value can be calculated for each outcome using the value function. A corresponding utility for the value is then found using the utility function, unless the decision is to be based on expected value. The tree is then "rolled back" by calculating the expected utility, or expected value, at each uncertainty node, moving from right to left. When a decision node is reached in the rollback, the alternative with the branch that has the highest expectation is chosen. This process continues until the initial decision node is reached.

Decision Policy

The rollback method indicates the optimal decision policy. The optimal decision policy is the alternative or set of alternatives that yield the highest

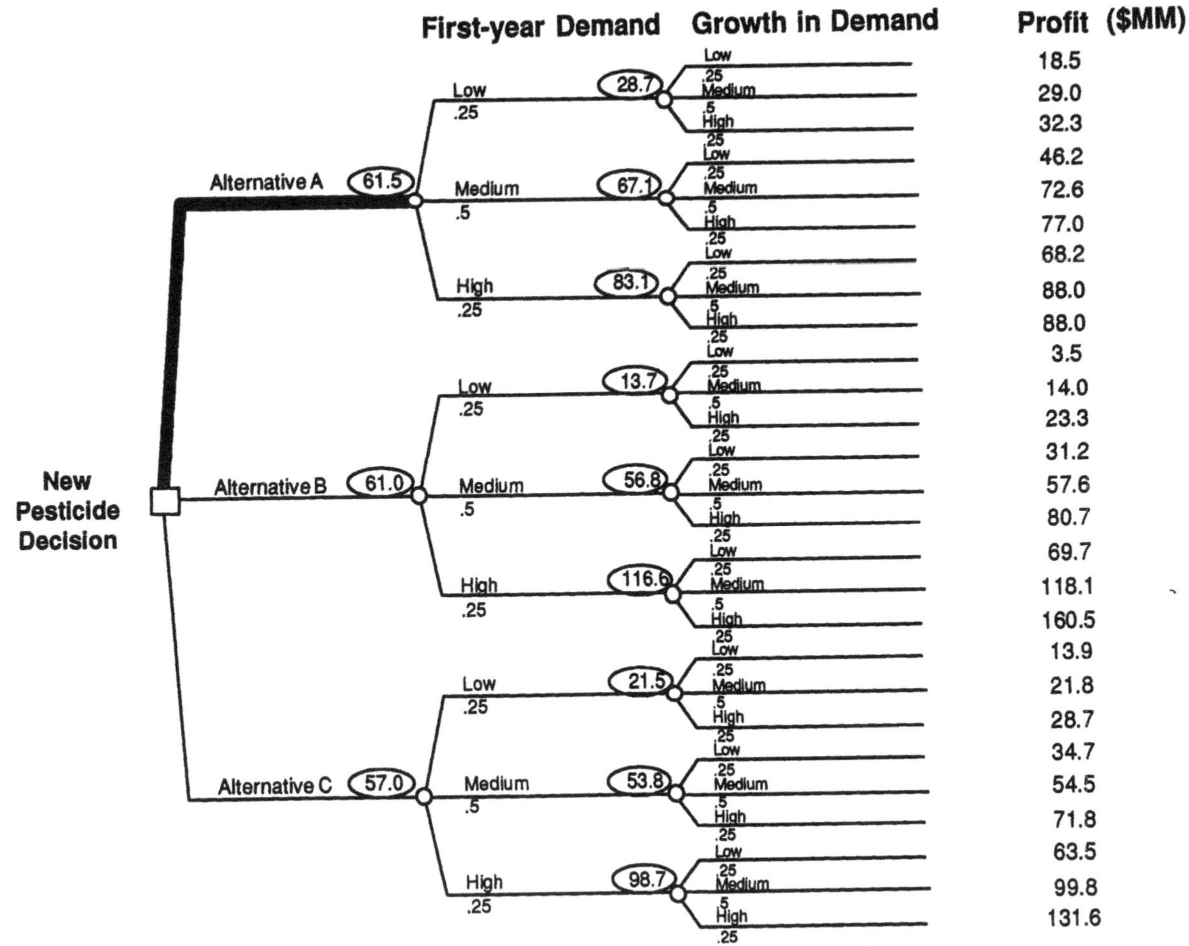

FIGURE 11.33. Rolled-back decision tree for Chemcorp example.

expectation. The decision policy will include one alternative for each decision in the problem. For instance, if there are two decisions—which product to develop and how to produce it—a decision policy will have two components: the recommended product and the recommended production process. The rollback of the Chemcorp tree is shown in Figure 11.33. In this example, the decision will be based on expected value, with risk aversion ignored. The expected profit at each node is circled.

To illustrate the expected-value calculation, consider Alternative A with an expected profit of $61.5 million. First, the expected profit at the end of each *Growth in demand* branch was calculated and placed at the end of the respective *First-year demand* branch. Then the expected profit of Alternative A was calculated from these as follows: $(0.25 \times 28.7) + (0.5 \times 67.1) + (0.25 \times 83.1) = 61.5$. The optimal decision policy, based on expected value, is to produce the pesticide in the existing facility with no partnership (Alternative A).

If risk attitude were a concern, a utility function would have been used to map the endpoint values (profits) to utilities. With these utilities, the expected utility would have been calculated using the identical approach as used with profit.

Probabilistic Sensitivity Analysis

The optimal decision policy indicates the best alternative(s) if the decision had to be made immediately. However, depending on the available time and resources, further analysis is often desirable. For instance, in the Chemcorp example, the expected profits of Alternatives A and B are so close that management might not be comfortable making a decision without further analysis. The analysis would continue with probabilistic sensitivity analysis. Probabilistic sensitivity differs from deterministic sensitivity. Recall that with deterministic sensitivity analysis, each variable is varied from its low to high setting, while keeping the other variables at their nominal setting. When performing a probabilistic sensitivity analysis, these other variables are allowed to vary probabilistically, rather than keeping them at their nominal settings. Each variable examined is adjusted slowly from an extreme low value to an extreme high value, and the points at which the optimal policy changes are recorded.

Figure 11.34 illustrates the results of such an analysis, for *Unit price* in the Chemcorp example. For each point on the curve, the setting of the variable under examination (*Unit price*) is given by the horizontal component, and the expected value (profits) of the optimal decision policy is given by the vertical component. As the variable is varied from its lowest setting to the highest, changes in the optimal policy are indicated on the chart by color changes. Figure 11.34 is sometimes referred to as a *rainbow diagram* because color changes can be used to represent decision changes. In this example, Chemcorp is examining the effect of *Unit price* on the optimal decision policy. Originally

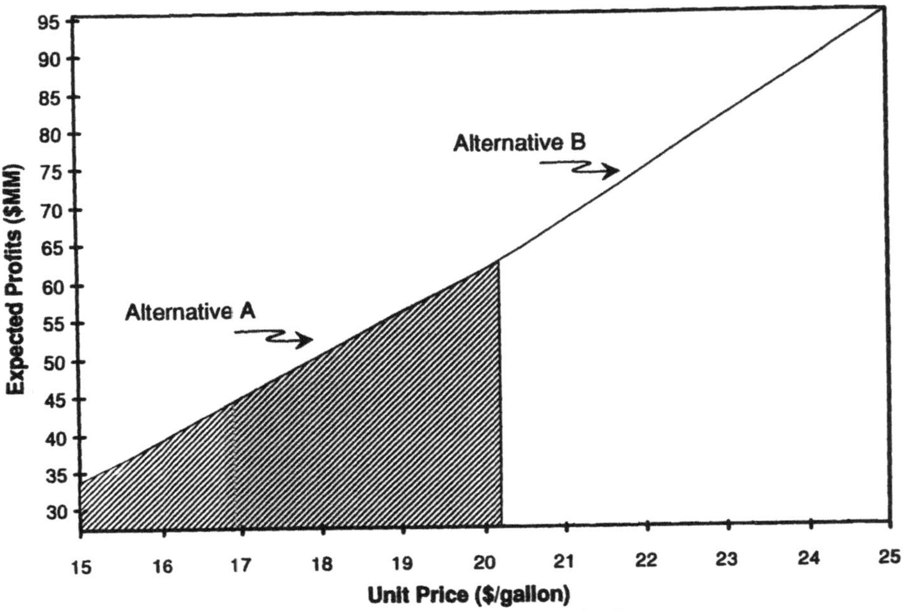

FIGURE 11.34. Probabilistic sensitivity analysis for Chemcorp's unit price.

the price was set at $20. From the figure, we see that the optimal policy changes from Alternative A to Alternative B as the unit price increases from $20 to $21 per gallon. Since the policy is highly sensitive near the unit price selected for the analysis, Chemcorp should reexamine the price assumption.

Sensitivity analysis will provide insight into the most important uncertainties and indicate which ones the analyst may want to assess more carefully. Sensitivity to the discount rate and risk attitude may also be examined to determine if variations make a difference in the optimal decision policy.

Value of Information

Whenever there is uncertainty in a decision, there is an implicit option to gather additional information to help reduce uncertainty, before making a final decision. When is it worthwhile to gather more information, and how much effort should one expend on information gathering? Decision analysis provides a way to calculate how valuable it is to the decision maker to obtain more information. This tells the decision maker whether the benefits from additional information exceed the cost of gathering it. The value of *perfect* information indicates the value of obtaining information from a source that knows for certain the true state of an uncertain quantity or future event. Perfect information is rarely available. However, the value of perfect information provides an upper bound on the amount of resources that should be allocated to gathering information. Imperfect information might include research, forecasts, experiments, models, or opinion polls.

To illustrate the value of perfect information, a schematic of the decision tree for the Chemcorp problem is shown in Figure 11.35. This schematic representation shows the relevant decisions and uncertainties, but not each possible path. The modified decision tree shown in Figure 11.36 indicates how the Chemcorp problem would be modeled to calculate the value of perfect information on the uncertainty *First-year Demand*. Note that the uncertainty is placed before the decision, since with perfect information the demand will be known before the decision is made. However, it remains an uncertainty, because demand is not known prior to deciding whether to buy the information. Value of information is calculated by comparing the expected value of the decision without the information to the expected value of the decision with the information. The expected value with perfect information is the difference between the expected value of the decision tree in Figure 11.36 and the expected value of the decision tree in Figure 11.35. The value of perfect information on first-year demand gives Chemcorp the upper bound on how much they should spend on market research. Similarly, the value of *imperfect* information can be obtained by adding an uncertainty to the tree that represents the uncertainty in the accuracy or completeness of the information. The reader should refer to a decision analysis text (e.g., Clemen 1991) for instruction on calculating value of information.

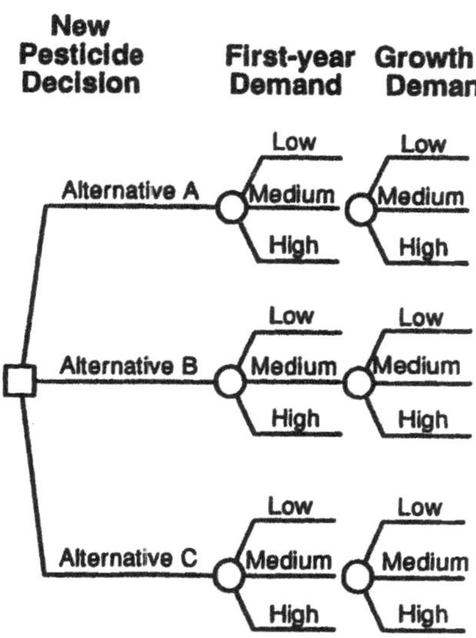

FIGURE 11.35. Schematic representation of Chemcorp decision tree.

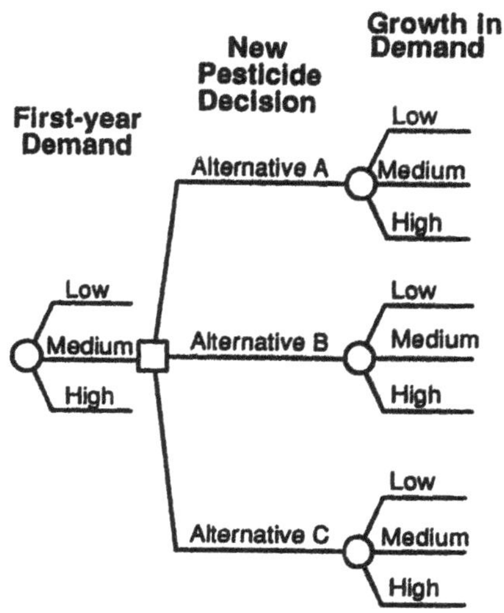

FIGURE 11.36. Tree used to determine value of perfect information on first-year demand.

Value of Control

Suppose we could take an action to force a particular state of an uncertainty. For example, suppose Chemcorp were able to sign contracts for all of its first-year production prior to making its decision. Doing this would eliminate uncertainty about *First-year demand*. The value of control indicates the benefit to the decision maker of being able to control an uncertainty to a desired state. To obtain the value of control on the *First-year demand* in Chemcorp's decision, the tree is restructured as shown in Figure 11.37 and the expected value recomputed. The expected value of the optimal decision with control on plant cost is compared to the expected value of the optimal decision under the original formulation. Notice in Figure 11.37 that control of demand means that it is no longer an uncertainty; demand has just the one state to which it was controlled. In our example, this value might give Chemcorp an estimate of the maximum amount to spend on marketing to promote high demand. The value of control is the maximum amount that should be paid to control a variable to the desired state. Note that the value of control for a variable will always be higher than the value of information because it is worth more to set a variable to a desirable state than to just know the true state without any control.

Next Steps

The evaluation phase provides a direction for action. This action may be to implement the optimal decision policy, to refine the model or the probability distributions, to gather more information in order to reduce the uncertainty,

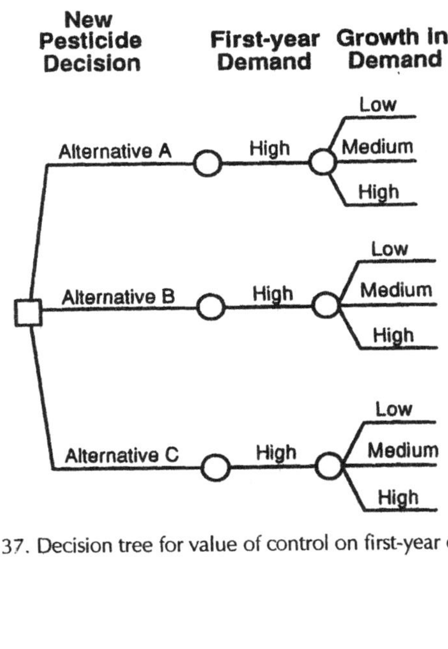

FIGURE 11.37. Decision tree for value of control on first-year demand.

or to attempt to control an uncertain event. In some cases, insights from the analysis will suggest entirely new alternatives. Part or all of the decision analysis cycle may be repeated at this point if refinements are necessary or additional information is gathered. If not, we are ready to communicate the results.

11.3.2.7 Communication Phase

Decision analysis facilitates communication of the decision to diverse audiences, including the technical team and the management team. The outputs of the formulation, modeling, and evaluation phases provide both insights and graphical tools for communicating these insights. For communication to an analysis team, the following items might be used in the presentation:

- A full influence diagram showing all of the variables considered and the relationships between them
- The value function (an equation, spreadsheet, or flowchart)
- Tornado diagrams showing the sensitivity of variables
- A full decision tree
- The probability distributions of the sensitive variables
- Risk profiles of the alternatives
- Diagrams showing the results of detailed sensitivity analyses
- Value of information and value of control.

For a management audience, the emphasis should be on the overall approach, key results, and recommendations. A presentation to management might include:

- A reduced influence diagram indicating only the most important factors
- Tornado diagrams showing the sensitivity of the variables
- A simplified decision tree
- A list of the experts used to assess the probabilistic variables (to indicate that several individuals contributed to the decision)
- Risk profiles of the alternatives
- Recommendations for future action.

In addition to these items, if risk aversion is a factor, the expected utility of each alternative can be presented to show what role risk considerations play in the decision. Also, "what if" scenarios can be addressed, using the results of various sensitivity analyses. For instance, someone may ask "What if demand is extremely low?" or "What if our manufacturing costs are higher than expected?" Decision analysis provides many options for presenting the results.

11.4 CASE STUDY: UNDERGROUND PIPELINE

An Application of Decision Analysis to Chemical Process Safety

11.4.1 Introduction

In this section, we use Case One, the underground pipeline, to illustrate a decision analysis. This case was introduced in Section 5.2; it was also used in the applications provided by Math Programming (Section 9.4) and Voting Methods (Section 6.4).

This case illustrates the decision analysis methodology. We present the key issues and steps, but do not elaborate on the mechanics. In a real problem with this level of complexity, decision analysis software would be used for the analysis. In this section we provide only a high-level view of the approach and solution to this problem. Therefore, there are some points in the case where we do not present sufficient detail for the reader to duplicate the results independently. We strongly recommend that the reader read Section 11.3 before proceeding with the case study.

This application addresses a decision regarding the risks of an aging section of an underground pipeline. The company experienced a leak of a highly flammable, toxic material from such a pipeline at another processing plant. Corporate management, concerned that the same risk may exist at their other facilities, sent a memo to plant managers at all of the facilities expressing its concern and asking the individual plant managers to "determine whether the potential for similar incidents exists at your facility, and if so, to undertake appropriate actions."

The plant manager of this particular facility received the memo from corporate management and consulted with the chief engineer. The chief engineer was concerned about an aging section of underground pipeline that is two miles long and delivers liquid feedstock to the facility. The plant manager agreed that there is a significant risk and asked the chief engineer to conduct an initial decision analysis.

Since the outcomes of the decision may result in costs on the order of a few million dollars, the plant manager assigned a fairly large staff to assist in the analysis over the next two months. The analysis team was led by the chief engineer. The other team members were a chemical engineer who led an earlier risk analysis of the pipeline, a financial analyst who reports to the business director, and a decision analyst from Corporate Planning to provide expertise on the methodology.

The remainder of this section summarizes the decision analysis that was performed. This is done by describing the results of each phase of the decision analysis cycle.

11.4.2 Formulation Phase

In its first meeting with the plant manager, the analysis team defined the decision problem. Specifically, the decision to be addressed was "What action, if any, should the company take to reduce the risk of failure of the aging pipeline?" The analysis team brainstormed to get an initial list of alternatives:

- Do nothing at this point in time
- Replace the current aging pipeline
- Insert automated shutoff valves
- Truck the feedstock to the plant.

The team determined that trucking the feedstock to the plant would cost too much, and therefore eliminated it from consideration. The analysis continued with the remaining three alternatives. The option to replace the pipeline would remove the risk of the current pipeline, but at a significant cost, both in terms of purchase and installation costs and downtime as the pipeline is replaced. Inserting valves just before the suspect section of pipeline would provide an additional measure of security. The valves would automatically stop flow of liquid whenever the pressure in the pipeline decreases precipitously. The valves would minimize damage from a large fault in the pipeline, but would have little effect on a small, extended leak.

The team also brainstormed to develop an initial list of uncertainties:

- The risk of the pipeline failing
- Whether a failure would result in a slow leak or a rupture
- The amount of chemical released due to failure
- The risk of ignition if there is a rupture
- The cost to clean up a chemical release from the pipeline.

The analysis team developed a preliminary list of concerns that affect the value the company will have for the various alternatives and their consequences. This list included:

- Cost of modifying the current pipeline or installing a new pipeline
- Cost of cleaning up a spill
- Disruption in operations
- Worker safety
- Public safety
- Public concern.

Worker safety and public safety were taken off of the list after the chemical engineer discussed the results of the earlier risk analysis, which showed that there is risk to property and the environment, but negligible risk to workers or the public. In addition, the public has not expressed concern about the pipeline. The analysis team settled on the net present value of the total cost (NPV Total Cost) as the value measure for the decision.

Additional important assumptions and simplifications made by the analysis team include the following:

- A 10-year time frame is adequate for the analysis. Costs beyond this horizon are not included in the NPV Total Cost calculation.
- The decision will be made by calculating an expected value for each alternative and choosing the alternative with the highest expected value. Since the net present value of the costs will be a small fraction of the company's total worth, incorporating risk aversion through a utility function is not necessary. (See "Assessing Risk Attitude" in Section 11.3.2.6 for more information.)
- At the end of the 10-year time frame the current pipeline will have no useful value.
- The probability of ignition of the released chemical and the costs due to a fire are the same regardless of how much chemical is released.

An influence diagram was constructed to illustrate the relationships between the various elements of the decision. As shown in Figure 11.38, there is one decision of what to do with the pipeline that is represented by the rectangular node labeled *Pipeline action*. (References in the text to the nodes in

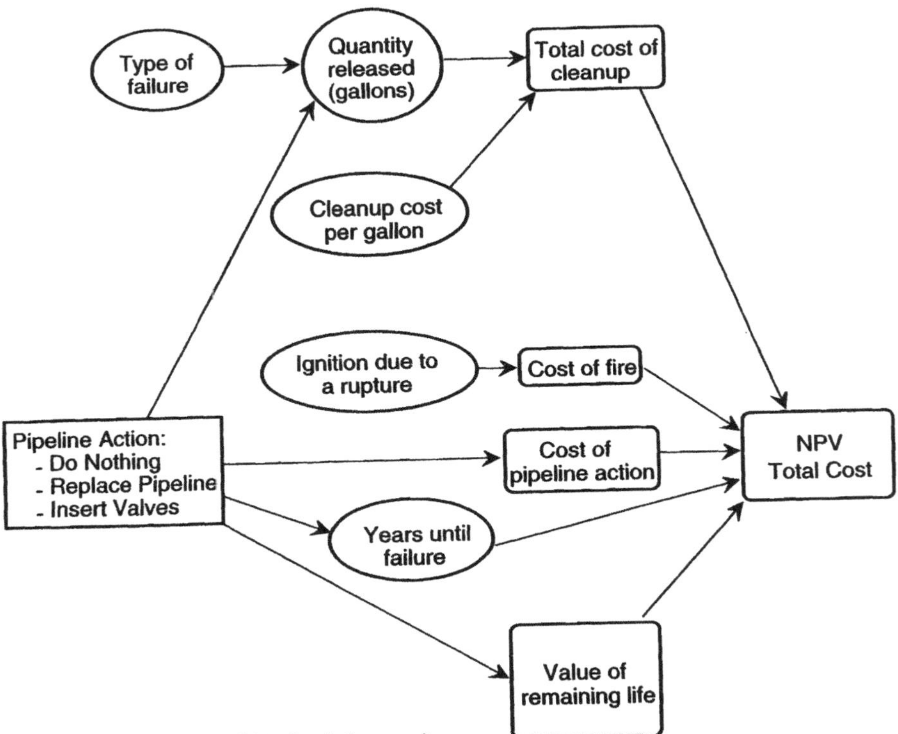

FIGURE 11.38. Underground Pipeline influence diagram.

the diagram are italicized.) *NPV total cost* is placed in a rounded rectangle to the far right in the diagram to represent the overall value measure. The factors contributing to *NPV total cost* are the *Cost of the cleanup*, the *Cost of fire* if one occurs, the *Cost of the pipeline action* taken, and the *Value of the remaining life* of the pipeline. The value of the remaining life accounts for the remaining life of the pipeline at the end of the 10-year horizon. The current pipeline would have no useful value in 10 years, but a new pipeline that replaces it now would still have a usable life remaining. The team put all of the factors contributing to *NPV total cost* in value nodes rather than uncertainty nodes. These factors are either known quantities or can be calculated given other variables in the diagram.

The *Type of failure* node refers to whether the leak is a slow leak or a rupture. The arrow from this node into the *Quantity released* indicates that the size of the leak depends on what type of failure occurs. The *Quantity released* also depends on the *Pipeline action* because the installation of automatic shutoff valves would drastically limit the quantity of chemical released from a pipeline rupture. Also, if a new pipeline were installed, it would have automatic shutoff valves. Arrows go from the *Quantity released* and the *Cleanup cost per gallon* to the *Total cost of cleanup* since this cost can be calculated given the number of gallons released and the cost per gallon released. *Total cost of cleanup* is a deterministic node (i.e., *determined* by quantity relased and cleanup cost per gallon), so it is placed in a rounded rectangle. Due to the frequency of minor spills, the company has cost data for the cleanup cost per gallon that are standard and fairly certain. However, the cost depends on the total number of cubic yards of soil contaminated. The number of cubic yards that would be contaminated by a pipe failure is uncertain. This is reflected in the influence diagram by making *Cleanup cost per gallon* an uncertainty. Alternatively, the team could have modeled this problem by having a separate uncertainty node for the cubic yards contaminated and having the *Cleanup cost per gallon* node represent a known quantity by using a rounded rectangle. There are many different ways to model this aspect of the problem. The team has chosen a simplified approach that will allow them to make a quick first pass through the cycle.[1]

The team has determined that if the failure is a slow leak, there is no chance of a fire because the liquid remains underground. In order to have a fire, there would have to be a rupture. There are two states for the *Ignition due to rupture* event—"ignition" or "no ignition." The arrow from *Ignition due to rupture* to *Cost of fire* represents the dependence of the cost on whether or not the liquid ignites. The cost of a fire is $0 if there is no ignition. The cost of a fire includes

1 For illustrative purposes, we have chosen to simplify this part of the model. In a more realistic model, the number of cubic yards contaminated would depend on which type of failure occurs and on the number of gallons released. The cleanup cost would then be treated as a cost per cubic yard of contamination.

the physical damage and the disruption in operations that would occur due to a power line located above the suspect section of the pipeline. The power line, which provides electricity to the plant, would likely be destroyed in a fire.

The *Cost of the pipeline action* and the *Value of remaining useful life* are represented by value nodes because each of these quantities is fairly certain. Both of these variables depend on which action is taken as indicated by the arrows into them from *Pipeline action*.

The *Years until failure* is relevant to the net present value calculation and depends on the action taken. A constant discount rate of 5% was chosen. The discount rate does not depend on which alternative is chosen and is not uncertain, so it is not shown in the diagram.

The influence diagram helped the analysis team define and scope the problem. It represents the way the team decided to model the real situation, including the assumptions and simplifications they made. Upon completion of the influence diagram, the team presented it to the plant manager for approval. They realized that the plant manager would not approve their analysis before she understood and agreed with the formulation.

11.4.3 Deterministic Phase

Moving on to the deterministic phase, the analysis team constructed the value function. Using total cost as the single outcome, the value function provides the value resulting from any combination of the aspects contributing to the outcome. The value function can be represented by the equations shown below.

$$\text{Total Cost} = C_A + C_C + C_F - V$$

$$C_C = C_G \times Q$$

where

C_A = Cost of the pipeline action
C_C = Cost of cleanup
C_F = Cost of fire
V = Value of the remaining life of the pipeline at the end of the 10-year horizon
C_G = Cleanup cost per gallon
Q = Quantity released due to failure

To get NPV total cost, the cost components and the value of the remaining life of the pipeline are discounted using the 5% discount rate and then combined using the Total Cost equation. The method for calculating net present value is described in Section 2.2.2.

Once the value function was constructed, each variable was assigned a numerical quantity. The inputs that are known for certain or can be estimated

TABLE 11.5
Inputs Treated as Certain

Cost of inserting valves	$80,000
Cost of replacing pipeline	$1.5 MM
Cost of fire	$2.0 MM
Discount rate	5%
Release quantity of rupture with present valves	40,000 gallons
Release quantity of rupture with additional valves	4500 gallons
Value of remaining life of current pipeline at 10 years	$0
Value of remaining life of new pipeline at 10 years	$500,000

with little uncertainty were given a single numerical quantity as shown in Table 11.5. All costs include the financial impacts from disruption in operations that would occur.

The inputs that are uncertain (shown as ovals in the influence diagram) were assigned ranges that represent their 10%, 50%, and 90% cumulative probability points. These ranges are shown in Table 11.6.

For each of the three alternatives considered, NPV total cost was calculated under the nominal scenario, i.e., using the value function with the inputs used at their nominal settings. As an illustration, the NPV Total Cost for Replace Pipeline was calculated as shown in Table 11.7

The results for each alternative are shown in Table 11.8. Do Nothing is called the *nominal* alternative since it is the optimal alternative (lowest cost) under the nominal scenario.

The analysis team performed a deterministic sensitivity analysis in order to establish how variations in the nominal inputs impact the model. They

TABLE 11.6
Inputs Treated as Uncertain

	Low (10%)	*Nominal (50%)*	*High (90%)*
Cleanup cost ($/gallon)	$300	$500	$800
Quantity released by slow leak (gallons)	2000 (gallons)	5000 (gallons)	10,000 (gallons)
Years until failure (current pipeline)	1 (year)	5 (years)	15 (years)
Years until failure (new pipeline)	1 (year)	10 (years)	35 (years)
Type of failure	Rupture	Slow leak	Slow leak
Ignition due to a rupture	Yes	No	No

TABLE 11.7
Example Calculation of NPV Total Cost for "Replace Pipeline" Alternative

		Year	*NPV*
Cost of action	$1,500,000	0	$1,500,000
Cleanup cost per gallon	$500		
Quantity released (gallons)	5000		
Cost of cleanup	$2,500,000	10	$1,534,783
Cost of fire (does not occur in nominal scenario)	0	—	0
Value of remaining life of pipeline	$500,000	10	$306,956
NPV TOTAL COST			$2,727,827

TABLE 11.8
NPV Total Cost for Each Alternative under Nominal Scenario

Do Nothing	$1.96 M
Replace Pipeline	$2.73 M
Insert Valves	$2.04 M

constructed the tornado diagram of Figure 11.39 to display the results of this analysis. (Refer to Section 11.3.2.4 for a discussion on constructing tornado diagrams.) The vertical line down the center of the diagram indicates the value of the nominal alternative. The endpoints of each bar indicate the NPV total cost when the given variable is at its extreme (10% and 90% cumulative probability) settings. The optimal alternative at each extreme is noted at the ends of the bars. For example, the *Years until failure* (of the current pipeline) was varied from 15 to 1. NPV total cost varies from $0 when there are 15 years until failure to $2.35 million when there is 1 year. (Fifteen years is beyond the 10-year time frame of interest.) The best alternative is Do Nothing under both scenarios.

Using the tornado diagram, the team chose the sensitive variables to be modeled probabilistically (i.e., to be assigned probability distributions instead of nominal values) in the next phase of the analysis. The sensitive variables are those for which variations from low to high settings cause a change in the optimal alternative or a large swing in NPV total cost. The change in color in the middle of a bar indicates the optimal alternative has changed with the

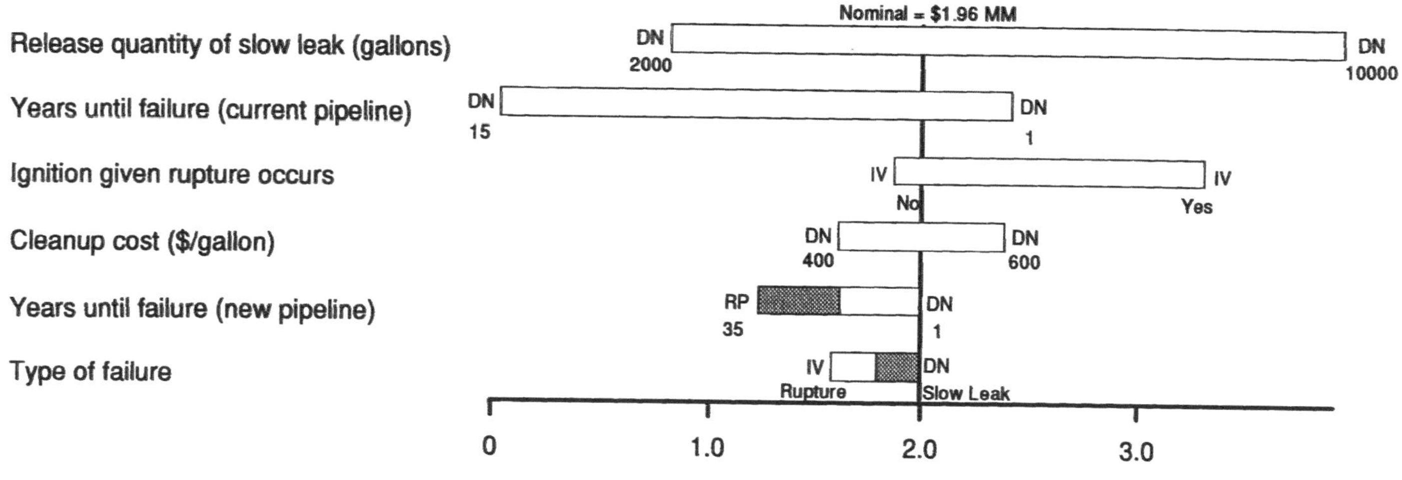

DN : Do Nothing
IV : Insert Valve
RP : Replace Pipe

<u>Notes</u> :
1. The shading change indicates that the optimal alternative differs from the nominal alternative at an extreme value. There is no significance in the location of the shading change; it is placed at the midpoint of the bar.
2. The *Ignition* variable was evaluated given that rupture occurs, since a fire will only occur with a rupture. However, the nominal state has no rupture.

FIGURE 11.39. Tornado diagram for Underground Pipeline case.

change in the value of the variable. As indicated on the tornado diagram, over the specified input ranges, the decision is sensitive to *Years until failure (new pipeline)* and to *Type of failure*. The optimal alternative changes with the swings in these variables. The team decided that all other variables except *Cleanup cost* would be considered sensitive variables based on the large swings in NPV total cost. Since varying *Cleanup cost* had no impact on the best decision and caused little swing in NPV total cost, it was used at its nominal setting in subsequent analysis.

11.4.4 Probabilistic Phase

Experts were found for each of the sensitive variables. The experts assessed probability distributions on these variables. For instance, the distribution shown in Figure 11.40 was assessed for the *Years until failure* of the current pipeline from the mechanical engineer who performed the original failure analysis on the pipeline. All distributions were then discretized and the discrete states and their probabilities were placed on the decision tree shown in Figure 11.41. The probability distributions more accurately express the uncertainty than the 10–50–90% ranges assessed in the deterministic phase.

11.4.5 Evaluation Phase

The team calculated the expected value of each alternative using the tree rollback method, as described in Section 11.3.2.6 in the paragraph headed

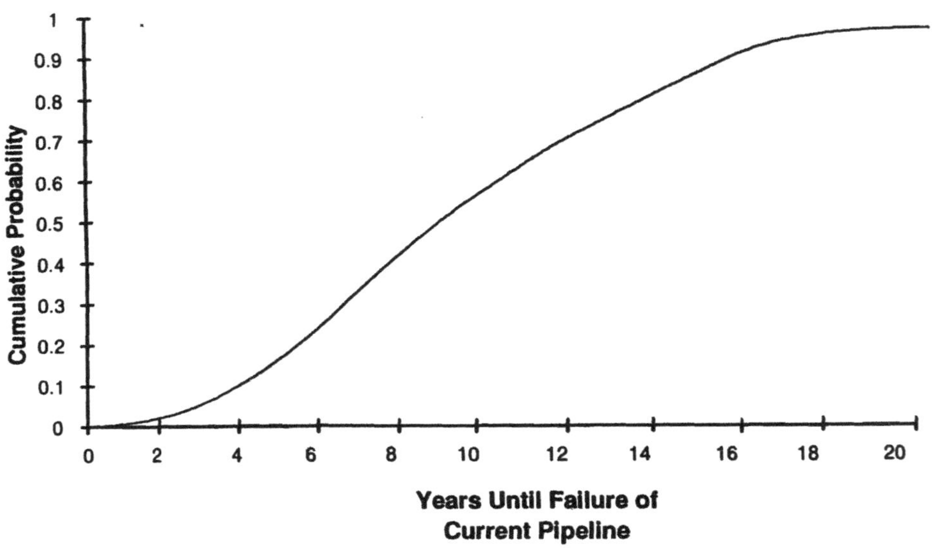

FIGURE 11.40. Assessment of years until failure of current pipeline.

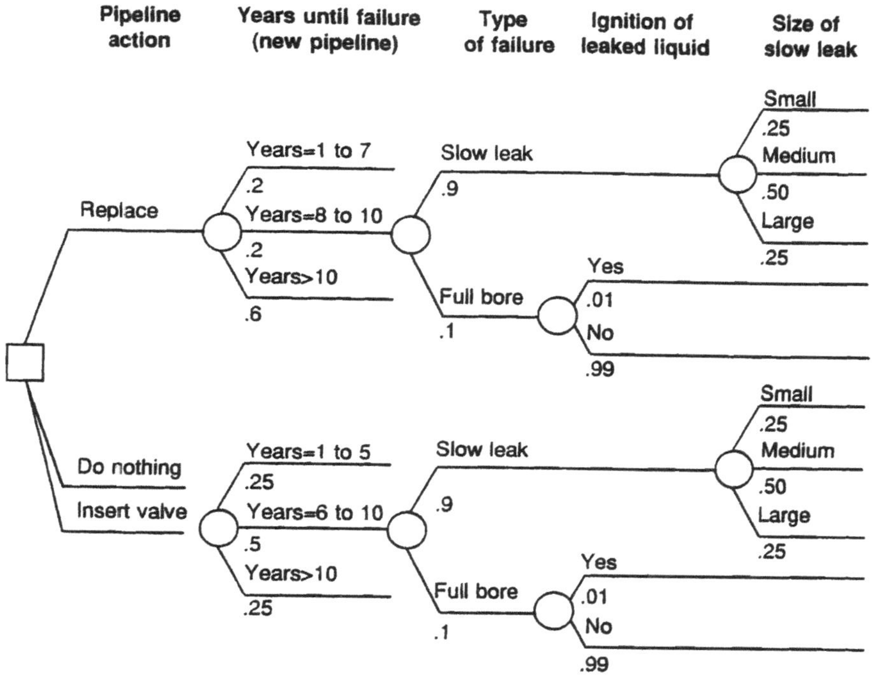

FIGURE 11.41. Decision tree for Underground Pipeline case.

"Rolling Back the Tree." They began by using the value function to calculate the values associated with each of the endpoints of the paths on the tree and then "rolled" these values back to find expected values for each alternative. The results are shown in Figure 11.42. The optimal alternative is to insert the valves. For each alternative, the risk profile, or cumulative probability distri-

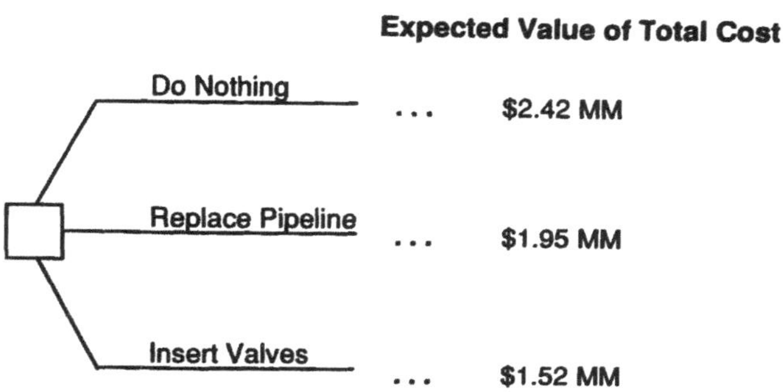

FIGURE 11.42. Expected values of the alternatives.

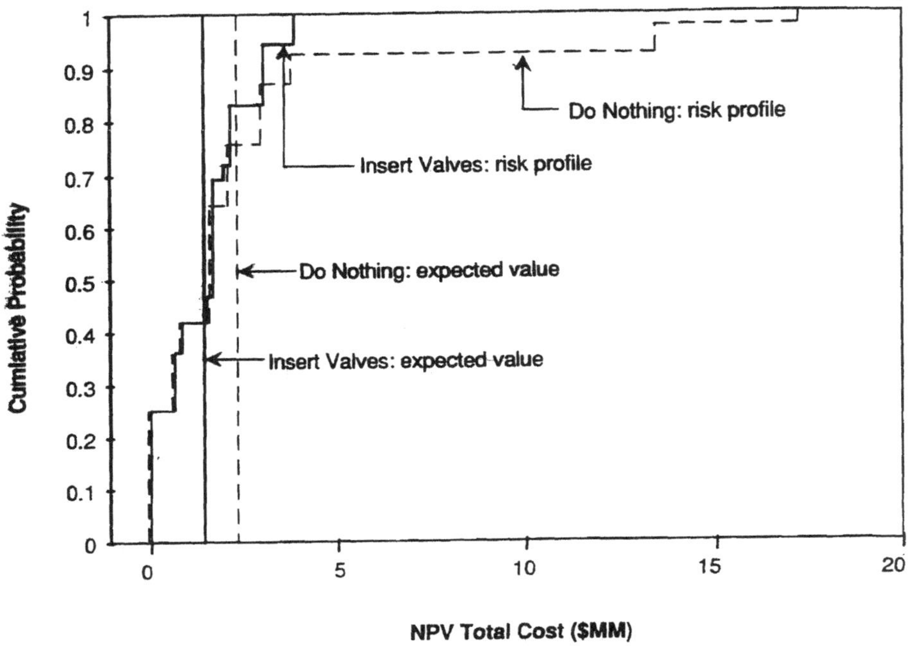

FIGURE 11.43. Risk profiles for Insert Valves and Do Nothing.

bution, was then generated using the values and probabilities at the end of the paths. Even though the team chose to make the decision on an expected value basis, they constructed risk profiles to check for concerns regarding extreme events. Figure 11.43 shows the risk profiles for Insert Valves and Do Nothing. The vertical lines are placed at the expected values. We can see from the diagram that Do Nothing has a higher expected value (NPV total cost). It also has a higher risk, since the highest-cost scenarios are significantly worse than those associated with doing nothing. From Figure 11.43, it is clear that Insert Valves is consistently lower in NPV total cost than Do Nothing. Figure 11.44 shows the risk profiles for Insert Valves and Replace Pipeline. All points except one on the Replace Pipeline profile lie to the right of the Insert Valves profile, indicating that Insert Valves comes very close to dominating Replace Pipeline. The profiles confirm that Iinsert Valves is the best alternative.

The decision analyst on the analysis team pointed out that *Cost of fire* was taken as a single, known input to the analysis. However, the down time and damage caused by fire are somewhat uncertain. The rest of the analysis team agreed, so sensitivity analysis was used to examine the impact of *Cost of fire* on the decision. As shown in Figure 11.45, *Cost of fire* was varied from $500,000 to $5 million, and the optimal policy and its expected value were determined at several points within this range. Varying this input over this range does not

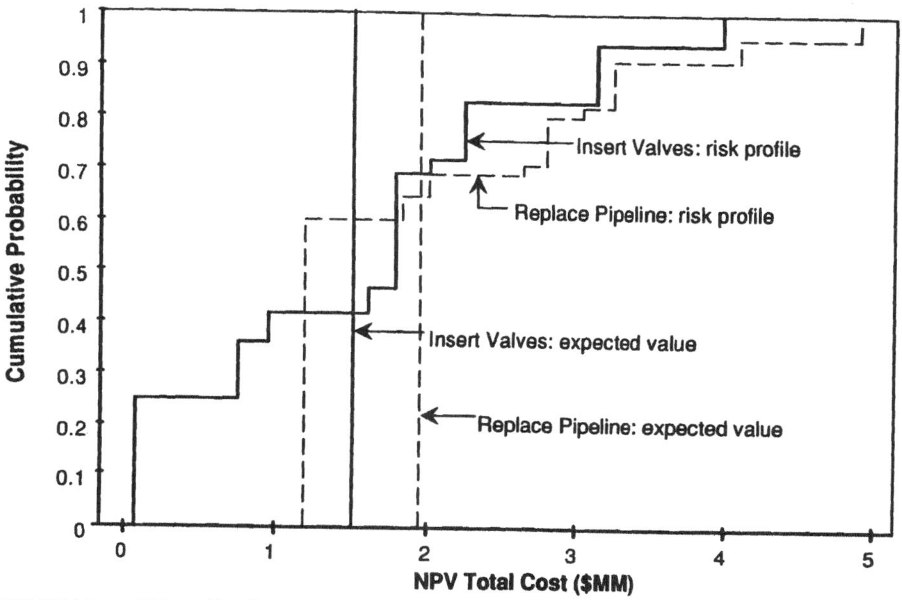

FIGURE 11.44. Risk profiles for Insert Valves and Replace Pipeline.

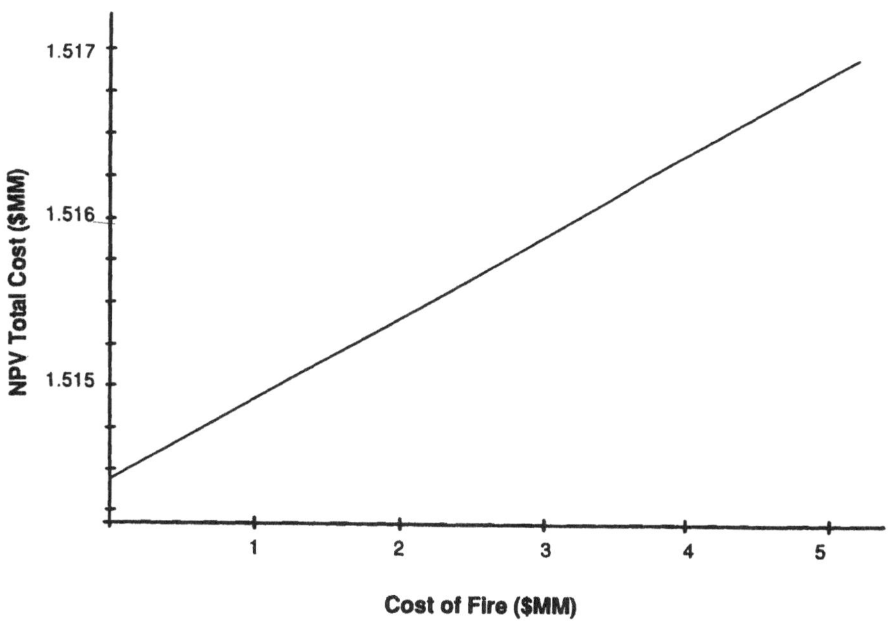

FIGURE 11.45. Sensitivity to *Cost of fire*.

change the optimal policy. It just increases the NPV total cost. Because *Cost of fire* is expected to be well within the range over which this sensitivity analysis was performed, the analysis team decided it was not necessary to change the *Cost of fire* node to an uncertainty.

Sensitivity analysis was also performed for *Cleanup cost*. Initially this input was used at a nominal setting of $500 per gallon. Recall that *Cleanup cost* varies depending on the number of cubic yards contaminated. A rupture would cause the greatest amount of contamination and *Cleanup cost* may be up to $800 per gallon. After examining the results shown in Figure 11.46, the teams saw that a policy change would not occur until *Cleanup cost* exceeds $800 per gallon. They felt comfortable with the nominal setting used in the analysis.

The final steps in the evaluation phase included calculating the value of information and the value of control. The chief engineer suggested conducting a study on the remaining life of the current pipeline but was unsure whether the cost of such a study was worthwhile. To help determine the benefit of such a study, the analysis team calculated the value of perfect information on *Years until failure (current pipeline)*. This value was found by subtracting the expected value of the decision if the remaining life of the current pipeline is known from the expected value of the decision without this information. The original decision tree provided the expected value without the information. The deci-

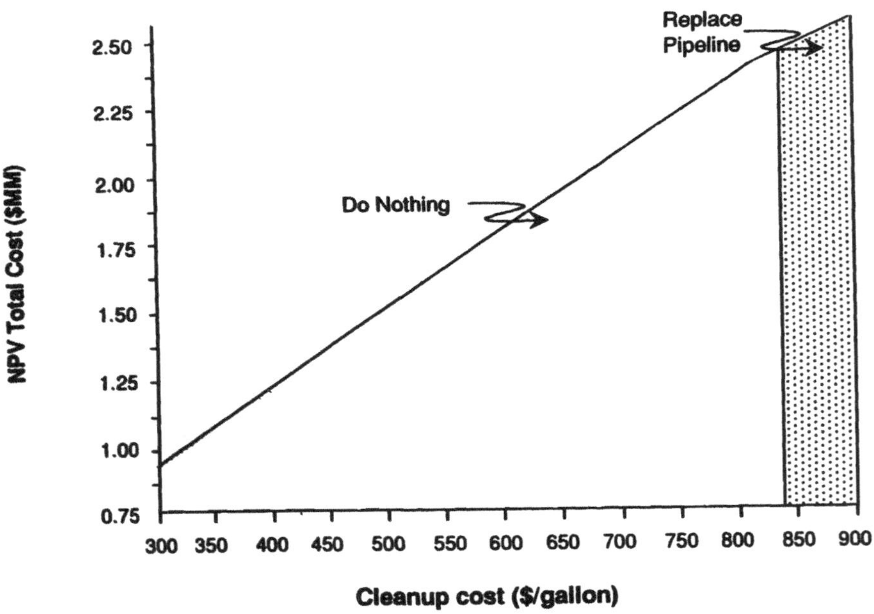

FIGURE 11.46. Sensitivity to *Cleanup cost*.

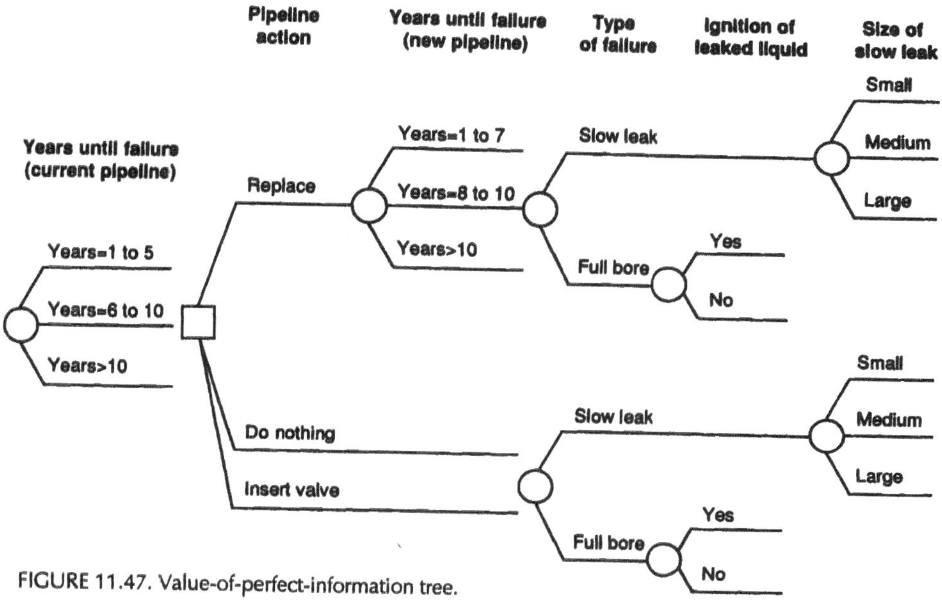

FIGURE 11.47. Value-of-perfect-information tree.

sion tree was restructured as shown in Figure 11.47 and used to calculate the expected value with perfect information. Note that this is the same tree constructed previously, except that the *Years until failure (current pipeline)* node is placed before the decision since, with perfect information, the years to failure would be known before the decision would be made. The value of perfect information on this variable was found to be $111,000. This is an upper bound on how much should be spent studying the remaining life of the current pipeline. Depending on the quality of information available from such a study, if the cost is less than $111,000, it might merit consideration.

The chemical engineer on the analysis team believes that a slow leak could be limited to a release of 250 gallons if a monitoring system were installed. A value of control calculation was performed to determine the maximum amount that should be spent on such a monitoring system. The value of control was calculated by subtracting the expected value of the decision if a slow leak was limited to 250 gallons from the expected value of the decision without having control of the size of the leak. The original tree provided the expected value without control. This tree was restructured as shown in Figure 11.48 and used to calculate the value of control. The value of control on the size of the slow leak was found to be $1.2 million. This is the maximum amount that should be spent on a monitoring system that could limit a slow leak to 250 gallons.

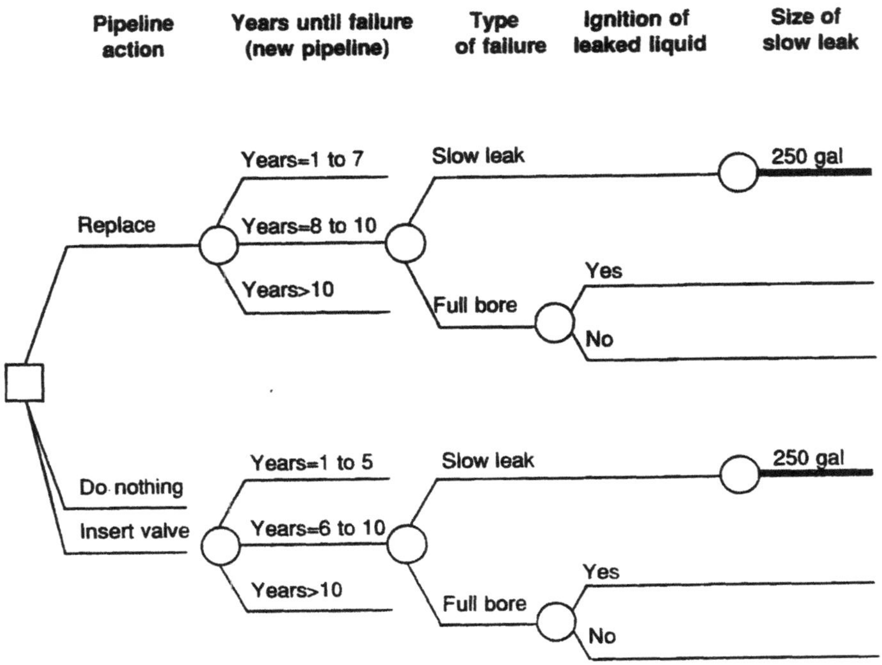

FIGURE 11.48. Value-of-perfect-control tree.

11.4.6 Communication Phase

In the communication phase, the analysis team presented the major factors influencing the decision, the important assumptions, and their conclusions to the plant manager. The analysis team provided the following conclusions:

- If immediate action with no further study is required, the decision should be to insert an additional valve.
- If immediate action is not required, the availability of more information on the number of years to failure of the current pipeline should be explored. $111,000 is the maximum amount that should be spent on such a study.
- A monitoring system that would limit a slow leak to 250 gallons is worth exploring. Such a system has a maximum value of $1.2 million to the decision maker; if it can be procured for less than $1.2 million it should be considered.

Since immediate action was not required, the plant manager arranged for a $20,000 ultrasonic study of the current pipeline. Any additional information gathered in the study will be incorporated into the decision model on the second pass through the cycle. In addition, the chief engineer will review costs for a monitoring system. If installation of a monitoring system is feasible, it may be added to the set of alternatives. Also, uncertainty regarding the effectiveness of the system and the amount of discharge prevented from early warning would be added to the model. Furthermore, the probabilities of the *Type of failure* will be verified and changed if necessary. An updated analysis will then be performed and recommendations made.

11.5 EXTENSIONS OF DECISION ANALYSIS

This section describes extensions of decision analysis in two different areas:

- Multiattribute utility analysis (MUA)
- Group interaction.

11.5.1 Multiattribute Utility Analysis

Multiattribute utility analysis has emerged from decision analysis as an alternative way to deal with decision outcomes that have multiple attributes. Instead of requiring the decision maker to combine multiple attributes into a single value measure, a utility function is created for each attribute. The individual utility functions are then aggregated into a single, multiattribute utility function. Multiattribute utility analysis and decision analysis share the same foundation in utility theory. Chapter 12 describes MUA in more detail.

11.5.2 Group Interaction

Decision analysis facilitates group interaction in some areas, but it was designed primarily for an individual decision maker. Generally it is necessary to involve many individuals in the analysis, but problems may arise when attempting to have a group make the decision. This section addresses these issues and ways of dealing with them in the following areas:

- Formulating the problem
- Procuring inputs and expertise
- Determining values, risk attitude, and discounting structure
- Making high-level strategic decisions.

One way to incorporate the knowledge of many individuals in the group is to have the group formulate the problem and develop the influence diagram. Knowledge of multiple experts may also be incorporated into the analysis, by having an expert most knowledgeable about each uncertain variable encode that variable's probability distribution. Another approach is to use multiple experts to encode a single probability distribution. This is significantly more complex than having a single expert, because a method of aggregating the various opinions is needed. Decision analysis research has attempted to provide guidance in this area (Genest and Zidek 1986). In addition to drawing on the opinions and knowledge of the group, multiple analysts can assist in developing the outcome model. The use of a common language and the explicitness of decision analysis facilitate these types of group interactions.

When it comes to analyzing the decision, one value function and one utility curve are needed. As in any group decision process, the question arises as to how to aggregate the knowledge and preferences of different decision makers into an adequate representation of the group's knowledge and preferences. With multiple decision makers, it is difficult to determine the value function, since different decision makers will have different preferences. For instance, a safety manager and a financial officer may feel differently about the relative importance of improved safety versus corporate financial impacts. Even when decision makers can agree on the various attributes of the outcome that are important, they may disagree on their relative importance. In addition to having different values, individuals may have different attitudes toward the time value of money and different attitudes toward risk taking.

Ideally, we would like the group of decision makers to come up with a common set of preferences for the value function, risk attitude, and discounting structure. One option is to get all decision makers together at one time and try to reach consensus through the exchange of information and the sharing of ideas. Another technique that attempts to get decision-makers' opinions or preferences to converge is the Delphi technique. Instead of open discussion, written feedback is given to an analyst who creates summary statistics and then passes the results back to the participants, enabling them to update their responses in light of feedback from the group. For more information on the Delphi technique, see Linstone and Turoff (1975). If the group cannot get their preferences to converge through group interaction, sensitivity analysis can be performed on each of the quantities about which there is disagreement. This is done by varying each quantity over the range of values desired by the individuals in the group. Sensitivity analysis may indicate that the disputed variable does not impact the results of the analysis over the range of values that have been assigned. In this case, getting the group to agree is not necessary.

The question of whose risk attitude to use in analyzing an organization's decision can also be addressed by identifying a corporate or group risk attitude. In addition to avoiding conflicts in individual decisions, a corporate risk attitude provides consistency in the use of risk in decision making throughout the organization. When an organization expresses a corporate risk attitude (e.g., through a corporate utility function for money), it avoids the need to revisit this issue with each new decision.

For high-level, strategic decisions, the principles of decision analysis have been combined with the principles of structured group problem solving and extended to a group decision-making process called CATALYST. CATALYST was developed by the Electric Power Research Institute to help organizations make decisions about complex, high-level problems in an uncertain environment (EPRI 1992). The CATALYST process is typically a 2- or 3-day, off-site meeting that aims to provide the participants with an understanding of the range of possible events that might occur and to facilitate the selection of a strategy that balances opportunities and risks. Chapter 10 (Payoff Matrix Analysis) provides a description of the CATALYST process.

11.6 IMPLEMENTATION NEEDS

Inputs to a decision analysis include descriptions of the alternatives, information, and preferences. Alternatives are the possible actions and are generated during the formulation stage. Information includes the certain and uncertain inputs that are identified in the course of developing the decision model. Data to characterize the certain inputs, and probabilistic assessments to characterize the uncertain ones are required. The preferences used in a decision analysis include the decision-maker's values (expressed by a value function), risk attitude (expressed by a utility function), and preferred discounting structure for cash flows (expressed by a discount rate).

For an initial decision analysis effort, the analyst should be trained in DA or be guided by a professional with experience in implementing DA. Although one can learn the basic concepts of decision analysis by reading a book, applying the concepts to complex problems involves details and nuances not easily communicated through text. A variety of textbooks, published research articles, and training courses are available. (See Section 11.7.2 for a recommended textbook.)

The time and human resources required to perform a decision analysis can vary, depending upon a number of factors. Needs increase with the complexity, amount of resources at stake, and the uniqueness of the decision to the individual or organization. The resources will also vary depending on the availability of information, experience of the analyst, amount of detail desired and the availability of decision analysis software. Repetitive decisions require minimal resources once a model has been developed, but requirements for unique decisions will depend on the complexity of the problem and the skills of the analyst. A typical decision analysis can range from a one-person effort for a few days to a three- or four-person effort over several weeks. If a high degree of refinement is desired and several passes through the cycle are needed, the DA may take up to several months. However, if time is a constraint, a quick initial pass through the decision analysis cycle often provides tremendous insight, especially in the formulation phase. Some analyses provide sufficient insights to be able to reach a decision after just one rough pass through the cycle. Decision analysis adds some benefit to the decision-making process at almost all levels of available resources.

Decision analysis software is available to decrease computation time, to aid in modeling, and to facilitate the generation of graphics, but software is not necessary in order to perform an analysis. In general, DA software is required when the number of uncertainties is large, resulting in a decision tree and computational burden that are unmanageable without software. The number of endpoints on a tree grows exponentially with the number of variables treated as uncertainties. Software is also necessary if a thorough sensitivity analysis is desired. Depending on the size and complexity of the problem, spreadsheets are often used alone or along with DA software. Simple problems can be done using paper, pencil, and a calculator. Several DA software packages are listed in the software appendix at the end of this book.

11.7 SUMMARY

11.7.1 Strengths and Weaknesses of Decision Analysis

Decision analysis provides a rigorous, structured process and a number of useful tools for decision making. The structured process and its axiomatic basis ensure logically sound decisions that can be documented and explained. As drawbacks, decision analysis is less suited to situations where there are multiple decision makers or when resources are greatly limited. Decision analysis requires that "soft" issues be quantified precisely and that multiple attributes of outcomes be combined into a single value measure. This places an information burden that does not occur when using less quantitative decision aids.

Decision analysis's unique features can be extremely useful in certain decision situations. For instance, the ability to calculate value of information is important in decisions where available studies or tests require substantial resources. In addition, the explicit representation of risk attitude via a utility function is a strength when many important decisions are being made and a consistent approach towards risk taking by the organization is desired. Decision analysis also handles multiple or sequential decisions, allowing the analyst to look at how downstream decisions will be impacted by earlier decisions and by events that will occur before later decisions are made.

One of its major strengths is that decision analysis provides tools to address uncertain and complex relationships among aspects of the decision. The systematic decision analysis process helps the decision maker consider all relevant factors, data, and value judgments. Another strength is that decision analysis guides the decision maker in thinking about the interactions between various factors in the problem. Through the DA process, the decision maker gains insights, not simply an indication of the best alternatives. The results help the decision maker determine specific issues on which to focus.

An additional strength of decision analysis is that it facilitates communication both among the participants in the decision analysis and with their audiences. Influence diagrams aid in formulating and communicating the structure of the problem, both during and after the analysis. Influence diagrams in and of themselves have proven to be valuable in structuring problems. Other graphic devices such as decision trees, cumulative probability distributions, and tornado diagrams are easily generated and aid in communication. In addition, decision analysis provides a precise, common "language." Not only is it easy for people to communicate about a complex decision by focusing on alternatives, uncertainties, and values, but ambiguity can be avoided by using mathematics and probability theory to express important aspects of the decision.

One weakness of decision analysis is that it may not adapt well to situations that involve multiple decision makers. Value judgments may vary with individuals, but the analysis requires that a single value function, utility function, and discounting structure be used. On the other hand, the formal structure of DA can help those who disagree on the best alternative to understand the dimensions on which they disagree. It will become apparent through a decision analysis whether the source of disagreement is in their understanding of the outcomes, assessment of probabilities for outcomes, values, risk attitudes, or a combination of the above. Similarly, when there is disagreement on a numerical input, sensitivity analysis may show that the decision is not sensitive to the input, making it unnecessary to agree on a precise numerical quantity for this input.

A second weakness of decision analysis is the potential difficulty in obtaining probability distributions and value measures. The process of assessing probability distributions is not familiar to most people. It may also be difficult to get access to decision makers to assess their values and risk preferences. When there are several outcomes of concern, the decision maker needs to express judgments about the relative importance of these outcomes. He may be reluctant to state these tradeoffs explicitly. Likewise, some people or organizations may hesitate to make their risk attitudes explicit.

Decision analysis may be less suited to major decisions that also have tight constraints on time, staffing, and other resources, particularly if the organization does not have the analytic capabilities already in place at the time of the decision. Those performing the analysis must first take time to learn DA or must hire a professional to help in the analysis, unless the decision is a repetitive one and a model has previously been developed. It should be kept in mind, however, that the depth of the analysis can be adapted to the time and resources available. Nevertheless, if the decision is complex and/or the magnitude of the possible outcomes large, a sizable commitment of time and/or money may be required. However, the training and initial time spent on learning decision analysis can be viewed as an investment. A sizable commitment of resources is often cost-effective for large, complex decisions. Each additional application of decision analysis becomes easier, and systems can be implemented to analyze recurring decisions more efficiently.

11.7.2 Suggestions for Further Information/Study

For further information on decision analysis, several textbooks are available. *Making Hard Decisions* by Clemen (1991) is recommended for a good explanation of basic concepts and illustrative examples. Other textbooks include *Applied Decision Analysis* (Bunn 1984), *Decision Making Under Uncertainty* (Holloway 1979), and *Decision Analysis and Behavioral Research* (von Winterfeldt and Edwards 1986). Covello (1987) provides additional reading on the strengths and weaknesses of DA and a comparison to cost–benefit analysis. Winkler (1990) provides a more technical comparison of DA to the analytic hierarchy process.

Prior to performing a decision analysis, it is important to be exposed to applications of the concepts. Unless training is obtained through a course or a seminar, DA should initially be applied to a small-scale decision. To speed up the development of a solid foundation in the basic principles, a seminar or course in decision analysis is recommended, along with the use of a textbook as a guide. Many decision analysis software packages provide tutorials with the program, but none of these packages is designed to enable the analyst to use the technique without prior knowledge of the fundamentals.

11.8 REFERENCES

Bohn, M. P. (1992). "Decision Making Under Uncertainty: An Investigation into the Application of Formal Decision-Making Methods to Safety Issue Decisions" (NUREG/CR-5906). Division of Safety Issues Research, Office of Nuclear Regulatory Research, U.S. Nuclear Regulatory Commission.

Boyd, D. W., D. Cohan, D. S. Wilson, E. V. Niemeyer, and W. C. Trench (1985). Managing PCB Risks from Transformer Fires. Proceedings of the American Power Conference, Electric Power Research Institute.

Braunstein, T. J. (1983). Decision Analysis of a Petrochemical Expansion Study, pp. 269–284 in Vol. I of R. A. Howard and J. E. Matheson (eds.) *Readings on the Principles and Applications of Decision Analysis* (2 vol.s). Menlo Park, CA: Strategic Decisions Group (1989).

Brown, R. V., and J. W. Ulvila (1988). Does a Reactor Need a Safety Backfit? Case Study on Communicating Decision and Risk Analysis Information to Managers," *Risk Analysis* 8(2):271–282.

Bunn, D. W. (1984). *Applied Decision Analysis*. New York: McGraw-Hill.

Clemen, R. T. (1991). *Making Hard Decisions: An Introduction to Decision Analysis*. Boston: PWS-Kent.

Covello, V. T. (1987). Decision Analysis and Risk Management Decision Making: Issues and Methods. *Risk Analysis* 7(2):131–139.

EPRI (1992). "CATALYST—A Group Process for Strategic Decision Making: Facilitator's Guidebook." Prepared by Applied Decision Analysis and The Brattle Group. TR-100583, Research Project 2631, Final Report. Palo Alto, CA: Electric Power Research Institute (September).

Genest, C., and J. V. Zidek (1986). Combining Probability Distributions: A Critique and an Annotated Bibliography. *Statistical Science* 1(1):114–148.

Hanna, R. C. (1979). Apply Decision Theory to Hazard Evaluation. *Hydrocarbon Processing* (December):133–150.

Holloway, C. A. (1979). *Decision Making Under Uncertainty: Models and Choices*. Englewood Cliffs, NJ: Prentice Hall.

Howard, R. A. (1968). The Foundations of Decision Analysis. *IEEE Transactions on Systems Science and Cybernetics* SSC 4(3):211–219.

Jensen, D. D., A. E. Tome, and W. P. Darby (1989). Applying Decision Analysis to Determine the Effect of Smoke Detector Laws on Fire Loss in the United States. *Risk Analysis* 9(1):79–89.

Keeney, R., and H. Raiffa (1975). *Decisions with Multiple Objectives, Preferences and Value Tradeoffs*. New York: Wiley and Sons.

Linstone, H. A., and M. Turoff (eds.) (1975). *The Delphi Method: Techniques and Applications*. Reading, MA: Addison–Wesley.

Luce, R. D., and H. Raiffa (1957). *Games and Decisions*. New York: Wiley.

Mauney, D. A. (1993). Economic Optimization of Multiple Component Replacement/Inspection in the Power System Environment. *Reliability and Risk in Pressure Vessels and Piping*, Vol. 251, ASME.

Merkhofer, M. W. (1987a). *Decision Science and Social Risk Management: A Comparative Evaluation of Cost-Benefit Analysis, Decision Analysis, and Other Formal Decision-Aiding Approaches*. Dordrecht, Holland: D. Reidel (Boston: Kluwer Academic Publishers Group).

Merkhofer, M. W. (1987b). Quantifying Judgmental Uncertainty: Methodology, Experiences, and Insights. *IEEE Transactions on Systems, Man, and Cybernetics* SMC-17(5):741–752.

Miller, A. C., and T. R. Rice (1983). Discrete Approximations of Probability Distributions. *Management Science* 29(3):352–362.

Newendorp, P. D. (1975). *Decision Analysis for Petroleum Exploration*. Tulsa, OK: PenWell Publishing Company.

Noonan, F., and C. A. Vidich (1992). Decision Analysis for Utilizing Hazardous Waste Site Assessments in Real Estate Acquisition. *Risk Analysis* 12(2):245–251.

North, W. (1968). A Tutorial Introduction to Decision Theory. *IEEE Transactions on Systems, Science, and Cybernetics* SSC 4:211–219.

Owen, D.L. (1981). The Design of Hazardous Products. *IEEE Transactions on Systems, Man, and Cybernetics* SMC–11(10).

Paté-Cornell, E. (1985). Fire Risks in Oil Refineries: Economic Analysis of Camera Monitoring. *Risk Analysis* 5(4):277–288.

Pratt, J., H. Raiffa, and R. Schlaifer (1965). *Introduction to Statistical Decision Theory*. New York: McGraw-Hill.

Savage, L. J. (1954). *The Foundations of Statistics*. New York: Wiley.

von Winterfeldt, D., and W. Edwards (1986). *Decision Analysis and Behavioral Research.* Cambridge: Cambridge University Press.
Winkler, R. L. (1990). Decision Modeling and Rational Choice: AHP and Utility Theory. *Management Science* 36(3):247–248.

12

MULTIATTRIBUTE UTILITY ANALYSIS

The real power and beauty of mathematics isn't in the way it can describe reality—it's in the way it can be used to improve reality.
—Michael Schrage, Los Angeles Times, 1991

12.1 PURPOSE OF CHAPTER

Chemical process safety decisions often involve difficult value tradeoffs among competing objectives or concerns. In choosing from among available alternatives, the decision maker expresses value judgments, either explicitly or implicitly. How should a corporation trade off the welfare of its employees against the viability of the company? How much weight should it place on public safety versus the safety of the company's employees? How should it choose between two alternatives that simultaneously vary on a number of important objectives?

Most of the decision aids discussed up to now have simplified such problems in one of several ways: by not *explicitly* addressing the various objectives (as with voting methods), by focusing on a *single* objective or collapsing objectives onto one objective (as with decision analysis), or by using a simplified scheme for weighting objectives and/or scoring alternatives on each objective (as with weighted scoring methods). These simplifications can streamline the analysis, but may result in decisions that do not completely reflect the preferences and concerns of the decision maker.

Multiattribute utility analysis (MUA) builds upon the principles and methods of decision analysis (DA) to address explicitly the value judgments associated with multiple, competing objectives. The fundamental problem addressed by MUA is how to determine the overall value of an alternative, and compare this alternative with others, based on how well the alternatives perform on multiple objectives. The process used for an MUA ensures that the multiple objectives are defined appropriately, that the assessment of the value of each alternative on each objective is accurate, that the uncertainties are factored in correctly, and that the computation of the overall value of each alternative based on its value for each objective is done properly. The axioms of decision analysis apply just as well to decisions with multiple objectives as they do to decisions with a single objective. However, the decision analysis axioms do not tell you when, for example, a simple "weight and add" function is appropriate to aggregate different objectives. The decision analysis axioms also do not tell you how to quantify a decision-maker's tradeoffs between different objectives. Multiattribute utility analysis extends decision analysis by filling in these gaps.

This chapter describes the inputs required to develop an MUA, but does not address in detail the techniques for eliciting decision-maker preferences. There is an extensive body of literature on this subject (referenced later in this chapter), and to apply these techniques usually requires specialized training and/or an experienced facilitator.

In what follows, we provide an overview of MUA and a discussion of the decisions and situations in which this decision aid is most appropriate. A detailed explanation of the aid is then provided, including key concepts, the steps in the analysis, and the types of results obtained. Multiattribute utility analysis is illustrated using a chemical process safety example. We describe similarities and differences between MUA and other decision aids such as weighted scoring methods, and discuss extensions to and variations of MUA, including applications to social decisions. We discuss requirements for implementing an MUA, including data needs and staffing requirements. Finally, this chapter concludes with a summary of the strengths and weaknesses of MUA and a guide to readers who want to learn more about this decision aid.

12.2 OVERVIEW OF MULTIATTRIBUTE UTILITY ANALYSIS

12.2.1 Introduction to Multiattribute Utility Analysis

Decision analysis and multiattribute utility analysis both provide a formal process for structuring and quantifying complex decisions that involve uncertainty. As does decision analysis, MUA incorporates what you can do (alternatives), what you know (uncertainties), and what you care about (values), into a quantitative analysis. Multiattribute utility analysis focuses on developing a *value model* when there are multiple concerns. In many cases, the associated objectives are often difficult to measure or quantify explicitly. Multiattribute utility analysis provides the logical and mathematical foundation for defining measurement scales in general. In addition, MUA explains how to mathematically combine multiple measures of value into a single measure, and thereby make possible the direct comparison of different alternatives. In this way, MUA ensures that there is a logical, rigorous, and defensible basis for the resulting decision for situations involving multiple stakeholders and competing value tradeoffs. A drawback of MUA is that the method generally requires more resources than other approaches, including the time and expertise necessary to perform such an analysis. Often, a major benefit from conducting an MUA is to make explicit the several underlying motivations for a decision, and to create detailed, objective criteria (in the form of measurement scales) for achieving success with respect to different goals of the decision.

12.2.2 History of Multiattribute Utility Analysis

Multiattribute utility analysis was developed relatively recently, in comparison with decision analysis. The theory behind decision analysis spans two centuries, beginning with the work of Pascal and Bernoulli in the 1700s. Section 11.2.2 of Chapter 11 describes the history of decision analysis and the utility theory that underlies these techniques. During the 1920s, the logician Frank Ramsey developed the mathematical concepts that would become the foundation of decision analysis. In the 1940s, von Neumann and Morgenstern developed axioms for utility theory, which show how decision-maker preferences may be explicitly combined with probabilistic uncertainty in a *utility function*. Some key results from this theory are that:

- A decision-maker's preferences can be represented by a mathematical utility function that associates outcomes with numbers called *utilities*
- The utility function incorporates the decision-maker's attitude toward uncertainty
- Preferences for complex alternatives can be calculated using the *expected utility* of each alternative, meaning the sum of utilities associated with individual outcomes multiplied by their probability of occurrence.

No refinement to the theory is required as a justification for MUA. However, utility theory (which basically forms the axioms for decision analysis) does not indicate how a utility function based on multiple values may be built up from simpler single-value components. Multiattribute utility analysis was developed in the 1960s to fill this gap and to make decision analysis more readily applicable to real-world, multiple-valued problems.

In a brief history of MUA by von Winterfeldt and Edwards (1986), Raiffa is credited with many of the key ideas behind the MUA approach (see Raiffa 1969). Keeney and Raiffa's text on MUA (1976) is considered a seminal work on practical approaches to assessing utilities and applying MUA techniques. Several early applications of MUA are described in this book, including decisions on airport development, air pollution control, instructional programs, fire department operations, corporate planning, computer system selection, and nuclear plant siting and licensing. Brief descriptions of other applications address aircraft safety, a hospital blood bank, sewage sludge disposal, job selection, transportion of hazardous substances, treatment for cleft palate, water quality, foreign policy, and forest pest management (Keeney and Raiffa 1976).

A prominent recent application of MUA addresses the Department of Energy's recommendation of a site for disposal of nuclear waste. Articles by Merkhofer and Keeney (1987) and Gregory and Lichtenstein (1987) provide a detailed description of this case, which is also discussed later in this chapter.

Other published applications of MUA address nuclear plant safety (Bohn 1992), ambient air quality standards (Keeney, Sarin, and Winkler 1984), ranking of hazardous waste cleanup sites (Merkhofer, Cotton, and Longo 1988), health programs (Torrance, Boyle, and Horwood 1982), offshore oil drilling (von Winterfeldt 1987), and ranking of hydroelectric power projects (Wenstop and Carlsen 1988). Because MUA provides so many insights into the decision-making process, some authors have come to see MUA as the natural starting point for applied decision analysis (Keeney 1992).

12.2.3 Decision Situations Most Appropriate for Multiattribute Utility Analysis

Multiattribute utility analysis is designed explicitly to treat problems that involve difficult value tradeoffs. While MUA can be used to analyze problems with both multiple values and complex uncertainties, the principal use of MUA is in value modeling, more than in uncertainty analysis. Multiattribute utility analysis is also useful for decisions involving difficult-to-quantify objectives. Outcomes such as ecological impacts, health impacts (e.g., acute vs. chronic), and social impacts may not have an obvious benefit measure for many applications. Multiattribute utility analysis indicates how to construct appropriate measurement scales (see Keeney 1992). Finally, MUA is well-suited to problems that involve multiple stakeholders or shared decision responsibility (Merkhofer 1987). Competing perspectives can be combined through a single MUA. Technical issues regarding the degree to which objectives can be attained can be separated from value-laden issues of the tradeoffs between such benefits and their associated costs.

12.3 EXPLANATION OF MULTIATTRIBUTE UTILITY ANALYSIS

12.3.1 Key Concepts of Multiattribute Utility Analysis

The building blocks of a multiattribute utility analysis are the same as those described for a single-attribute decision analysis: alternatives, uncertainties, and values. Alternatives are the possible actions that the decision maker can take. Uncertainties are aspects that influence the decision, but which are known only imperfectly. Values are quantitative measures that the decision maker associates with each possible outcome of the decision; the outcomes will depend upon which alternative is chosen and on what actually occurs or is achieved as a result of that action. These outcomes are expressed in terms of an attribute (such as health impacts) and an associated measurement scale. Referring again to the decision analysis cycle described in Section 11.3.2 of the preceding chapter, MUA is primarily used to develop a value model in the deterministic phase of the decision analysis, as shown in Figure 12.1.

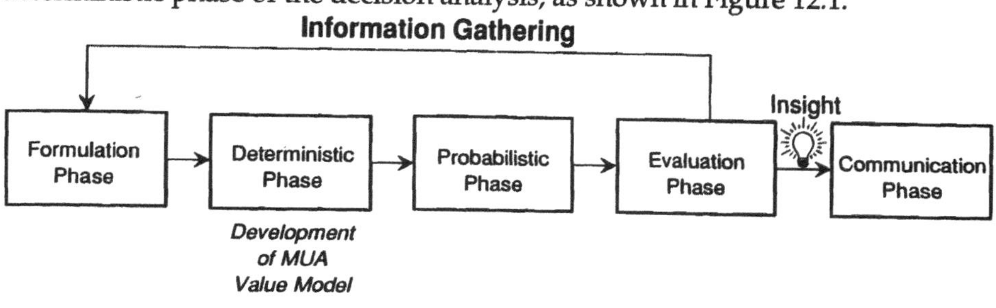

FIGURE 12.1. Role of MUA within the decision analysis cycle.

The remainder of this section focuses on the development of the value model, captured through a multiattribute utility function. The reader is directed to the preceding chapter for details on the other steps of the decision analysis not covered here. In the discussion that follows we use the commuter's decision example from Section 11.3 to illustrate the MUA approach.

Figure 12.2 depicts the key steps required to develop a multiattribute utility function:

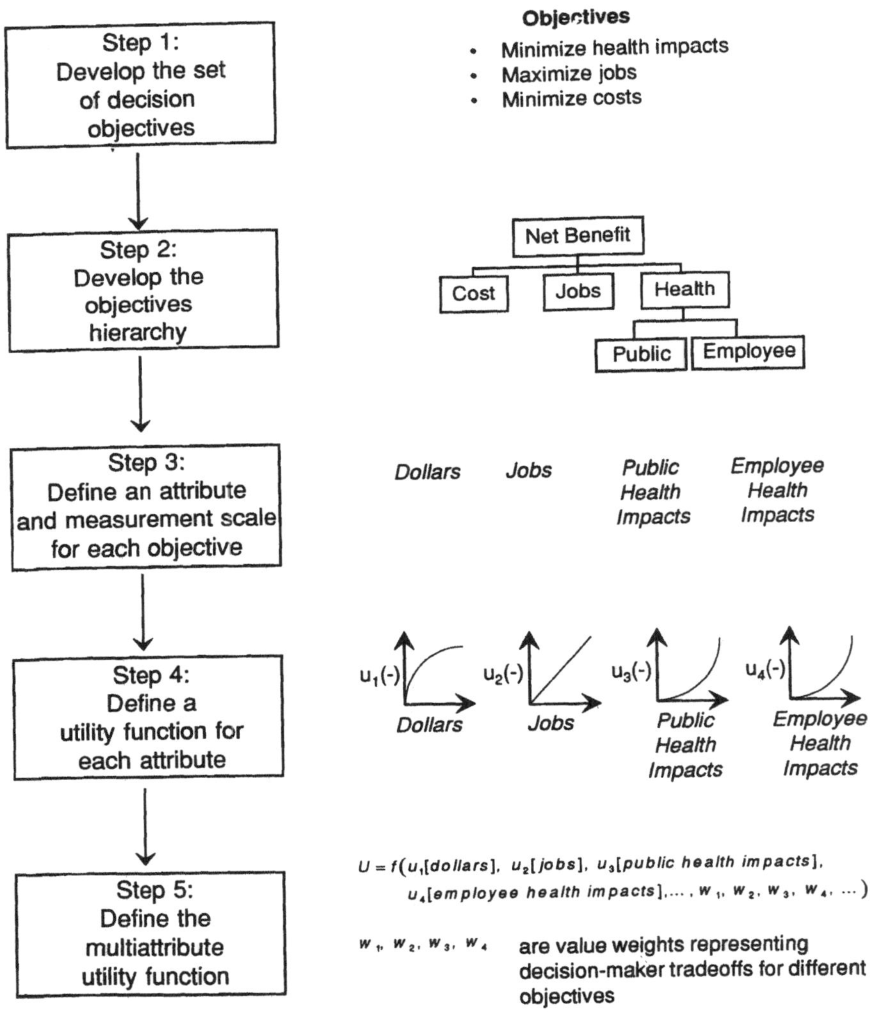

FIGURE 12.2. Key steps in defining the multiattribute utility function.

- *Step 1:* Develop the set of decision objectives.
- *Step 2:* Develop an objectives hierarchy. The objectives hierarchy (or value hierarchy) organizes the set of objectives defined in Step 1 in a logical structure. From the theory of MUA, when the objectives hierarchy has certain "independence" properties, the multiattribute utility function will be a simple linear function of single-attribute utilities.
- *Step 3:* Define an attribute and measurement scale for each lowest-level objective. For each objective, an attribute must be identified and a measurement scale given that allows the measurement of performance on that objective.
- *Step 4:* Define a utility function for each attribute. A single-attribute utility function represents the decision-maker's preferences and risk attitude for a single objective, such as health impacts.
- *Step 5:* Define the multiattribute utility function. In Step 2, the form (linear, multiplicative, etc.) of the multiattribute utility function will be determined by the type of value hierarchy developed. In this step, the multiattribute function is completed by assessing required parameters, called *value weights,* which represent decision-maker tradeoffs for different objectives.

Step 1. Develop the set of decision objectives.
The set of objectives should represent what may be achieved by making the decision: *Minimize health impacts, Maximize revenues, Minimize environmental impacts, Maximize public acceptance,* etc. An objective has a "direction" of better or worse, and an "object" of value, such as health or environmental impacts. In addition, the objectives are defined relative to the decision context. *Minimize health impacts* may refer only to choices such as process safety options, but not, for example, to whether an employee goes rock climbing on weekends.

The set of objectives should be both complete and nonredundant. All objectives relevant to the decision and, typically, decision stakeholders, should be included for the MUA to be defensible. However, if an objective makes no difference in comparing alternatives, it may be dropped. For example, if all relevant decision alternatives provide the same level of environmental protection, then that objective may be dropped from the analysis. The set of objectives should also not count the same objective twice, for instance by first counting employee injuries and then employee health impacts. Such double-counting will result in miscalculation of utilities.

The commuter's decision described in the last chapter (Section 11.3.1) involved a choice between driving to work and riding a train to work, given uncertainty about the freeway congestion that might be encountered. This example involved a relatively simple set of objectives. The primary objective was to *Maximize* [the commuter's] *overall satisfaction.* This objective was broken into two subobjectives: *Minimize commute time* and *Minimize commute cost.* The

Maximize overall satisfaction objective must be broken down because there is no obvious attribute by which to measure its performance.

Step 2. Develop the objectives hierarchy.
The second step is to develop an objectives hierarchy for the decision. An *objectives hierarchy* structures the list of objectives to indicate how more specific objectives contribute toward broader objectives. This helps to ensure consideration of all objectives, and to identify an appropriate form for the multiattribute utility function. This is one point at which the MUA theory is important. If the objectives hierarchy is structured properly, the multiattribute utility function will have a simple form, such as linear or multiplicative. One cannot merely assume a simple form for the aggregation function—it depends upon a properly structured objectives hierarchy. A property that is often used is called "preferential independence," and is explained in detail in MUA texts (e.g., von Winterfeldt and Edwards 1986). Figure 12.3 provides the objectives hierarchy for the commuter's decision.

Step 3. Define an attribute for each lowest-level objective.
The third step in assessing the utility function is to define an attribute and measurement scale for each of the objectives that appear at the lowest level of the hierarchy. An attribute is simply the indicator used to measure the degree to which an objective is satisfied. For example, health impacts may be measured by number of acute injuries, individual risk of fatality, societal risk of fatality over 15 years, etc. Each objective will have its own attribute, although some of these may employ the same units of measure. For example, construction cost and operating cost are different attributes that are both expressed in dollars. For some objectives, no obvious attribute exists by which the objective can be measured, and in this situation, it is necessary for the analyst to develop a "constructed scale" especially for the decision at hand. Examples of such objectives are *Corporate image*, *Public concern*, and *Customer value*. (A sample constructed scale is presented in Table 12.5 in Section 12.3.2.3.)

For the commuter's decision, the following attributes may be used to represent the two lowest-level objectives:

OBJECTIVE	ATTRIBUTE
Minimize commute time	Commute time (minutes)
Minimize commute cost	Commute cost (dollars)

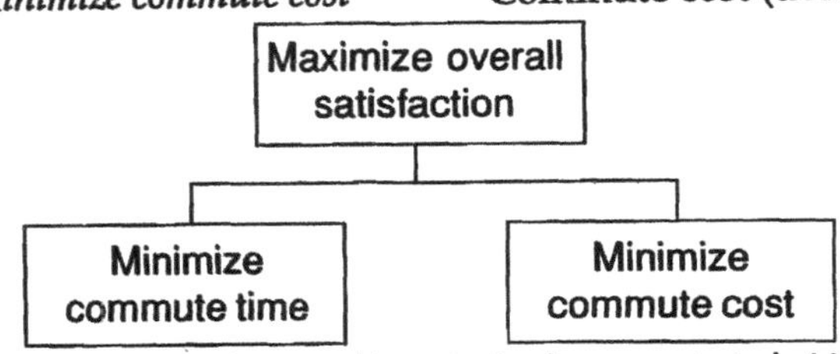

FIGURE 12.3. Objectives hierarchy for the commuter's decision.

Once the attributes are defined, we can measure the performance of each alternative on each objective. For some or all alternatives, it may be uncertain what performance will be achieved with respect to one or more of the objectives. A probabilistic assessment of the possible performance outcomes, or of the other uncertainties that affect performance, is used to quantify this uncertainty. When a multiple-attribute decision involves uncertainty, the MUA theory gives additional conditions that must be true of the objectives hierarchy (and the decision-maker's preferences over these objectives) so that a simple linear or multiplicative multiattribute function is appropriate. It is often possible to ignore these conditions in building the MUA model, meaning that the analyst can develop a multiple-attribute *value* model. ("Utility" strictly means that there is at least one uncertainty.) We will make this assumption also for the remainder of the chapter and sometimes use "value" and "utility" interchangeably.

In the commute decision example, Table 12.1 provides a complete characterization of the two alternatives.

Step 4. Define a utility function for each attribute.

The fourth step is to specify a utility function for each attribute. This *single-attribute utility function* expresses the decision-maker's relative desire for each of the possible results that could occur for an attribute. For example, you may be indifferent between all commutes of between 0 and 10 minutes, feel slightly worse about commutes in the 10- to 15-minute range, and be quite dissatisfied with commutes over 15 minutes. In an MUA, you must account for relative preferences *within* an attribute, as well as the tradeoffs *between* attributes. A single-attribute utility function can have any mathematical form, depending on the attribute and the decision-maker's preferences. Determining the form of the utility function involves assessing the decision-maker's preferences for the attribute as described later in this chapter.

In the commute decision example, the single-attribute utility functions are as shown in Figure 12.4. Each is assumed to be a linear function of the attribute measured. The scales are normalized so that values range from 0 to 1, but any other minimum and maximum is acceptable. The units in such scales are called *utiles*.

TABLE 12.1
Characterization of the Alternatives in the Commuter's Decision

Alternative	*Commute time (minutes)*	*Commute cost (dollars)*
Drive	5, 10, or 40 minutes with probability 0.3, 0.5, and 0.2, respectively	$2.00
Ride Train	15 minutes	$1.50

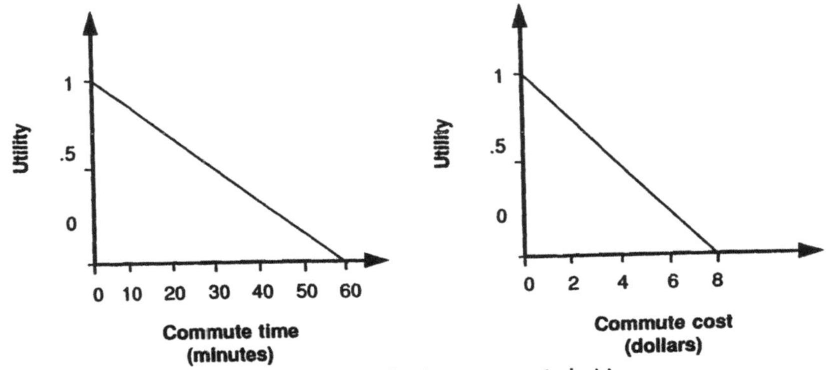

FIGURE 12.4. Single-attribute utility functions for the commuter's decision.

Step 5. Define the multiattribute utility function.
The final step is to combine the single-attribute utility functions with a set of tradeoff or value weights into a multiattribute utility function. The value weights express the decision-maker's preferences for tradeoffs between different objectives. These weights may be thought of as "rates of exchange" between objectives, expressing the value of one objective in terms of any other. As mentioned earlier, the form of the multiattribute function (e.g., a linear weighted sum of the single-attribute utility functions) depends on the objectives hierarchy and the single-attribute utility functions.

In the commute decision example, the multiattribute utility function expresses tradeoffs between commute time and commute cost. Discussion with the decision maker revealed equal preference (indifference) among the following options:

Combination 1: 5-minute commute time, \$6.25 cost
Combination 2: 10-minute commute time, \$5.00 cost
Combination 3: 30-minute commute time, zero cost

For illustrative purposes, we selected combinations that imply a value of \$15.00 per hour of avoided commute time, consistent with assumptions used for this example in the previous chapter. Based on the value of avoided commute time, and the single-attribute utility functions from Figure 12.4, the decision-maker's preferences can be characterized by the following multiattribute utility function:

$$U(\text{time, cost}) = 1.0u(\text{cost}) + 0.5u(\text{time})$$

Using this function and the single-attribute utility functions, the utility of each combination above will be 1.0. For example, Combination 1 involves a 5-minute commute time and \$6.25 cost. From Figure 12.4, a 5-minute commute has a utility of 0.92, and a $6.25 cost has a utility of 0.17. Using the function *U* above,

$$U(\text{Combination 1}) = (1.0 \times 0.92) + (0.5 \times 0.17) = 1.0.$$

The same approach can be used to compute the utility of any combination of commute time and commute cost.

The coefficients in this multiattribute utility function must have a 2 : 1 ratio, but otherwise are arbitrary. For example, weights of 2 and 1 or 10 and 5 would define two other multiattribute utility functions, but represent exactly the same preferences as the one above. The coefficients and calculated utilities should not be interpreted in isolation or in absolute terms.

The multiattribute utility analysis can now be completed using the approach described in the preceding chapter on decision analysis. This analysis will account for the uncertainties and value tradeoffs in determining the expected utility for each alternative, with the recommended alternative being the one with highest expected utility. Sensitivity analyses, value-of-information analyses, and other analyses that are typically included in a decision analysis can be performed in a multiattribute utility analysis.

Figure 12.5 provides the decision tree for the commuter's decision example. (For comparison with the DA treatment of this example, see Figure 11.16 in the preceding chapter.) The utilities for time and cost shown in the next to last column are obtained from Figure 12.4, while the overall utility shown in the last column is obtained from the function *U* above.

Using the approach described for decision analysis, we compute the expected utility of each alternative. Expected utility is calculated by summing the products of the probabilities of each outcome times the corresponding

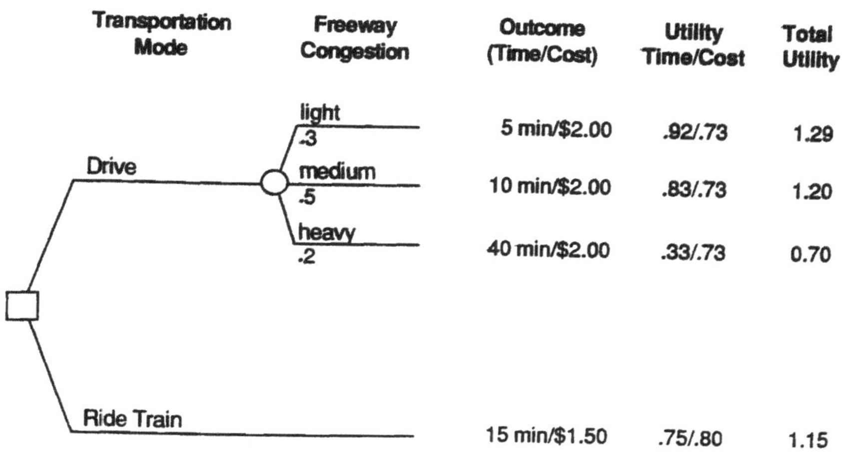

FIGURE 12.5. Decision tree showing utility for outcomes of commuter's decision (MUA approach).

utilities. For the Drive alternative there is a 0.3 chance of receiving a utility of 1.29, a 0.5 chance of receiving a utility of 1.20, and a 0.2 chance of receiving a utility of 0.70. Multiplying the utilities by the respective probabilities, we get:

For Drive alternative:

$$\text{Expected utility} = (0.3 \times 1.29) + (0.5 \times 1.20) + (0.2 \times 0.70) = 1.13$$

The expected utility of the Ride Train alternative is the single utility of 1.15, which is greater than the 1.13 expected utility for driving. In terms of the MUA, the best alternative for the commuter is to Ride the Train. (The difference in expected utility, 0.02, is equal to the utility for a reduction of 1–2 minutes in the commute time.) The result in this example is consistent with the result obtained through the decision analysis treatment of the problem, in Section 11.3.1.

12.3.2 Description of Multiattribute Utility Analysis

The steps involved in developing a multiattribute utility function are described in the preceding section. This section describes each of these steps in more detail.

12.3.2.1 Developing the Set of Decision Objectives

An objective expresses one motivation underlying the decision-maker's interest in the decision. More precisely, an objective is defined by a context, an object, and an orientation of preference. For example, in a decision involving chemical process safety, one of the objectives might be expressed as follows:

Example objective: *Fatalities from accidental chemical releases should be minimized.*

In this example, the context is "accidental chemical releases," the object is "fatalities," and the orientation is to "minimize." Each of these three elements is required for an objective to be well defined. The context identifies the range of alternatives and events considered (for example, fatalities from other causes are excluded); the object identifies what is valued; and the direction of preference indicates that the decision maker can always make judgments of "better," "worse," or "indifferent."

Table 12.2 illustrates possible objectives. The objective does not tell us the specific alternative to adopt, only one dimension on which to compare alternatives. Thus the third objective in Table 12.2, *Minimize costs for corporate safety improvements*, does not mean that the lowest-cost alternative will be chosen, only that this is one of several competing (possibly even contradictory) goals.

Typically, exercises that help stakeholders consider all possible costs and benefits associated with the decision are used to develop the list of objectives. For example, participants may think into the future, and consider all outcomes

TABLE 12.2
Illustrative Decision-Making Objectives

Objective	*Context*	*Object*	*Orientation*
Minimize fatalities resulting from accidental chemical releases	Accidental chemical releases	Fatalities	Minimize
Maximize profits from operations in the Western Region	Operations in the Western Region	Profits	Maximize
Minimize costs for corporate safety improvements	Corporate safety improvements	Costs	Minimize

that could make the decision maker or other stakeholders happy (or unhappy) that a particular alternative was chosen. Keeney (1992) discusses many approaches to defining objectives. The analyst should ensure that individual objectives are well-defined and have a convenient attribute by which performance can be measured. A vaguely defined objective leads to confusion or disagreement about what the objective means, and will make it difficult to obtain unbiased assessments of the performance of an alternative on that objective, and a value weight. In the earlier example, it would be important that all participants in the analysis agree on what counts as a fatality caused by the accident (e.g., are indirect casualties resulting from a traffic accident during evacuation included?), on what is meant by a "chemical release" (e.g., are accidents during transportation of materials to the facility included?), and on other crucial aspects that define the objective.

Each single objective should be measurable. Eventually the analysis will require a quantitative attribute for each objective, and the performance of each alternative will be specified using this attribute. An objective is measurable if it is possible to know exactly how an alternative performs on that objective if all uncertainties were resolved. Someone with perfect knowledge of the future should be able to express unambiguously the actual performance that will be achieved on each objective. If not, the objective probably does not meet the measurability criterion. For example, using a rating of "Public satisfaction with level of safety investment" is not by itself sufficient as an objective, since it cannot be determined exactly how well an alternative performs on this objective.

The set of objectives as a whole should also fulfill key criteria. Specifically, the group of objectives should be comprehensive, nonredundant, and independent.

Comprehensive means that the set of objectives should address all outcomes of interest, for all decision alternatives. One way to check for comprehensiveness is to list the pros and cons of each alternative, and determine whether the set of objectives incorporates all of these aspects.

Nonredundant means that no two objectives should overlap, thus causing the same objective to be counted twice. Redundancy will result in giving double-credit (or double-penalty) to an alternative on the basis of essentially the same objective. Partial overlap of objectives must also be avoided. If objectives are redundant, the objectives should be redefined to eliminate the overlap, or define a combined objective that incorporates those that overlap. For example, the objectives "maximize profit" and "minimize cost" are redundant, as both include costs. Instead, the analysis could use "maximize revenue" and "minimize cost" as nonredundant objectives.

Independent (or preference-independent) means that the decision-maker's value for the performance on one objective should not depend on the performance on another objective. For example, "minimize accident exposures" and "minimize potential hazard severity" (e.g., fatal/nonfatal/permanent/temporary) are not preference-independent. The value placed on reducing potential hazard severity (e.g., by changing a chemical process) will likely be greater when exposures are high than when exposures are low. This preference dependence results from the fact that risk is typically a multiplicative relationship between severity and exposure, and that decomposition is not appropriate in an MUA. When two objectives are preference-dependent, the analyst should develop a new objective that incorporates the two. Alternatively, the analysis can and should be modified to address the dependency explicitly, but the analysis and assessment are more difficult. In practice, it is almost always possible to structure a set of objectives so that the objectives are preference-independent. Often, such restructuring reveals hidden or implicit objectives.

Finally, the number of objectives should be kept small, say, to about a dozen at most. If an objective will be satisfied almost equally well by all of the alternatives, it may be eliminated from the analysis. One may also exclude objectives for which the decision-maker's value for the differences in performance among the alternatives is known in advance to be too small to alter the decision. To further reduce the number of objectives, it may be practical to combine two or more objectives into a higher-level, more aggregated objective (e.g., "cost" and "revenue" might be combined as "profit"). A disadvantage of eliminating objectives, however, is that you may lose the ability to demonstrate (in)sensitivity of the decision to the objective or its value weight. The "option" of the value hierarchy may also not be acceptable to decision stakeholders, regardless of what is shown by analysis.

12.3.2.2 Developing the Objectives Hierarchy

Figure 12.6 provides an example of an objectives hierarchy that might apply to a decision involving chemical process safety. This example is illustrative only; each corporation would need to develop an objectives hierarchy that is appropriate to its own decision-making context. A possible overall objective might be to *Maximize corporate viability*, recognizing that fulfilling this objective will depend on aspects internal to the organization, such as financial strength and employee welfare, and on external aspects such as public welfare and corporate image. In the figure, four objectives contributing to corporate viability are shown in the second level of the hierarchy.

Each lower level of the hierarchy includes objectives that contribute to fulfilling objectives in the level just above it. For example, the third level of the hierarchy in Figure 12.6 includes *Maximize public confidence* and *Minimize public concern*. These two lower-level objectives reflect key aspects that the decision maker in this example considered important influences on maximizing the corporate image. Public confidence has to do with the ongoing public perception of the company, while public concern relates to the public's reaction in the event of an accident. It is plausible that a decision maker would need to decide among alternatives that differ strongly from one another on these two aspects. One alternative might involve significant investments towards preventing an accident, which may provide a payoff in enhanced public confidence. Another alternative might involve investing more heavily towards reducing the consequences of an accident, and hence contribute more toward the public concern objective. At the lowest level of the objectives hierarchy, the distinctions among the objectives can be subtle, but crucial.

Each lower-level objective answers the question, "What aspects of the higher level objective are important to achieving this objective?" There is no restriction on the number of objectives in each level of the hierarchy, the number of levels, or in the relationships specified among the objectives. Figure 12.6 shows that the lower-level objective that seeks to *Maximize employment of workers* also contributes to two of the second-level objectives: *Employee welfare* and *Public welfare*. There are really two lower-level objectives associated with employment—one for employees and one for the public.

Often, one develops the list of objectives and builds the objectives hierarchy at the same time. In some situations it is easiest to build the objectives hierarchy starting from the top, by redefining objectives in more specific terms, rather than assembling a list of objectives.

In developing the objectives hierarchy, when does one stop? As with other decision aids, the depth of the analysis should be commensurate with the importance of the decision and the availability of time and resources. The number of assessments required as inputs to the analysis will increase linearly

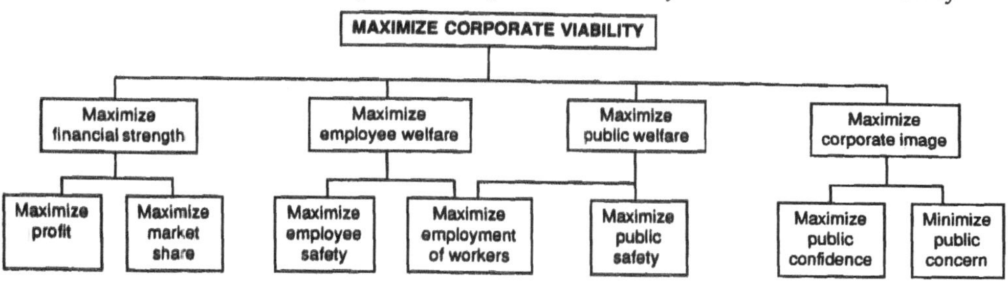

FIGURE 12.6. Sample objectives hierarchy.

with the number of objectives, since each alternative must be evaluated with respect to each objective. For some alternatives and objectives, the performance may be uncertain, and these uncertainties need to be assessed as well. In addition, defining the multiattribute utility function may become more complex, since dependencies among objectives need to be checked for and incorporated into the analysis if they are present.

Figure 12.7 illustrates an objectives hierarchy that was developed as part of the Department of Energy's evaluation of alternative sites for the high level nuclear waste repository (Merkhofer and Keeney 1987). In this study, conducted from Fall 1985 to Spring 1986, the Department of Energy (DOE) evaluated five alternative sites. According to DOE guidelines, the decision needed to address objectives both for the time period prior to closure of the repository (first 10,000 years) and for the period following closure (10,000 to 100,000 years). Figure 12.7 shows the preclosure objectives hierarchy, which has as its top-level objective to "minimize adverse preclosure impacts". This objective was broken into a total of 23 contributing subobjectives.

Although the objectives hierarchy may include a large number of objectives at many levels, only the objectives at the lowest level are relevant to the subsequent steps of the analysis. These objectives will be used to define the multiattribute utility function, and the performance of each alternative must be measured for each of these objectives. For example, the objectives hierarchy in Figure 12.7 includes 14 lowest-level, or "fundamental," objectives. The MUA therefore involved measuring the performance of each of the five alternatives on 14 attributes.

12.3.2.3 Defining the Attributes

An attribute is a feature or characteristics associated with an objective which can be used to measure the degree to which an objective is met. The analyst must identify or define an attribute for each fundamental objective (at the lowest level of the hierarchy).

An attribute should provide a convenient and natural means by which to evaluate how well an alternative performs with respect to an objective. Sample attributes include those shown in Table 12.3.

TABLE 12.3
Examples of Objectives and Corresponding Attributes

Objective	*Attribute*
Minimize costs of transporting hazardous chemicals	Average single round-trip transportation cost in dollars
Minimize acute health risks associated with the transport of hazardous chemicals	Expected number of fatalities per year (probabilistic)

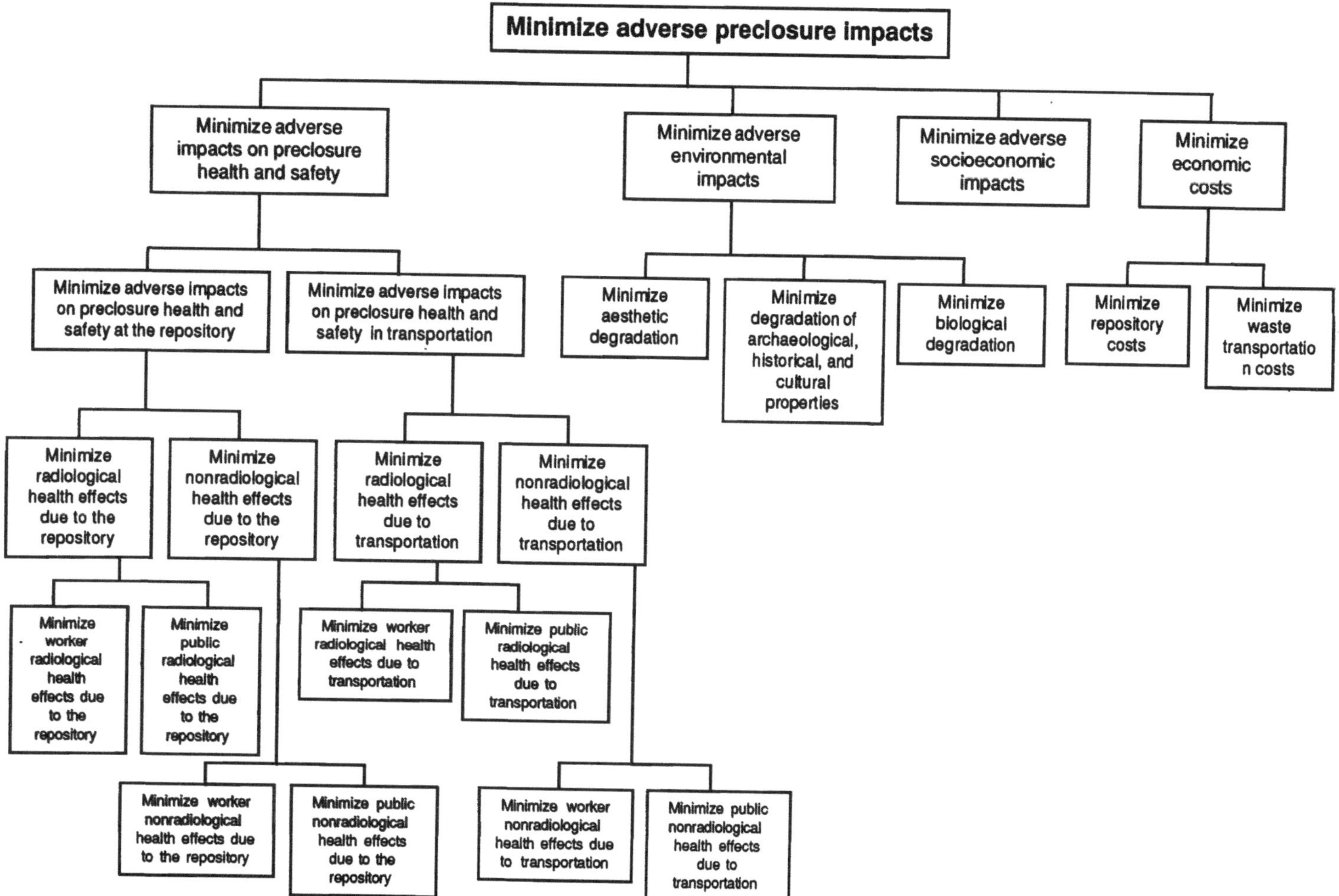

FIGURE 12.7. Objectives hierarchy used in an MUA of alternative sites for the disposal of nuclear waste (Merkhofer and Keeney 1987, p.175).

TABLE 12.4
Examples of Direct and Proxy Attributes

	Attribute	
Objective	*Direct*	*Proxy*
Minimize public nonfatal health effects from catastrophic releases	Number of hospital admissions per year	Volume of toxic material released per year
Maximize employment	Number of new jobs provided per year	Pounds of finished product manufactured per year

Attributes may be either *direct* or *proxy* measures. A *direct* measure focuses on the degree to which an objective is met. A *proxy* measure indirectly gauges the fulfillment of the objective based on related factors. Sample direct and proxy measures for two objectives are provided in Table 12.4. For example, SO_2 emissions do not directly measure environmental impacts, and most decision makers do not value lowered SO_2 emission levels *per se* . Lowering SO_2 emissions is a means toward achieving a fundamental objective, which may be achieved through other means as well. Put otherwise, if you could reduce environmental (or other relevant) impacts, through means other than changing SO_2 levels, you would not then still value SO_2 reductions, since they have no intrinsic relationship to the fundamental motivations for decision making. Understanding and clarifying the roles of means and end objectives and their attributes are two of the most important aspects of an MUA.

Once an attribute has been chosen, a measurement scale or mathematical model must be defined by which performance will be explicitly evaluated. The scale used to measure performance may be either *natural* or *constructed.* A *natural* scale is one that enjoys common usage and interpretation. For example, a natural scale for cost is dollars. A natural scale for a quantity of toxic vapor released might be the volume in millions of cubic feet. A *constructed* scale is one developed when no natural scale exists for the problem at hand. Table 12.5 provides a simplified description of a constructed scale for impacts on historical properties that was used in the DOE study discussed above (Merkhofer and Keeney 1987).

In this example, the meaning associated with each scale point is carefully defined, thus minimizing ambiguity and facilitating the scoring of each alternative. Vaguely defined scales, using only descriptors such as "desirability" and "satisfaction," are easily misused and subject to misinterpretation. Behavioral research shows that people provide a range of interpretations for such simple terms, hence the need for detail and rigor in constructed scales (von Winterfeldt and Edwards 1986). The specificity of a well-designed constructed

TABLE 12.5
Example Constructed Scale (Simplified Description)

Impact level	*Impacts on historical properties in the affected area*[a]
0	There are no impacts on any significant historical properties
1	One historical property of major significance or 5 historical properties of minor significance are subjected to minimal adverse impacts
2	Two historical properties of major significance or 10 historical properties of minor significance are subjected to minimal adverse impacts
3	Two historical properties of major significance or 10 historical properties of minor significance are subjected to major adverse impacts
4	Three historical properties of major significance or 15 historical properties of minor significance are subjected to major adverse impacts
5	Four historical properties of major significance or 20 historical properties of minor significance are subjected to major adverse impacts

[a] A historical property of major significance meets the criteria for the National Register of Historic Places (e.g., first town hall in a community; cave sites representative of an Indian people at one stage of their history; a Civil War battlefield). A historical property of minor significance has local significance (e.g., a homestead or miner's cabin, an archaeological site that is representative of a period for which there are many examples). Major impacts change the integrity or significance of the historical property. Minimal impacts may alter the historical property, but will not change its integrity or its significance.

Source: Merkhofer and Keeney 1987, p.177.

scale also facilitates defining the multiattribute utility function, because the meaning of a specified difference between two alternatives will be clear.

12.3.2.4 Defining the Single-Attribute Utility Function

A utility function, whether single- or multiattribute, is a mathematical function that represents the subjective, relative desirability of consequences as described by the attributes. This concept of a utility function is identical to the concept introduced in the preceding chapter on decision analysis. In that chapter we described the axiomatic basis for utility functions and the rule of expected utility maximization in decision problems.

A single-attribute utility function represents the desirability or preferences of possible outcomes expressed with a single outcome measure, such as dollars, health impacts avoided, environmental impacts avoided, etc. Figure 12.8 illustrates that single-attribute utility functions may be linear or nonlinear, discrete or continuous. Virtually any mathematical function is possible, and the specific form will depend on the decision-maker's preferences for each attribute.

Various techniques are used to elicit the specific form of the utility function. No single technique is uniformly best, and usually a suite of assessments

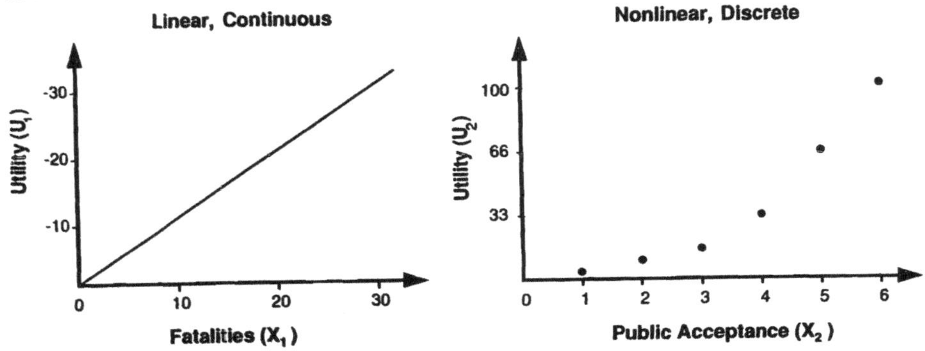

FIGURE 12.8. Sample single-attribute utility functions.

will be used to "triangulate" the final function. One approach is described in Section 11.3.2.6, and involves posing choices among hypothetical risky options. Keeney and Raiffa (1976) and von Winterfeldt and Edwards (1986) describe a variety of approaches to utility assessment.

12.3.2.5 Defining the Multiattribute Utility Function

A multiattribute utility function represents the desirability of possible combinations of outcomes that can occur with a set of objectives. Figure 12.9 illustrates a multiattribute utility function that combines utilities for the two attributes in Figure 12.8. In general, a multiattribute utility function is difficult to assess: it must express the relative value that the decision maker has for any possible combination of outcomes on the attributes. Even if we limited ourselves to two or three points on each performance scale, the number of possible combinations that could occur is very large, even with only two or three

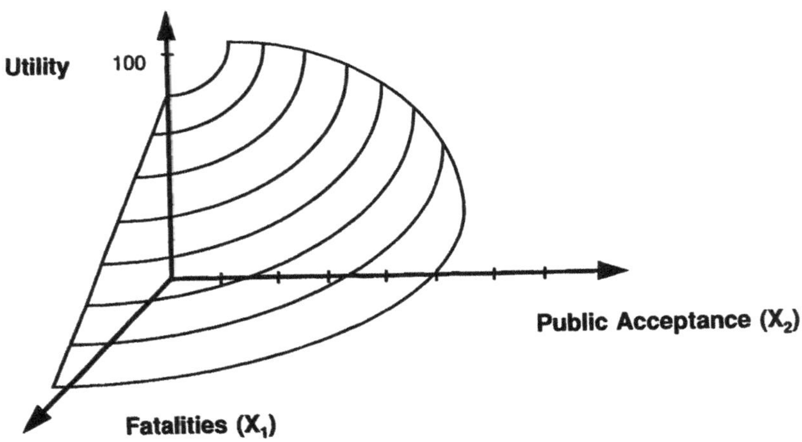

FIGURE 12.9. Sample multiattribute utility function.

objectives. To elicit the decision-maker's value for each combination would be an overwhelming task, and subsequent analyses would reference extensive sets of data. This is why it is important to design an MUA so that the multiattribute utility function has a form that is simple to assess and calculate.

The multiattribute utility functions used most often are simple aggregations of single-attribute utility functions like the ones discussed in the preceding chapter. The additive form is used frequently:

$$U(x_1, x_2, \ldots, x_n) = k_1u_1(x_1) + k_2u_2(x_2) + \cdots + k_nu_n(x_n)$$

where

- x_i is the outcome associated with a given alternative on the *i*th objective, expressed in terms of that objective's attribute (dollars, jobs, etc.)
- $u_i(x_i)$ is the single-attribute utility function for the *i*th attribute
- k_i is a weighting factor that applies to the *i*th utility function

In this formulation, each single-attribute utility function expresses the relative desirability of outcomes on each attribute. Each k_i rescales a single-attribute utility function to reflect the relative desirability of utility increases on one objective relative to other objectives.

A multiplicative utility function is sometimes used. Using the same notation as before:

$$U(x_1, x_2, \ldots, x_n) = (1/k)\{[1 + kk_1u_1(x_1)][1 + kk_2u_2(x_2)] \cdots [1 + kk_nu_n(x_n)] - 1\}$$

where k is a constant satisfying the equation:

$$1 + k = (1 + kk_1)(1 + kk_2) \cdots (1 + kk_n).$$

The additive form is a special case of the multiplicative form in which the scaling constants sum to 1. Keeney (1992) and von Winterfeldt and Edwards (1986) give detailed guidance on developing MUAs so that a linear or multiplicative model is appropriate.

Both the additive and the multiplicative forms of the utility functions allow the utility function to be decomposed into single-attribute utility functions, and this greatly simplifies the task of assessing the parameters of the utility function. However, *preferential independence*[1] conditions among the objectives are required to justify the use of either of these two forms. Preferential independence for the additive form means that the relative desirability of outcomes on one objective does not depend upon the outcome on another objective. For example, if two objectives of a chemical process safety decision are to minimize cost and to minimize fatalities, preferential independence is violated if the decision maker feels that the value of a specified reduction in fatalities differs depending upon the cost that was incurred for the project.

1 "Preferential independence" may refer to other conditions as well. We use it in the sense of "additive difference independence" as in von Winterfeldt and Edwards (1986).

Preferential independence is a crucial requirement for the use of the relatively simple forms of multiattribute utility functions described above. Keeney and Raiffa (1976) describe straightforward procedures to test for preferential independence. If two objectives are found to be dependent, it may be possible to combine them in order to eliminate the dependency. Alternatively, approaches are available for incorporating the dependency explicitly into the formulation.

A difficult issue in making chemical process safety decisions is how to trade off the value of reducing risks to human life against the value of financial aspects of the decision. Chapter 2, Section 2.5, includes a discussion of *value-of-life* issues. A multiattribute utility function can be developed without requiring the decision maker to directly express the equivalence between fatalities and another attribute such as dollars. Some organizations find this appealing, but it should be recognized that with MUA and most other quantitative decision aids, the dollar value placed on reduced fatality risk, even if only an implicit value, can usually be determined from the analysis.

Despite the relative simplicity achieved through the additive or multiplicative utility functions described above, the burden of assessing the complete set of single-attribute utility functions and multiattribute parameters can be substantial. Often, key stakeholders or the decision makers will spend an entire day with a decision analyst to develop the complete set of parameters. (The exercise is much more than a mechanical "question and answer," and decision makers often find that it gives them considerable insight into the decision-making process.) The assessment burden depends on the complexity of the single-attribute utility functions, the number of objectives, and the dependencies among objectives. To reduce the assessment burden, the MUA model can be simplified by combining related objectives, or by using a single-attribute utility function, such as an exponential function, which requires assessment of only one parameter. Assessing the weighting factors is also an important task, and arbitrary schemes such as point allocations may fail to capture important dependencies. An effective approach is to infer the weights by having the decision maker express preferences among alternatives that vary on two or more attributes. Indirect assessment of this kind is more accurate and resistant to bias than "direct assessments" that ask for the weights explicitly.

12.3.2.6 Incorporating the Multiattribute Utility Function into the Analysis

Once the multiattribute utility function has been developed, the remainder of the analysis follows the same approach as the one described for decision analysis in the preceding chapter. Since the multiattribute utility function introduces additional structure, and parameters, into the analysis, some of the outputs of an MUA may differ from those of a decision analysis. For example, sensitivity analysis is typically performed on the attribute value weights. This may involve placing all weight on a single objective, or setting the weight on a single objective to zero, and seeing how the optimal decision changes.

12.4 CASE STUDY: DISTILLATION COLUMN

An Application of Multiattribute Utility Analysis to Chemical Process Safety

This application involves Case Three, the distillation column, which was introduced in Section 5.4. The example was also used in the application provided for Weighted Scoring Methods (Section 7.4).

The case study involves an existing plant at which a distillation column is used to separate hexane and heptane. Close to the facility is a residential area of about 200 people. In response to management concerns, a risk analysis was undertaken. The risk analysis focused on the safety of the residential community, and excluded possible concerns about employee safety that had been addressed when the distillation column was originally designed.

After reviewing the results of the risk analysis, management wishes to consider various alternatives that might be undertaken to reduce the risk. Because of the magnitude of the expenditures that might be required, and the potential risks posed to the community, the decision will be made at the corporate level. A *corporate team* has been formed that has final decision-making authority. This team includes the:

- Business director
- Director of engineering
- Member of a capital projects review team
- Operations general manager.

An *analysis team* has been given responsibility to conduct the analysis. This team includes staff that can guide the analysis and facilitate the gathering of relevant expertise both inside and outside of the company. The analysis team consists of three persons: an individual who led the earlier risk analysis and who reports to the safety manager, an individual who reports to the business director and who can help with modeling the financial aspects of the decision, and a consultant with expertise in multiattribute utility analysis.

Figure 12.1 showed the phases of a multiattribute utility analysis. The remainder of this section summarizes the activities and results of each phase.

12.4.1 Formulation Phase

The formulation phase involves defining the decision problem and developing an understanding of the alternatives, uncertainties, and values relevant to the decision. In its first meeting, the analysis team agreed on the decision to be addressed and on a preliminary list of alternatives. Specifically, the problem addressed is what action the company should take, if any, to reduce the risks to the residents located near the distillation column operation.

With input from the review team and others, the analysis team developed three alternatives to be analyzed:

- Do Nothing
- Use Two Smaller Columns
- Install Water Curtain.

With input from safety engineers and from plant and process engineers, some important uncertainties were identified during the formulation phase. These include:

- The risk of failure and resulting releases from the distillation column
- The risk of ignition of released gases resulting from column failure
- Uncertainty about the wind speed and direction that will prevail at the time any releases occur
- Uncertainty about whether the water curtain will function properly (applies only to Install Water Curtain alternative)
- Uncertainty about the costs associated with each alternative
- Uncertainty about the number of fatalities that will result from a column failure.

The analysis team also developed a preliminary list of concerns that affect the value the company will have for the various alternatives and their consequences. This list included:

- Capital cost, including both construction and relocation of people
- Operating cost
- Revenues from the operation
- Public safety, including both fatalities and injuries
- Employee safety, including both fatalities and injuries
- Concern of the community over risks from the column
- Public image.

After further discussions with the corporate team, it was agreed that employee safety would be eliminated as a separate issue: an appropriate level of employee safety would be "engineered" into any solution, and public safety is viewed by the company as a far more important issue to be addressed in this decision. The team also agreed that safety could be characterized in terms of fatalities alone; even though injuries are a concern, far greater importance is placed on the risk of fatality. Moreover, the team felt that it was unlikely that any of the alternatives would vary significantly from others in the risk of injury as a proportion of the risk of fatality under that alternative. The same was believed to be true of employee safety. Thus ignoring injuries and employee safety would be unlikely to bias the analysis in favor of any alternative.

It was also determined that revenues from the operation could be ignored since none of the alternatives appeared to significantly affect company revenues. Finally, at the suggestion of the Business Director, it was agreed that the analysis should use a 20-year time frame, and that operating costs should be discounted at 8%. This allowed the capital cost and operating cost for each alternative to be expressed as a present-value discounted cost, consistent with other investment decisions of the organization.

As a result of these discussions, the list of concerns were distilled into three key objectives, as illustrated in the simple objectives hierarchy shown in Figure 12.10:

- Minimize the present-value discounted cost (capital + operating expenses) to the company
- Minimize public fatalities
- Maximize community goodwill.

To represent the decision graphically, an influence diagram was developed that expresses the relationship among the uncertainties, alternatives, and values. This diagram is shown in Figure 12.11. The italic text on the figure indicates the possible states of the uncertainties. *Column failure* refers to the type of release from the column, with possible states "instantaneous," "continuous," or "none." The *ignitable quantity* is determined by the action taken, the column failure, and the water curtain effectiveness. For instance, if no action is taken, an instantaneous failure would release 28,000 kg with a potential for ignition. A continuous failure would release 11 kg/second. If the water curtain is installed, and it works, it will reduce the ignitable quantity to 14,000 kg in the case of instantaneous release and 5 kg/second in the case of continuous release. Similarly, using two columns instead of one would reduce the instantaneous release quantity to 14,000 kg, and the continuous release rate to 5 kg/second, but there would be twice the probability of failure. *Ignition* represents the type of explosion or fire resulting from a release. The states are "immediate ignition" (resulting in a BLEVE for instantaneous releases and a jet fire for continuous releases), "delayed ignition resulting in a UVCE," "delayed ignition resulting in a flash fire," and "no ignition".

Wind direction is the direction from which the wind blows. The directions toward the residential community are southwest, west, and northwest. The problem was simplified by assuming one average wind speed and weather stability.

The three components of the value model are shown with arrows leading to *Total Utility*. *Public fatalities* are influenced by the type of column failure, the

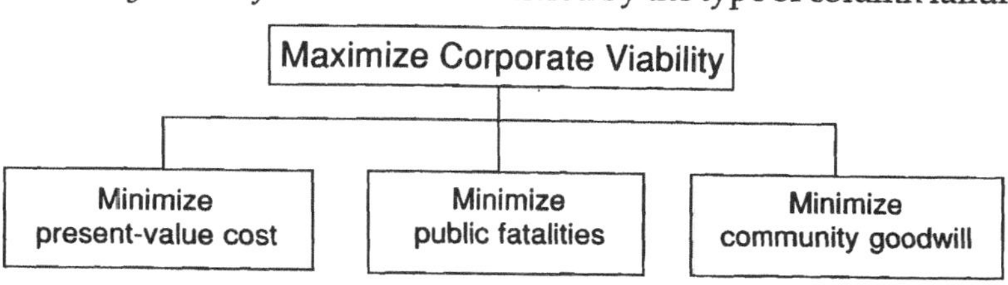

FIGURE 12.10. Objectives hierarchy for Distillation Column case.

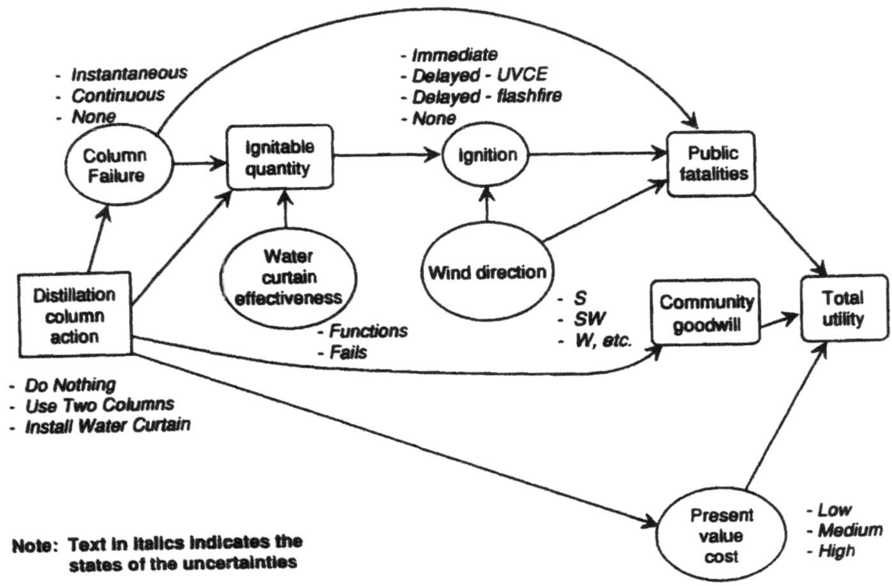

FIGURE 12.11. Influence diagram for Distillation Column case.

type of ignition, and the wind direction. *Community goodwill* represents the concern the community has over the problem. The action taken affects the company's reputation. *Present-value cost* includes the capital and operating costs. Total utility will be defined by combining public fatalities, community goodwill, and present-value cost through a multiattribute value function in the next phase of analysis.

The influence diagram contains attributes associated with fundamental objectives (health impacts, community goodwill, costs); means objectives (e.g., ignitable quantity) which can be used to influence the outcome of alternatives with respect to fundamental objectives; and other variables (e.g., wind direction) that are associated with neither end nor means objectives, since they are uncontrollable.

At the end of the Formulation phase, the analysis team met with the corporate team to present the framework for the study. The team's primary goal for this meeting was to gain agreement on the problem statement, the alternatives to be analyzed, and the criteria used to evaluate alternatives. The influence diagram provided an orientation to issues addressed in the analysis. Finally, the analysis team reviewed and received approval for the project schedule and summarized the key tasks, staff time, outside experts, and costs anticipated for the remaining phases of the analysis.

12.4.2 Deterministic Phase

With the problem structured, the next step is to construct a deterministic decision model, that is, a model that calculates the value for a given alternative and a given set of future states of uncertainty variables. A deterministic model maps each alternative and each possible occurrence or setting of each variable into a single (set) of outcome(s) that results under that scenario. It also determines the overall value the decision maker has for these outcomes with their resulting costs, fatalities, and community goodwill. For an MUA, a key portion of the deterministic model is the value function. Other portions of the deterministic model specify relationships among variables that influence inputs to the value function.

The part of the model representing public fatalities was developed through the quantitative risk assessment conducted earlier. Specifically, the quantitative risk assessment provides an estimate of the number of public fatalities that will result from any combination of ignitable quantity, ignition, wind direction, and distance. The second attribute, present-value cost, requires a simple financial model that applies the appropriate discount rate to the stream of 20-year operating costs and combines this estimate with the capital cost in the first year.

The third attribute, community goodwill, has been modeled using the scale shown in Figure 12.12. Each score is described in terms of the public's concerns about the safety of the distillation column. The bullets below each score give the scale its content and are essential. Often, intermediate scores (such as a "3.5") are used with an interpolation (linear or nonlinear) appropriate for the single-attribute utility function. Each alternative receives a score on the community goodwill scale corresponding to the level of community concern that is most likely if the alternative is chosen. The analysis team agrees to use the following values for community goodwill:

ALTERNATIVE	SCORE
Do Nothing	4
Use Two Smaller Columns	2.5
Install Water Curtain	2

The desirability of possible outcomes is provided by single-attribute value (utility) functions for the individual attributes and a multiattribute value (utility) function for overall value. Developing these single- and multiattribute value functions constitutes the core of the MUA approach, and the form used for this value function distinguishes MUA from DA.

Tentatively, an additive form was used to combine the cost, fatality, and community goodwill measures into a single multiattribute utility function:

$$U(\text{Cost,Fatality,Goodwill}) = k_1 u_1(\text{Cost}) + k_2 u_2(\text{Fatalities}) + k_3 u_3(\text{Goodwill})$$

0 = Community concern is very low or nonexistent.
- Area residents demonstrate almost no knowledge of or interest in the risks due to the chemical plant.

1 = Community concern is low.
- The chemical plant occasionally (i.e., several times per year) appears in local news stories.
- There is little significant discussion of the distillation column at community meetings.
- Public trust and company credibility are not seriously threatened.

2 = Community concern is moderate.
- There is occasional (i.e., several times per year) negative coverage of the plant and its hazardous materials in media serving localized areas within the surrounding area.
- Occasional small-scale public protests are common.
- Some area residents perceive the company response to be somewhat misguided. Public trust and company credibility are somewhat damaged.

3 = Community concern is high.
- Local and regional news coverage is frequent (i.e., greater than monthly) and typically negative, and there is occasional national coverage.
- Local government bodies initiate action regarding the handling of hazardous materials (such as proclamations or resolutions) and/or single-purpose public groups form to pressure action.
- Many area residents perceive the company response to be seriously inappropriate. Public trust and company credibility are seriously and permanently damaged.

4 = Community concern is very high.
- Local and regional news coverage is frequent (i.e., greater than monthly) and quite negative, and there is occasional negative national coverage.
- Small-scale public protests are common and local referenda related to possible releases and ignition from the distillation column addressed by the action plan are held; state officials become involved.
- A large segment of the public perceives the company response to be a blatant disregard for public health and safety, or a reneging on a promise or commitment of fundamental importance. Public trust and company credibility are damaged irreparably.

FIGURE 12.12. Community goodwill scale.

where u_1, u_2, and u_3 are single-attribute utility functions for present-value cost, fatalities, and community concern, respectively. The coefficients k_1, k_2, and k_3 represent the decision-maker's tradeoffs between the utilities for each attribute.

Using the assessment approaches outlined in Chapter 11, Decision Analysis, the single-attribute utility functions were assessed. These assessments were performed with members of the corporate team who could express the company's values for possible outcomes. The results of these assessments are provided in Figures 12.13, 12.14, and 12.15. Each of these curves expresses the desirability to the company of various outcomes for each attribute.

The curve for present-value cost shown in Figure 12.13 is linear within the region from \$0 to \$12 million since each additional dollar spent is valued equally by the company:

$$\text{Cost} = \text{present-value cost in \$ million}$$

$$u_1(\text{Cost}) = -\frac{\text{Cost}}{12}$$

Utilities were assessed for fatalities, with the resulting curve shown in Figure 12.14 represented by the following function:

$$u_2(\text{Fatalities}) = -\frac{(\text{Fatalities})^2}{40{,}000}$$

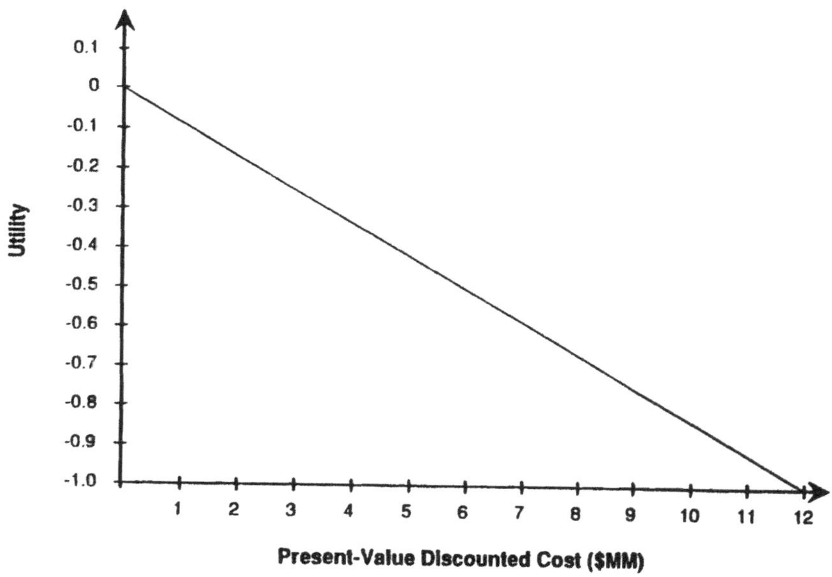

FIGURE 12.13. Single-attribute utility function for present-value cost.

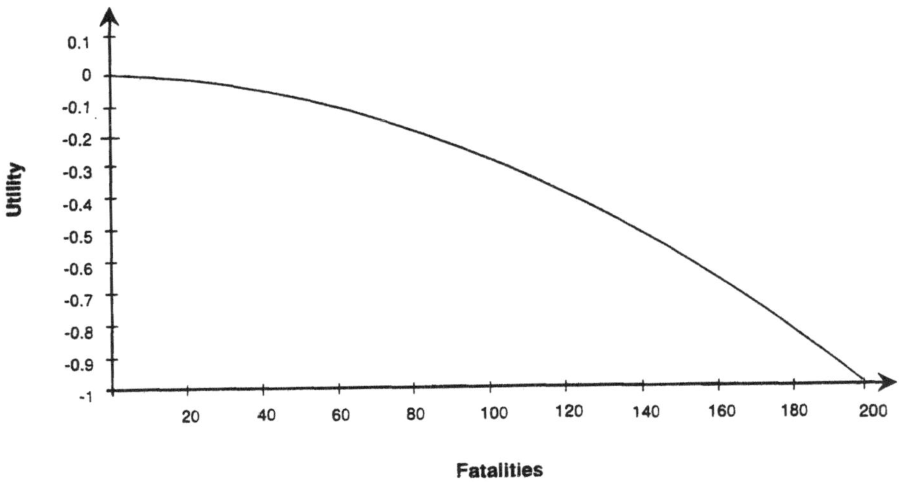

FIGURE 12.14. Single-attribute utility function for number of fatalities.

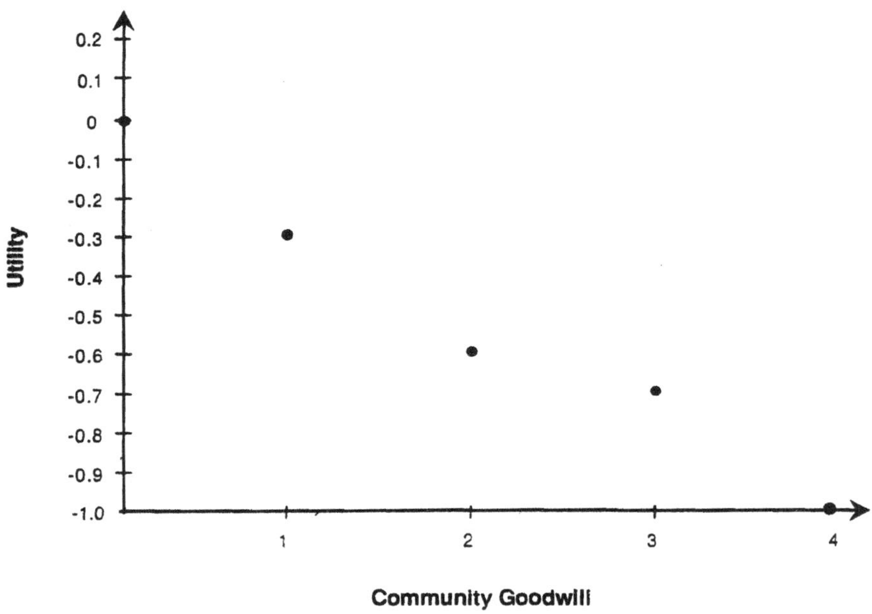

FIGURE 12.15. Single-attribute utility function for community goodwill.

The curve has an increasingly steep slope, since the disutility for each additional fatality increases. For example, the utility for 200 fatalities is 10 times greater per fatality than the utility per fatality for 20 fatalities.

The utility function for community goodwill shown in Figure 12.15 is discrete, with a utility assigned to each score on the community goodwill scale. However, utilities for intermediate scores (such as the score of 2.5 assigned to the two columns alternative) are estimated by interpolation.

To determine the tradeoffs among attributes (k's), exercises were conducted with members of the corporate team choosing among hypothetical combinations of cost, fatalities, and score on the community goodwill scale. The values of the k's were inferred from the tradeoffs implied by the corporate team members' choices. Meetings were conducted to resolve differences of opinion. Subsequent validation exercises confirmed that the linear-additive form of the multiattribute utility function was appropriate. The weighting factors that resulted appear in Table 12.6.

Implications about the corporate team's values can be determined by comparing the utilities for incremental changes for each attribute. For example, the value function implies that the following outcomes are about equally desirable to (hence equally valued by) the corporate team:

- A cost decrease from \$8MM to \$0MM
- A decrease from 16 expected fatalities to 0 expected fatalities
- A decrease from 4 to 2 on the community concern scale.

Although significant time and effort were required at this stage of the analysis, the corporate team found this worthwhile given the importance of the decision and how critical the value tradeoffs are to the decision. The value function could be further expanded to address other objectives, such as maximizing employment and minimizing environmental effects. To do so would involve defining attributes for each of these objectives, assessing associated utility functions, and assessing the corresponding value weights.

The final step in the deterministic phase is to conduct a sensitivity analysis on the inputs to the problem to determine if it can be simplified. This deterministic sensitivity analysis involves varying each of the uncertain values from

TABLE 12.6
Value Function Weighting Factors

Attribute	*Weighting Factor*
Present-value Cost	1.0
Fatalities	100.0
Community Goodwill	1.5

one extreme value to another, one at a time. For example, the state of the uncertainty *Water curtain effectiveness* was varied from "functions" to "fails." If it fails, Do Nothing is the alternative with the highest expected value. If it functions, Install Water Curtain is the best alternative. The decision is said to be *sensitive* to the uncertainty. The decision was found to be sensitive to all of the variables. Although varying *Present-value cost* from one extreme to the other did not change the preferred alternative, it did have a significant impact on the utility. Therefore, all of the uncertain variables will be described in more detail by assessing probability distributions in the next phase.

Figure 12.16 provides a schematic of the decision tree resulting from the Deterministic Phase. The full tree is too large to display. The analysis team met with the corporate team to describe the results of the deterministic analysis and to decide on experts to provide probabilistic assessments.

12.4.3 Probabilistic Phase

Key steps in the probabilistic phase are to identify an expert to assess each of the uncertain variables, conduct the assessments, and construct a final decision tree including probabilities.

Working with appropriate experts, the analysis team obtained judgmental uncertainty estimates that express the degree of uncertainty about the variables in the analysis. Figure 12.17 shows the probabilities assessed for *Column failure* and *Water curtain effectiveness*. The probabilities of column failure depend on the action taken and, specifically, on whether or not a water curtain is installed. If a water curtain is installed, there is a 0.1 probability that the column will not function adequately.

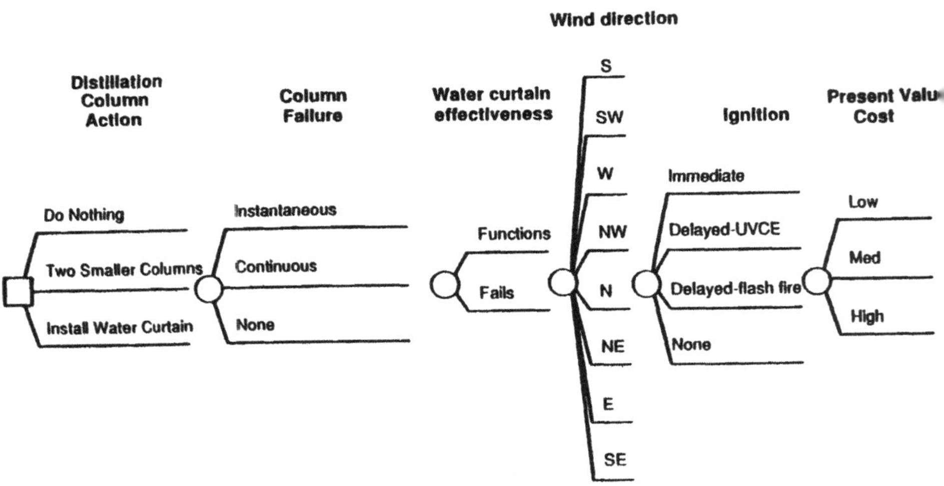

FIGURE 12.16. Schematic decision tree for Distillation Column case.

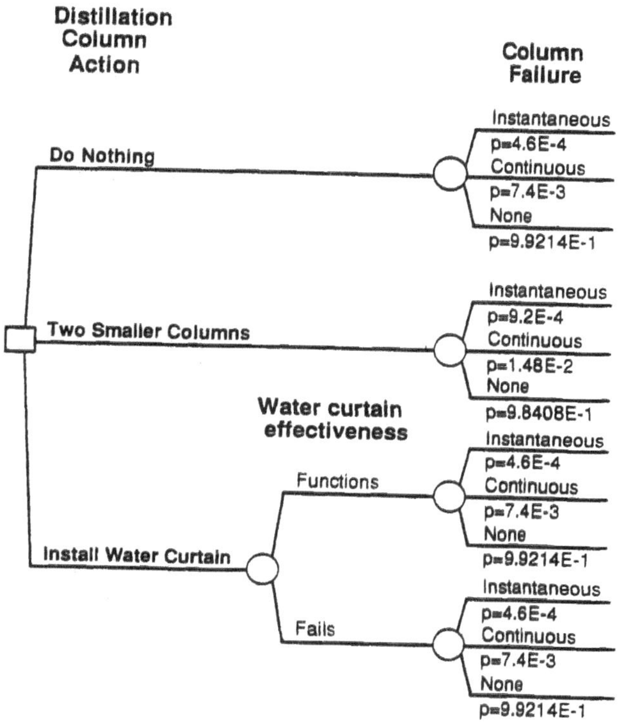

FIGURE 12.17. Probability assessments for column failure (20 years).

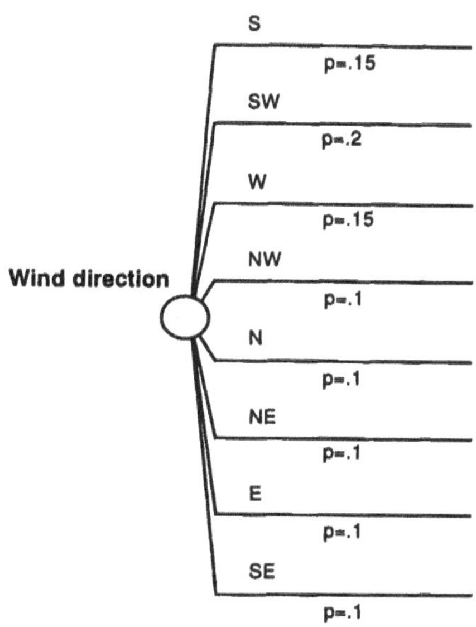

FIGURE 12.18. Probability assessments for wind direction.

Probabilities were also assessed for *Wind direction* (as shown in Figure 12.18) and for *Ignition*. Tables 12.7 and 12.8 show the probabilities of the different states of *Ignition* for each type of column failure. Table 12.7 is for the scenarios when the wind is from the southwest, west, and northwest, in which cases a release would be directed toward the residential area and be more likely to ignite. The probabilities for the wind from the other directions are shown in Table 12.8.

The expected number of fatalities under each possible scenario was estimated using the earlier risk analysis. The uncertain events of wind direction, column failure, and ignition imply a probability distribution for the number of fatalities. This distribution is calculated by assigning the number of fatalities associated with each of the possible outcomes, or scenarios, that result from the combination of the states of these events. The probability of each outcome can be calculated from the probabilities of the states leading to the outcome. F–N curves for the number of fatalities in 20 years are shown in Figure 12.19 for each alternative. The curves are based on all of the possible outcomes. The expected fatalities for each alternative appear in Table 12.9.

TABLE 12.7
Probability of *Ignition* with Wind Direction toward Community (from SW, W, and NW)

Ignition	*Instantaneous (28,000 kg)*	*Instantaneous (14,000 kg)*	*Continuous (11 kg/s)*	*Continuous (5 kg/s)*	*None*
Immediate	.25	.25	.10	.10	0
Delayed—UVCE	.34	.20	0	0	0
Delayed—flash fire	.34	.20	.68	.40	0
None	.07	.35	.22	.50	1

TABLE 12.8
Probability of *Ignition* with Wind Direction away from Community (from N, NE, E, SE, and S)

Ignition	*Instantaneous (28,000 kg)*	*Instantaneous (14,000 kg)*	*Continuous (11 kg/s)*	*Continuous (5 kg/s)*	*None*
Immediate	.25	.25	.10	.10	0
Delayed—UVCE	.08	.06	0	0	0
Delayed—flash fire	.08	.06	.09	.06	0
None	.59	.63	.81	.84	1

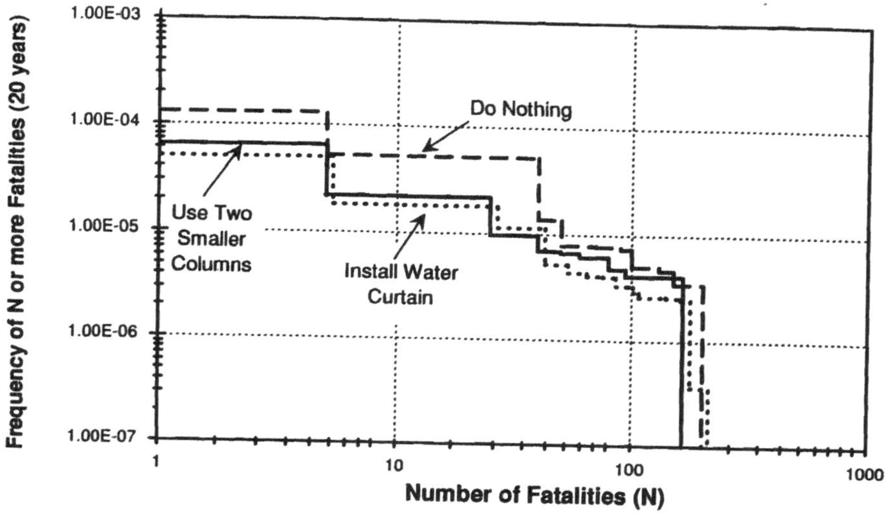

FIGURE 12.19. F-N curves for Distillation Column alternatives.

TABLE 12.9
Expected Fatalities for Distillation Column Alternatives

Alternative	*Expected Fatalities (20-year period)*
Do Nothing	.07
Use Two Smaller Columns	.03
Install Water Curtain	.02

The analysis team also assessed probability distributions for the present-value cost for each alternative. The discrete points shown in Figure 12.20 were chosen to represent the assessed probability distributions. Assessment and discretization of probability distributions are discussed in Section 11.3.2.5.

12.4.4 Evaluation Phase

Key steps in the evaluation phase are to develop a risk profile (the probability distribution over possible values) for each alternative, determine the optimal alternative, conduct sensitivity analysis, and estimate the value of gathering further information.

The expected utilities were calculated as described in Section 11.3.2.6 and are shown in the last column of Table 12.10. Install Water Curtain has the highest expected utility and is therefore preferred. The total utility includes a

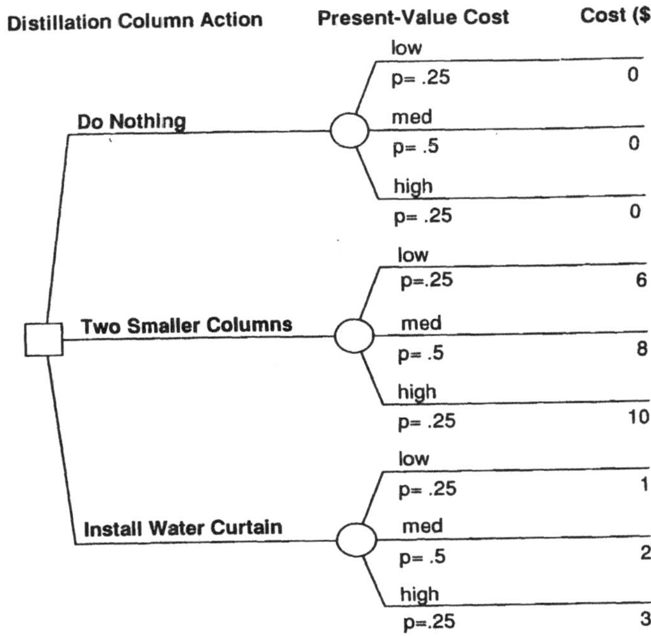

FIGURE 12.20. Probability assessments for *Present-value cost* ($MM).

contribution from each of the three attributes, as shown in Table 12.10. Install Water Curtain outperforms the other alternatives on all three attributes.

In Section 11.3.2.6 we described how risk profiles help characterize the degree of uncertainty around the expected utility for each alternative. The risk profile indicates the likelihood that any undesirable outcome will occur. Figure 12.21 provides the risk profile for each alternative. In this figure, only the crucial portions of the risk profiles are shown (i.e., where the least desirable outcomes occur). The risk profiles are extremely close for the three alternatives. However, the risk profile for Install Water Curtain confirms that choosing this alternative helps reduce the likelihood of extreme outcomes. With all three alternatives there is a probability of .002 or less that the utility will be below –5.0.

TABLE 12.10
Expected Utility for Distillation Column Alternatives

Alternative	*Expected k_1u_1(Cost)*	*Expected k_2u_2(Fatalities)*	*Expected k_3u_3(Goodwill)*	*Total Expected Utility*
Do Nothing	0	–0.014	–1.500	–1.514
Use Two Smaller Columns	–0.667	–0.007	–0.975	–1.649
Install Water Curtain	–0.167	–0.005	–0.900	–1.072

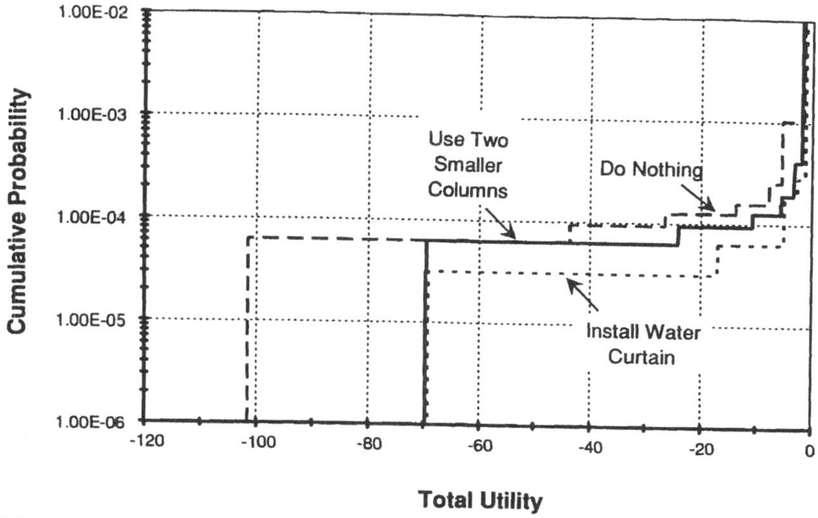

FIGURE 12.21. Risk profiles for utility of the alternatives.

Sensitivity analysis can be performed on key inputs, including parameters of the multiattribute utility function, uncertainties, and cost estimates. These analyses can help support the recommendations and provide backup detail that can be used to answer questions about the assumptions underlying the analysis. Figure 12.22, for example, shows how the expected utility changes as the cost of Install Water Curtain changes. This output helped respond to a

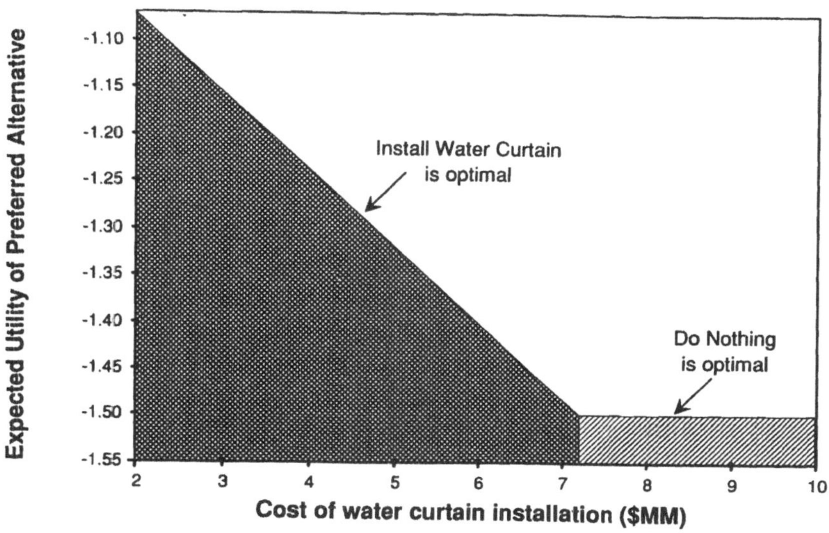

FIGURE 12.22. Sensitivity of optimal decision alternative and expected utility to cost of water curtain installation.

reviewer's concern that the cost of the disruption in operations caused by installing the water curtain may have been underestimated. The sensitivity analysis indicated that Install Water Curtain is the best alternative unless its cost exceeds $7.3 million. In this case, to Do Nothing would be optimal. Similar results could be developed for other key inputs to this decision.

Value-of-information analysis can be performed on key uncertainties. The results can help indicate whether it is worthwhile to obtain additional information to help reduce uncertainties prior to making a final decision. For example, we might want to know how much to pay for a study that would reduce the uncertainty on the ignition of a release. Value-of-information calculations are illustrated in Section 11.4.5.

12.4.5 Communication Phase

The final phase involves communicating the assumptions, results, and implications of the analysis and obtaining support for implementing the decision. In this case, the analysis team presented to the corporate team an overview of the structure of the analysis in the form of a simple influence diagram. The analysis team presented the expected value of each alternative and the risk profile for the recommended alternative. The presentation concluded with a recommended action plan. Backup charts and figures like those shown earlier were used to respond to questions that arose during the final presentation.

12.5 EXTENSIONS OF MULTIATTRIBUTE UTILITY ANALYSIS

Multiattribute utility analysis is a flexible technique that is equipped to address a broad range of problems. Much of the attention on extending MUA focuses on improving and broadening the scope of application. Important issues include simplifying applications and determining whether and how MUA might be implemented as part of the public participation process that is often essential for the solution of large-scale public policy problems.

Simplification is needed to enable the approach to be used successfully by those other than highly experienced experts in the technique. Many of the necessary steps to properly apply MUA are more difficult than they might first appear. For example, applying the tests necessary to determine the proper functional form of the utility function is sufficiently difficult that many applications simply assume a linear, "weight and add" form without verifying that this form is correct. Although the weight-and-add form is nearly always intuitively pleasing, assuming the weight-and-add form when it is not appropriate can cause the analysis to produce incorrect conclusions. Such errors of application are subtle and may go undetected unless and until the analysis is subjected to technical peer review.

Applying MUA in analyses with stakeholder participation is a challenge, because the stakeholders to a decision may hold very different values and objectives. Obtaining accurate value inputs without overwhelming the person(s) providing these inputs, is a key concern. For example, Keeney, von Winterfeldt, and Eppel (1990) describe an approach they call the "public value forum" that combines elements of group discussion techniques (such as focus groups) and of value elicitation techniques (such as swing weighting [von Winterfeldt and Edwards 1986]). Merkhofer (1987) and Howard (1975) have addressed societal decision making and group decision processes employing MUA. Benefiting from the growing use of MUA in the public sector, practitioners are testing ways to ensure that MUA effectively and accurately captures the judgments of opinion leaders and stakeholders involved in the decision.

Dyer et al. (1992) provide a perspective on this field with suggested directions for future development.

12.6 IMPLEMENTATION NEEDS

Key inputs to a decision analysis are the alternatives, uncertainties, and preferences. Alternatives are the possible actions that the decision maker can take, and are generated through brainstorming or other techniques during the formulation phase. Uncertainties affect the possible outcomes from the decision, and are identified and structured during the deterministic phase. At the same time, other deterministic inputs to the outcome model are developed and the outcome model specified.

The preferences used in the decision analysis are represented by a multiattribute utility function. This function incorporates the values for outcomes associated with each of the attributes, the relative weights on the various objectives, the preferential dependencies among objectives, risk preference, and time preferences or discounting. To develop this function requires identification of (1) the objectives and an attribute measuring performance on each objective, (2) a single-attribute utility function for each attribute, and (3) weighting factors for the attributes. More complex formulations of the utility function may require assessment of additional parameters, such as those capturing the cross-impact of attributes that are preferentially dependent. In some applications, uncertainties may be incorporated into the analysis.

The training required to conduct a multiattribute utility analysis is similar to what was described for decision analysis. Initially it is suggested that the analyst(s) be trained in decision analysis and/or work with a professional who has experience in decision analysis. Additional training is required in how to assess a multiattribute utility function and incorporate it into the decision analysis. This training may be a part of an introductory course in decision analysis, but is more typically covered in a self-contained course. A variety of textbooks, journal articles, and training courses is available, as described in Section 12.7.2.

A multiattribute utility analysis will typically involve more time and staff resources than a single-attribute decision analysis, and these decision aids are generally more resource-intensive than others. A primary reason for this resource intensity is the effort required to formulate and customize the approach to individual problems. The information and resource requirements depend on the complexity of the problem and the depth of analysis desired. A "back-of-the-envelope" decision analysis can be performed in a few minutes or hours. The multiattribute utility function in such an analysis could take the form of a simple weighted sum of a few key attributes on which the alternatives are scored, as used in weighted-scoring methods. A typical multiattribute utility analysis can range in scope from a 2- to 3-person team working for several weeks, to efforts involving several persons for up to a year (as in the DOE study described earlier). However, a simplified MUA can be performed in a matter of days if warranted by the problem.

Decision analysis software is available to help with structuring, model-building, analysis, and reporting of results. Most decision analysis software can handle multiattribute utility functions. Spreadsheets and other computational models can be employed for aspects of the problem. Software also exists to develop the multiattribute utility function and assess the defining parameters.

12.7 SUMMARY

12.7.1 Strengths and Weaknesses of Multiattribute Utility Analysis

Multiattribute utility analysis is especially designed for problems involving complex value tradeoffs among several objectives with incommensurate attributes, such as dollars and fatalities. Factors that are difficult to quantify, such as "corporate image" and "public concern," can also be addressed.

As with decision analysis, MUA incorporates both uncertainties and values explicitly into the analysis. Other methods may not distinguish between uncertainties about a possible outcome and the value for the outcome. Instead, uncertainties are ignored, treated as a separate concern, or combined implicitly with the outcomes. This can result in decisions that do not correctly reflect the decision-maker's knowledge and preferences.

As with a DA, there is a sound theory behind the MUA approach. The results of the analysis have a logical, rigorous basis that is documented and defensible. The procedures systematically elicit the relevant information about value tradeoffs and risk aversion. Sensitivity analyses can be performed that explicitly identify the role of uncertainties and value judgments.

As disadvantages, MUA involves more time and resources than do other approaches. Training in decision analysis and in multiattribute utility assessment are required. The training needed to practice decision analysis may lead to more specialized skills than many organizations need to have in-house, and thus outside decision analysis professionals are often brought in to help with larger problems.

Multiattribute utility analysis is not specifically designed to address problems involving multiple decision makers. In fact, one can prove mathematically that no simple procedure exists for aggregating the value weights of different individuals. The problem is analogous to the impossibility of a voting rule that consistently ranks all candidates in a runoff with more than two choices (see Chapter 6 on voting paradoxes). However, multiattribute utility analyses have been successfully applied in group settings. The overall value can be separated into underlying objectives, with appropriate individuals involved in specifying values for each objective. The explicitness of the assumptions and logic underlying the analysis can facilitate group decision making, since the analysis can easily be reviewed by all participants and refined as necessary.

12.7.2 Suggestions for Further Information/Study

Textbooks, journals, and courses are available for those who wish to further explore the theory and application of multiattribute utility analysis. Clemen (1991) provides a good introduction to decision analysis, and Keeney and Raiffa (1976) provide an introduction to multiattribute utility analysis and describe a number of applications. Von Winterfeldt and Edwards (1986) also provide a useful introduction to decision analysis and multiattribute utility analysis. Keeney (1992) targets decision makers with some quantitative background.

Seminars are available on both decision analysis and multiattribute utility analysis. These courses are an effective way to learn the techniques since they are usually taught by experienced practitioners and incorporate practical considerations and insights. Relevant journals include *Multi-criteria Decision Analysis, Management Science, Operations Research,* and *Risk Analysis*. The journal *Interfaces*, published by The Institute for Management Sciences (TIMS), provides applications-oriented articles.

To gain familiarity with the practical application of MUA, we suggest reading Bell, Keeney, and Raiffa (1977); Keeney (1992); Merkhofer and Keeney (1987); von Winterfeldt and Edwards (1986); or other applications referenced in Section 12.2.2.

12.8 REFERENCES

Bell, D., R. Keeney, and H. Raiffa (eds.) (1977). *Conflicting Objectives in Decisions*. New York: John Wiley and Sons.

Bohn, M. P. (1992). "Decision Making under Uncertainty: An Investigation into the Application of Formal Decision-Making Methods to Safety Decisions" (NUREG/

CR-5906). Division of Safety Issues Resolution, Office of Nuclear Regulatory Research, U.S. Nuclear Regulatory Commission.
Clemen, R. T. (1991). *Making Hard Decisions: An Introduction to Decision Analysis*. Boston: PWS-KENT.
Dyer, J. S., P. C. Fishburn, R. E. Steuer, J. Wallenius, and S. Zionts (1992). Multiple Criteria Decision Making, Multiattribute Utility Theory: The Next Ten Years. *Management Science* 38(5):645–654.
Gregory, R., and S. Lichtenstein (1987). Review of the High-Level Nuclear Waste Repository Siting Analysis. *Risk Analysis* 7(2):219–223.
Howard, R.A. (1975). Social Decision Analysis. *Proceedings of the Institute of Electrical and Electronics Engineers* 63(3):359–371.
Keeney, R. L. (1992). *Value-Focused Thinking*. Cambridge, MA: Harvard University Press.
Keeney, R. L., and H. Raiffa (1976). *Decisions with Multiple Objectives: Preferences and Value Tradeoffs*. New York: John Wiley and Sons.
Keeney, R. L., R. K. Sarin, and R. L. Winkler (1984). Analysis of Alternative National Ambient Carbon Monoxide Standards. *Management Science* 30(4):518–528.
Keeney, R. L., D. von Winterfeldt, and T. Eppel (1990). Eliciting Public Values for Complex Policy Decisions. *Management Science* 16(9):1011–1030.
Merkhofer, M. W. (1987). *Decision Science and Social Risk Management: A Comparative Evaluation of Cost–Benefit Analysis, Decision Analysis, and Other Formal Decision-Aiding Approaches*. Dordrecht, Holland: D. Reidel (Boston: Kluwer Academic Publishers Group).
Merkhofer, M. W., T. A. Cotton, and T. P. Longo (1988). A Program Optimization System for Aiding Decisions to Fund the Cleanup of Hazardous Waste Sites at Department of Energy Defense Facilities. Presented at *Superfund 1998, Proceedings of the Ninth National Conference*, sponsored by the Hazardous Materials Control Research Institute, November 28–30, 1988, Washington, D.C.
Merkhofer, M. W., and R. L. Keeney (1987). A Multiattribute Utility Analysis of Alternative Sites for the Disposal of Nuclear Waste. *Risk Analysis* 7(2):173–194.
Raiffa, H. (1969). "Preferences for Multiattributed Alternatives" (RM-5868-DOT/RC). Santa Monica, CA: Rand Corporation.
Torrance, G. W., M. H. Boyle, and S. P. Horwood (1982). Application of Multi-Attribute Utility Theory to Measure Social Preferences for Health States. *Operations Research* 30(6):1043–1069.
von Winterfeldt, D. (1987). Value Tree Analysis: An Introduction and an Application to Offshore Oil Drilling. In P. Kleindorfer and H. Kunreuther (eds.), *Insuring and Managing Hazardous Risks: From Seveso to Bhopal and Beyond*, pp. 349–377. New York: Springer-Verlag.
von Winterfeldt, D., and W. Edwards (1986). *Decision Analysis and Behavioral Research*. Cambridge: Cambridge University Press.
Wenstop, F. E., and A. J. Carlsen (1988). Ranking Hydroelectric Power Projects with Multicriteria Decision Analysis. *Interfaces* 18(4):36–48.

13

REVIEW OF CASE STUDIES

Who shall decide when doctors disagree,
And soundest casuists doubt, like you and me?
—Alexander Pope, Moral Essays, Epistle iii

13.1 PURPOSE OF CHAPTER

In this chapter, each case study is summarized briefly with respect to the key differences in the way that the selected decision aids were applied to the decision problem. The three case studies used in the book were the Underground Pipeline, the Chlorine Rail Tank Car Loading Facility, and the Distillation Column. (The background of the case studies was presented in Chapter 5 and will not be completely reviewed here.)

Chapter 4 presented key distinguishing characteristics that can be used to evaluate decision aids. These are:

- Resource requirements
- Depth of analysis/Complexity
- Logical rigor
- Group focus
- Quantitativeness
- Track record.

The case studies served to illustrate some differences among the decision aids on the first five of these characteristics, as we point out in this chapter.

The results obtained with one decision aid should not be compared with those of another. In some chapters the case study description was modified so that a particular aid would be better suited for the case. Although a case is used with several decision aids, the assumptions and inputs were varied to suit each decision aid.

13.2 CASE ONE: UNDERGROUND PIPELINE

The Underground Pipeline case was used as an illustration for three decision aids:

- Voting Methods (Chapter 6)
- Mathematical Programming (Chapter 9)
- Decision Analysis (Chapter 11)

This case involved concerns over aging underground pipelines. For the voting and decision analysis applications, a single plant manager was faced with a decision of what to do with an aging section of an underground pipeline. In the mathematical programming application the case was modified. A director of safety was faced with a decision of how to allocate a budget among six plants with aging underground pipelines. Each plant submitted one or more action plans for reducing risks from sections of pipelines and the director of safety needed to decide which activities to fund.

The various decision aids met different objectives. In the voting application, the plant manager was concerned with incorporating as many perspectives as possible into the decision. Voting provided a group decision-making method and allowed all group members to see the views of others. In the math programming example, the director of safety faced a resource allocation problem. Math programming was an appropriate aid because the goal was unambiguous (i.e., select action plans so as to maximize total benefits) and the factors that constrain the set of alternatives were clear (i.e., the budget was limited and at least one plan from each facility was to be funded. The decision analysis application focused on the uncertainty around a potential failure, including the type of failure, the size of a release, the years until failure, and the clean-up costs.

Math programming and decision analysis provided much more depth of analysis, but had greater resource requirements and involved a greater degree of quantitativeness. Voting had less depth and quantitativeness, so the decision was reached in a short time frame. A key example of the difference in depth is that the decision analysis allowed for a calculation of a monetary value of conducting tests on the pipeline. In voting, metallurgy and ultrasound tests were included as alternatives, but the measure of their worth was given by the number of votes they received, not a dollar value. As another example, sensitivity analyses were performed with both the math programming and decision analysis applications.

The decision analysis and math programming applications are logically rigorous and analytically sound. The voting application, on the other hand, was less rigorous. The results were subject to possible manipulation or "gaming." In addition, the criteria upon which the alternatives were evaluated in voting varied by individual. This can result in a choice that does not truly reflect the preferences of the group.

Voting facilitated having many perspectives incorporated into the process. Each group member was involved in the decision. The mathematical programming and decision analysis applications did not have a group focus, but the opinions and knowledge of others were incorporated as inputs to the analyses.

13.3 CASE TWO: CHLORINE RAIL TANK CAR LOADING FACILITY

This case concerned a decision regarding an existing chlorine loading facility situated close to a populated residential area. The company had conducted a quantitative risk analysis and was deciding what actions, if any, should be taken to reduce the risk associated with the facility. The two decision aids applied to the problem were:

- Cost–Benefit Analysis (Chapter 8)
- Payoff Matrix Analysis (Chapter 10)

For the cost–benefit analysis, an analysis team was formed to evaluate many alternatives in detail. Input was obtained as required from operations, maintenance, plant process engineering, logistics, and corporate-level legal, engineering, and business groups. The payoff matrix analysis had more of a group focus. A less detailed analysis of the problem was made by assembling a cross-section of plant staff. The result was a plan of action after a 2-day off-site session.

The cost–benefit study provided greater depth of analysis, at the cost of greater resource requirements. It involved a thorough evaluation of the decision alternatives while the payoff matrix analysis provided insights into the key issues and a plan for next steps. In the cost–benefit analysis, a broad range of factors including community goodwill and property values were addressed in the analysis. An extensive amount of data was gathered, and the alternatives were analyzed from both corporate and societal viewpoints. The highly quantitative approach allowed the analysis team to perform various sensitivity analyses. The payoff matrix analysis focused strictly on the impacts to the company. Although the payoff matrix analysis identified the key factors in the analysis, the company was not comfortable in selecting an alternative after the 2-day session. However, clear recommendations for further data gathering and a detailed risk analysis were developed.

13.4 CASE THREE: DISTILLATION COLUMN

This case involved an existing plant that employs a distillation column to separate hexane and heptane. The operation is located near a residential area. A risk analysis was performed and the company was faced with a decision on what action to take, if any, to reduce the risk from the column. The two decision aids applied to this case study were:

- Analytic Hierarchy Process (AHP) (Chapter 7)
- Multiattribute Utility Analysis (MUA) (Chapter 12)

The analytic hierarchy process was used in this case to do a quick analysis to examine the relative attractiveness of the decision alternatives. The same alternatives were analyzed in the MUA case study, but in greater depth. A primary difference in the two analyses was that MUA treated uncertainty quantitatively by defining possible future outcomes and estimating their likelihoods of occurrence. In addition, the MUA separated outcomes and data from value judgments, while AHP did not. AHP involved making assessments of how the alternatives perform on each of three decision criteria, without explicitly considering possible outcomes and their likelihoods of occurrence. Because of the separation of outcomes from values, MUA had greater logical rigor but larger data and analysis requirements.

The multiattribute utility analysis was more logically rigorous in its treatment of the decision-makers' relative preferences for the three decision criteria—cost, risk to the community, and community goodwill. In using weighted scoring methods, a single weight for each criterion was developed. This weight was applied over the entire range of values for the criterion, assuming a constant marginal utility. In MUA, the decision-makers' preferences were quantified in a utility function, which allowed the marginal utility to vary over the range of values for a given criterion. In addition, in the AHP application, we saw that the criteria were not tied to concrete measures upon which the judgments could be made. Therefore, the numerical judgments were subject to individual interpretation and inconsistencies.

14

IMPLEMENTING IMPROVEMENTS IN RISK DECISION MAKING

Competing pressures tempt one to believe that an issue deferred is a problem avoided; more often it is a crisis invented.
—Henry Kissinger

14.1 PURPOSE OF CHAPTER

The decision aids described in this book can benefit many organizations. However, the novelty of the decision aids and lack of corporate experience with them may hamper their use. The purpose of this chapter is to provide some guidance on how to encourage the use of these tools to improve risk decision making.

To promote better decision making, four elements are essential:

- *Responsibility:* The responsibilities for effective decision making must be understood and accepted by appropriate individuals. Responsibilities include making the decision, analyzing the decision, and providing input to the decision.
- *Commitment:* There must be a commitment to making decisions that are logical, rigorous, and defensible; to establishing clear guidance for these decisions; and to consistently allocating sufficient resources to support quality decisions.

- *Knowledge:* The corporate philosophy must be understood, and participants in decisions must have adequate expertise in the decision aids that support quality decision making.
- *Experience:* Quality decision making must be made routine throughout the organization, and experience must be accumulated in a wide range of decision contexts.

The following section elaborates on how these elements can be established.

14.2 KEYS TO IMPLEMENTATION

Various approaches can be used to implement better approaches to risk decision making. Some organizations simply hope that it happens as a natural result of good management. Some organizations commit themselves to rapid and massive change, organized from the top, and backed by significant resources. Others empower specific groups to experiment, and then foster those approaches that work best.

Broad corporate initiatives for change can create a positive atmosphere for implementing higher quality approaches to risk decision making. Many organizations in the chemical process industry are undertaking significant initiatives to change how they conduct business and deal with chemical process risks.

Because good risk decision making requires strong cooperation among different parts of the organization, leadership is needed to improve risk decision making throughout the organization. Without management's active support, change will occur slowly. Leadership is needed to instill into the organization the four elements that support good decision making: responsibility, commitment, knowledge, and experience.

14.2.1 Responsibility

The responsibilities for creating quality decisions are depicted in Figure 14.1. A single individual or different individuals may fulfill these responsibilities; however, quality decisions require that the right person takes on each responsibility. The *decision-making* responsibilities include clearly stating the specific organizational goals and the values that are to be maximized through the decision. These responsibilities must be assumed by the decision maker or decision-making body and cannot be delegated without harming the decision-making process. Second, there is a responsibility for *expertise*. Within the time and resource constraints imposed by the problem, the most knowledgeable experts inside and outside the firm should be responsible for supplying unbiased information and evaluations. Finally, there is responsibility for *analysis* of the decision. The individual or individuals doing the analysis are responsible for using the inputs in a logical and unbiased fashion and for gaining support for proposed solutions.

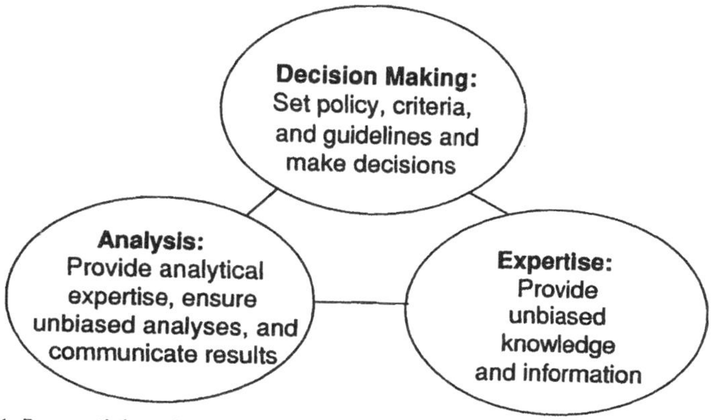

FIGURE 14.1. Responsibilities for making quality decisions.

14.2.2 Commitment

A first step toward commitment is often for the organization to support pilot projects or educational efforts that introduce new decision aids and decision-making approaches. Management's direct involvement is invaluable in early pilot projects. When pilot projects prove successful management can then confidently expand the organization's commitment.

A more substantial and long-run commitment to improved decision-making approaches is established through clear policies and given substance through guidelines and standards. *Policies* express the principles to use in decision making, *guidelines* provide a target for implementing a policy, and *standards* control design and operation in order to meet guidelines. In setting policies and guidelines, there is a tradeoff between flexibility and control. If a very general policy is chosen and not specified through guidelines and standards, management must be prepared to accept a variety of interpretations and a level of inconsistency in decision making.

Another important part of commitment is providing sufficient support for quality decision making. This commitment includes training, well publicized access to internal and external expertise, and provision of resources for analysis and decision making.

Finally, commitment is demonstrated by support for quality decision making. Management should emphasize the importance of understanding and documenting the rationale behind decisions, and should expect that a significant decision will be underpinned by a thorough analysis.

14.2.3 Knowledge

The quality of risk decision making is improved by more widespread knowledge of decision aids and greater experience with their use. Table 14.1 illus-

TABLE 14.1
Knowledge of Decision Aids Needed in the Organization

Responsibility	*Required Knowledge*	*Why*
Decision Making	Overview	To understand when and what decision aids are useful in decision making and the role of decision makers in the decision process.
Expertise	Awareness	To understand how information will be used and how to most effectively support decision making.
Analysis	Detailed knowledge	To understand when and what decision aids are useful, to lead analysis teams, and to promote quality decision making.

trates the knowledge needed to assume different responsibilities in the decision process.

Many persons in every organization make decisions of such a scope and importance that formal decision aids may be useful. Thus, a very broad range of managers, senior support staff, and others who help to make or are significantly affected by risk decisions should have an overview of decision aids. The overview should encompass the types of decision aids that exist, the level of corporate support for the use of decision aids, the situations in which decision aids are useful, and where to go in the organization for decision-making support. Decision makers must also have access to information on the required staff, resources, time, and other inputs needed in the use of a decision aid. Finally, decision makers must understand that the use of some decision aids requires decision makers to specify decision-making guidelines and criteria.

For persons with frequent responsibility for providing expertise in decision studies, a somewhat higher level of awareness may be useful. When experts understand and support how their information will be used, they usually work harder and more effectively to supply useful information.

Persons responsible for analysis need to have an overview of decision aids and a more detailed understanding of the specific aids that they will use in an analysis. Their knowledge of the decision aids that they will use should cover the theoretical basis of the aid and its practical application. Knowledge is built through both training and experience. The analyst should also be effective in team leadership and project management.

14.2.4 Experience

Three steps to building experience are to:

- Demonstrate that formal decision aids can improve decision making
- Develop expertise in using the aids and adapt the decision aids to the specific organization

- Build a history of using decision aids and promote their use throughout the organization.

The initial step in gaining experience in quality decision making is to demonstrate that the use of formal decision aids does indeed help the decision maker and result in better decisions. One way to begin is to work with in-house or outside experts to create some winning projects that exemplify quality decision making. In sharing these success stories, emphasize the analysis that led to the right decision.

The second or developmental step in gaining experience involves refining the use of decision aids and making use of the lessons learned in earlier applications of decision aids. During this stage, policies will be redefined and restated as the organization learns what works in its unique situation. The organization will learn who can make the best use of training and support. Finally, the organization's confidence in formal decision aids will grow. During this stage, a core team of internal decision experts can be a focus and a repository for organizational memory regarding decision making.

The organization can continue to improve decision making over the longer term by preserving organizational memory. This ensures that its policies are stable and consistent, reinforces internal expertise, contributes to training, and allows successes and failures to be recognized. The organization should require that analyses of decisions be documented and should encourage their circulation both internally and externally. For example, some firms have a regular newsletter that is published by their decision support center. Organizations can have internal experts help teach others. Only internal experts can communicate the organizational memory and the unique constraints on decision making within the organization.

14.3 SUMMARY

Responsibility, commitment, knowledge, and experience are crucial to an organizations implementation of the decision approaches described in this book. Responsibility ensures that each participant contributes most effectively to the decision. Commitment ensures that new decision processes spread from pilot projects, demonstrations, and education, to organization-wide application. Knowledge ensures the quality and efficiency with which new decision approaches are executed. Experience ensures that the organization is equipped to handle new decision-making approaches, and that the entire organization learns from previous applications.

15
FUTURE DEVELOPMENTS

The basic texture of research consists of dreams into which the threads of reasoning, measurement, and calculation are woven.
—**Albert Szent-Györgyi**

15.1 PURPOSE OF CHAPTER

This chapter outlines directions for research and development aimed at improving understanding of decision making and methods to support decision making. The chapter begins with a description of the decision-making research field, including relevant journals and research-sponsoring agencies. This is followed by a discussion of specific areas for future research and development.

Many of the research ideas listed in this chapter are from Bouyssou et al. (1993), Dyer et al. (1992), and Simon et al. (1987). The interested reader should consult these articles and others listed in the topical bibliography under the category "Future Research" for more details on the state of current research.

15.2 THE FIELD OF RESEARCH ON DECISION MAKING

Relevant research is occurring in a number of academic areas. These include:

- Applied Mathematics
- Business Management
- Computer Science
- Economics
- Operations Research
- Psychology.

Much of the relevant work falls within a field currently referred to as the "Management Sciences" or "Decision Science" area. Professional organizations supporting work in this field include:

- Institute for Operations Research and the Management Sciences (INFORMS) (formerly the Operations Research Society of America [ORSA] and The Institute of Management Sciences [TIMS])
- International Institute for Applied Systems Analysis (IIASA)
- International Society on Multiple Criteria Decision Making
- Society for Judgment and Decision Making
- Society for Medical Decision Making.

Important sources of information on new theoretical and practical developments include the following journals and newsletters:

- *Decision Analysis Newsletter of the ORSA Special Interest Group on Decision Analysis*
- *Decision Sciences*
- *Decision Support Systems*
- *European Journal of Operational Research*
- *Group Decision and Negotiation*
- *IEEE Transactions on Systems, Man, and Cybernetics* (formerly the *IEEE Transactions on Systems Science and Cybernetics*, published by the Institute of Electrical and Electronics Engineers)
- *Interfaces*
- *Journal of Behavioral Decision Making*
- *Journal of Experimental Psychology*
- *Journal of Multi-Criteria Decision Analysis*
- *Journal of the Operational Research Society* (formerly *Operational Research Quarterly*)
- *Judgment and Decision Making Newsletter* (Society for Judgment and Decision Making)
- *Management Science*
- *MCDM WorldScan Newsletter*

- *Medical Decision Making*
- *Omega*
- *Operations Research*
- *Operations Research Letters*
- *Organizational Behavior and Human Decision Processes*
- *Risk Analysis.*

Research on decision making and on decision-aiding methodologies is being supported by:

- National Science Foundation (NSF)
- Advanced Research Projects Agency (ARPA)
- Electric Power Research Institute (EPRI)
- Office of Naval Research
- National Institutes of Health.

At the NSF, projects are sponsored by the Social Sciences Division; the Decision, Risk, and Management Science (DRMS) Program; the Division of Behavioral and Neural Sciences; and the Directorate of Computer Science and Engineering. As examples of the types of work supported by these agencies, NSF research addresses decision analysis and other decision aids, probability theory, statistical and other mathematical techniques, decision-making behavior, and expert judgment. ARPA supports research in computer science, artificial intelligence, decision-support systems, and decision analytic methods. NIH research addresses medical decision making, expert judgment, management of cumulative experience, and analyses of specific categories of medical decisions. EPRI sponsors research on risk perceptions, decision analysis methodologies, and the CATALYST approach to payoff matrix analysis, with particular focus on R&D and investment decisions in the industry.

Research that more specifically addresses risk decision making in the chemical industry is supported by the American Institute of Chemical Engineers (AIChE), particularly through its Center for Chemical Process Safety (CCPS), and by the Chemical Manufacturers Association (CMA). CCPS supports work to develop improved techniques, written guidelines, and training programs to support many aspects of risk management including hazard evaluation, risk analysis, consequence modeling, practices for storage and transport of hazardous materials, mitigation strategies, risk analysis, and risk decision making. CMA-supported research addresses risk analysis, risk management, and risk communication.

Within Europe, IIASA has been a focal point for research on decision making. IIASA seeks to bring multi-attribute and multiobjective decision aids to bear on complex problems such as those involving natural resources, environmental quality, and population issues (Dyer et al. 1992).

15.3 SPECIFIC AREAS OF RESEARCH

This section provides brief descriptions of current and future research areas:

- Behavioral or descriptive research
- Ways to facilitate real-time decision making and negotiation
- Methods to facilitate problem definition and problem structuring
- Techniques to help generate and screen alternatives
- Methods to ease data gathering and improve data quality
- Improved ways to communicate results
- Extensions to address specific issues in risk decision making
- Improving the credibility/acceptance of formal decision making
- Improvements to existing decision aids.

What follows is a discussion of each of these research areas.

15.3.1 Behavioral or Descriptive Research

This field aims to understand how people process information and reach decisions, which can support development of more practical methods reflecting the decision-making processes of individuals and groups. Some areas of inquiry include research on:

- How the human mind processes information and solves problems
- How individuals make decisions, and how this varies with characteristics of the individual such as wealth, attitudes towards risk activities, and cultural background
- The attitudes that individuals and organizations have toward risk and uncertainty, either as demonstrated through previous decisions and actions, or as elicited through experimental research
- The values that people have for various objectives, either as implied by past decisions or as elicited experimentally
- How groups make decisions, and how this depends upon aspects such as the type of problem addressed and the roles of the individual participants
- How organizations make decisions, and how this can change with factors such as the level of the organization in which the decision is made, its size, its attitude toward making risky investments, and corporate culture
- The hidden agendas and biases that individuals or groups may have.

15.3.2 Ways to Facilitate Real-Time Decision Making and Negotiation

Many of the decision aids discussed in this book involve modeling, data collection, and analysis that is done by an analyst or analytic team separated somewhat from the decision maker. CATALYST (Section 10.3.3) and various extensions to voting methods (Section 6.5) are examples of techniques designed to put the analysis closer to the decision maker, ideally as a group process in which the problem is analyzed and a decision reached in a session or series of sessions without the need for separate analytic and data-gathering steps. Work in this area includes:

- Improving upon existing decision aids such as decision analysis and weighted scoring methods to streamline the modeling, data-gathering, and analysis steps and to facilitate real-time use of the technique
- Developing other type of decision aids or processes to facilitate decision making, especially real-time group decisions and negotiations
- Designing approaches that allow the problem to be addressed progressively with the depth and direction of analysis indicated by the results of earlier iterations.

15.3.3 Methods to Facilitate Problem Definition and Problem Structuring

Particularly difficult are decisions that address poorly specified problems or situations in which the goals are not well-defined. Research in this area is aimed at helping individuals and groups in the early stage of analyzing a decision, when the problem is first being scoped out. Some areas of research include:

- Approaches to agenda setting and problem framing to best focus the initial work on solving the problem
- Ways to represent problems visually or in words so as to facilitate communication and subsequent analysis
- Tools for building models of the problem, such as influence diagrams and decision trees
- Methods that more tightly integrate the visual representation of the problem structure with the data-gathering and analysis steps.

15.3.4 Techniques to Help Generate and Screen Alternatives

Many of the decision aids assume that an appropriate set of alternatives has been identified prior to the application of the decision aid. Some decision aids include processes to facilitate the identification of appropriate alternatives. In some cases, there are too many alternatives to consider, and a screening procedure is required to reduce the number of alternatives initially evaluated. Some areas of research include:

- Ways to facilitate group brainstorming and other processes aimed at eliciting alternatives

- Methods for using the results from the initial analysis of the decision to identify new alternatives to evaluate
- Screening procedures to rule out alternatives that are dominated or likely to be dominated by other alternatives, or that are a suitably representative subset of the alternatives.

15.3.5 Methods to Ease Data Gathering and Improve Data Quality

For most decision aids, key inputs to the analysis of a decision include the preferences and knowledge of the decision maker. Some key areas of research include:

- Tools for eliciting and validating the judgmental uncertainty and preferences of a decision maker or expert
- Tools which can help estimate uncertainty or value inputs based on incomplete or imprecise data, such as fuzzy set theory
- Tools that address the findings of psychological research on judgmental biases and heuristics
- Methods for successively or even interactively eliciting these inputs as-needed, to help focus data gathering on the inputs that are crucial to selecting the best alternative
- Methods for combining objective data such as historical failure rates with subjective data such as judgmental uncertainty
- Methods such as Bayesian analysis that improve upon an initial uncertainty estimate as information is obtained
- Methods to incorporate organizational learning into the analysis, for example, techniques to infer organizational values from past decisions and to update these values as new decisions are made
- Techniques for combining the judgments of multiple experts or forecasts by multiple sources
- Processes that help groups resolve differences in judgmental uncertainty or conflicting value judgments
- Methods for simplifying or reducing the number of uncertainties addressed in an analysis, for example, by coalescing uncertainties into scenarios.

15.3.6 Improved Ways to Communicate Results

Some important areas of research include:

- Ways to efficiently communicate the key assumptions and logic underlying the analysis
- Graphical aids for communicating results of sensitivity analyses and conveying key inputs

- Tools to help communicate with the public and other stakeholders about the basis and results of the analysis; in particular communicating risks meaningfully to the public is an important issue.

15.3.7 Extensions to Address Specific Issues in Risk Decision Making

Decisions that address process safety involve some unusual aspects; research is attempting to improve the ability of decision aids to address these issues. As examples, some key aspects of risk decisions include:

- Extremely low-probability and high-consequence events, whose risks are difficult to estimate
- Short-term investments, with risks and rewards that occur over a very long time frame
- Sharply conflicting values between members of the public, the corporation, and other stakeholders
- Tradeoffs between meeting the public's desire for more information, and avoiding causing undue public concern
- Quantifying the "value of life" or the benefit of reduced fatality risk.

15.3.8 Improving the Credibility/Acceptance of Formal Decision Making

This book suggests the applicability of more formal decision-making processes using various decision aids. Affecting the acceptance of this approach is the concern organizations have about extending these processes to new applications. Some key research areas include:

- Empirical research to document the use and acceptance of various approaches in specific industries and public policy areas, and the range of applications addressed
- Documentation of specific examples or case studies in which various decision aids were applied in the past
- Experimental applications or case studies that demonstrate use of a decision aid on representative risk decision problems and help identify ways to improve the applicability of the method for these decisions
- Research on the organizational contexts and decision situations in which various decision aids are most likely to succeed or have been most successful
- Legal research on the risks of formal decision making, for example, whether revealing the explicit basis for a safety-related decision increases or decreases the organization's legal vulnerability
- Research on the roles that regulatory agencies could or should play in risk-related decisions of the industry, for example, the use of standards for the type of analysis, inputs to the analysis, and alternatives considered.

15.3.9 Improvements to Existing Decision Aids

Each of the decision aid chapters (6–12) describe specific extensions and enhancements to the methodology. As general improvements to decision aids, some areas for research include:

- Developing hybrid approaches that pick and choose among alternative decision aids and tools as required by the problem
- Improvements in the useability of software to help guide someone through the analysis who may have little or no training in a decision aid
- More powerful software to enable larger and more complex problems to be solved, and to better address real-time decision making
- Software that is customized to specific risk decision problems, such as a generic tool to help allocate resources among competing safety projects, or a generic tool for choosing among alternatives each characterized by their dollar cost and a resulting F-N curve
- Applications of artificial intelligence, expert system, learning system, or decision support system approaches to risk decision making
- Training programs that expose management to formal processes for risk decision making and that enable organizations to implement more sophisticated decision aids using in-house staff.

15.4 REFERENCES

Bouyssou, D., P. Perny, M. Pirlot, A. Tsoukias, and P. Vincke (1993). A Manifesto for the New MCDA Era. *Journal of Multi-Criteria Decision Analysis* 2:125–127.

Dyer, J. S., P. C. Fishburn, R. E. Steuer, J. Wallenius, and S. Zionts (1992). Multiple Criteria Decision Making, Multiattribute Utility Theory: The Next Ten Years. *Management Science* 38(5):645–654.

Simon, H. A., G. B. Dantzig, R. Hogarth, C. Plott, H. Raiffa, T. C. Schelling, K. A. Shepsle, R. Thaler, A. Tversky, and S. Winter (1987). Decision Making and Problem Solving. *Interfaces* 17(5):11–31.

APPENDIX

A

SOFTWARE

This appendix provides a list of commercial software packages used to implement some of the decision aids discussed in this book:

Cost–Benefit Analysis (Chapter 8)
Decision Analysis (Chapter 11)
ELECTRE (Chapter 3)
Mathematical Programming (Chapter 9)
Multiattribute Utility Analysis (Chapter 12)
Payoff Matrix Analysis (Chapter 10)
Weighted Scoring Methods (Chapter 7)

The list includes both packages with which we are familiar and other packages found in published software surveys. We have not conducted a comprehensive search, but only intend to give the reader a starting point in finding software. The references listed at the end of this section were used in compiling this Appendix.

We have not reviewed the software packages and make no claims about their capabilities. We caution the reader to examine the methodology used before selecting software, since terminology differences exist. For instance, the term "decision analysis" is used broadly and often applied to weighted scoring methods. The software packages vary in the amount of complexity they can handle and the depth of the analysis that can be performed with them. Please contact the companies listed for a demonstration and additional information. The reader may also want to refer to some of the articles listed in the reference section of the appendix for software reviews or further information.

COST–BENEFIT ANALYSIS

Bayesian Decision Support System
Ozier Peterse, and Associates USA
870 Market St., Suite 1001
San Francisco, CA 94102

DECISION ANALYSIS

Arborist
Texas Instruments Inc.
P.O. Box 100137
Atlanta, GA 30384-0001

Arizona State University Software
Department of Decision &
Information Systems
College of Business
Arizona State University
Tempe, AZ 85287-4206

DATA (Decision Analysis by TreeAge)
TreeAge Software Inc.
1 Post Office Square
23rd Floor
Boston, MA 02109

DAVID
Center for Academic Computing
Duke University
North Building, Research Drive
Durham, NC 27706

Decide
Software Inventions, Inc
7400 Hillside Way
Anchorage, AK 99516

Decision Analysis Techniques
Lionheart Press, Inc.
P.O. Box 4056
Allentown, PA 18105

Demos
Lumina Decision Systems
350 Cambridge Ave., Suite 390
Palo Alto, CA 94306

DPL (Decision Programming Language)
Applied Decision Analysis, Inc.
2710 Sand Hill Road
Menlo Park, CA 94025

FAST QM
McGraw Hill Publishing
Princeton Road
Heightstown, NJ 08520

Influence Diagram Processor (InDia)
Decision Focus Inc.
650 Castro Street, Suite #300
Mountain View, CA 94041

@RISK
Palisade Corporation
31 Decker Road
Newfield, NY 14867

SMLTREE
Dr. James P. Hollenberg
445 East 68th Street, Apt. 11P
New York, NY 10021

Stratatree
Decision Support Software Inc.
4922 Ellsworth Ave.
Pittsburgh, PA 15213

Supertree
SDG Decision Systems
2440 Sand Hill Road
Menlo Park, CA 94025

Treeval
Decision Analysis Group
University of Southern California
Los Angeles, CA 90089

ELECTRE

Decision Analysis Techniques
Lionheart Press, Inc.
P.O. Box 4056
Allentown, PA 18105

Decision Matrix Method
Armada Systems
P.O. Box 637, Station A
Downsview, Ontario M3M 3A9
Canada

ELECTRE
Professor Bernard Roy
University of Paris-Dauphine
Place du Marechal Del Lattre
de Tassigny
75775 Paris Cedex 16, France

MATHEMATICAL PROGRAMMING

AT&T KORBX
AT&T Advanced Decision Support
System
101 Crawfords Corner Rd.
Holmdel, NJ 07733

Best Answer
Heizer Software
P.O. Box 232019
Pleasant Hill, CA 94523

BLP88,LP88,MILP88,MXLP88,TSA88
Eastern Software Products
P.O. Box 15328
Alexandria, VA 22309

CPLEX
CPLEX Optimization Inc.
Suite 279
930 Tahoe Blvd, Building 802
Incline Village, NV 89451-9436

Decision Analysis Techniques
Lionheart Press, Inc.
P.O. Box 4056
Allentown, PA 18105

EQUITY
KRYSALIS LTD.
28 Derwent Drive
Maidenhead SL6 6LB, England

FortLP
Numerical Algorithms Group
1400 Opus Place, Suite 200
Downers Grove, IL 60515

GAMS
The Scientific Press, Inc.
651 Gateway Blvd., Suite 1100
So. San Francisco, CA 94080-7014

GINO, LINDO, What's Best
LINDO Systems Inc.
1415 N. Dayton
Chicago, IL 60622

HS/LP Linear Optimizer, OMNI LP Model Management
Haverly Systems Inc.
P.O. Box 919
Danville, NJ 07834

INCEPTA
Insight Logistics Ltd.
3 Robert Speck Parkway, Suite 270
Toronto, Ontario, Canada

IXL: The Discovery Machine
Intelligence Ware Inc.
5933 W. Century Blvd.
Los Angeles, CA 90045

LAMPS
Advanced Mathematical Software
#4 Yukon Road
London SW129PU, England

LAO1, LAO2, LAO5
AEA Technology, Building 424.4
Harwell 0XII ORA, England

LOQO
Robert J. Vanderbei
Dept. of Civil Engineering &
Operations Research
Princeton University
Princeton, NJ 08544

MathPro, XPress-MP
MathPro Inc.
1019 19th Street, NW
Washington, DC 20036

MS Excel Math Programming Add-in
Microsoft Corporation
One Microsoft Way
Redmond, WA 98052

MILP Linear Optimizer
Computer & Automation Institute
Victor Hugo 18-22
H-1132 Budapest, Hungary

MIMI/LP
Chesapeake Decision Science Inc.
200 South Street
New Providence, NJ 07974

MPL Modelling System
Maximal Software Inc.
1655 N. Fort Myer Drive, Suite 700
Arlington, VA 22209

MPS III, Optimization and Modeling Library
Ketron Management Science
2200 Wilson Blvd., Suite 220
Arlington, VA 22201

MSLP-PC
CompuEd
P.O. Box 35544, Station E
Vancouver, B.C. V6M 4G8 Canada

OB1, YLP
XMP Software Inc.
930 Tahoe Blvd., Suite 802-279
Incline Village, NV 89451-9436

OMP, PC-PROG
Computing & Systems Consultants
8205 Old Deer Trail
Raleigh, NC 27615

One-up
Comshare Inc.
3001 S. State St.
Ann Arbor, MI 48108

Optimal Manager
Transpower Corp.
1 Oak Drive
Parkerford, PA 19457

Optimization Subroutine Library
IBM Corp.
Neighborhood Rd.
Kingston, NY 12401

PIMS
Bechtel Corporation
3000 Post Oak Blvd.
Houston, TX 77056

SAS/OR Software
SAS Institute Inc.
SAS Campus Dr.
Cary, NC 27513

Sciconic
SD-SCICON UK Ltd.
Wavendon Tower
Wavendon,
Milton Keynes MKH 8LX, England

VMP/6000,VMP/PC,VMP/OSL
Sundown Software System Inc.
P.O. Box 10417
Silver Spring, MD 20914

XIA
Sunset Software Technology
1613 Chelsea Rd., Suite 153
San Marino, CA 91108

MULTIATTRIBUTE UTILITY ANALYSIS

Arizona State University Software
Department of Decision &
Information Systems
College of Business
Arizona State University
Tempe, AZ 85287-4206

Continuous Multi-Attribute Risk
NEVADA Simulations
P.O. Box 32272
Phoenix, AZ 85064-2272

Decision-Analyst
Executive Software Inc.
Two North State Street
Dover, DE 19901

Decision Map
Softstyle
Suite 205
7192 Kalanianaole Hwy.
Honolulu, HI 96825

DPL Advanced Version
Applied Decision Analysis, Inc.
2710 Sand Hill Road
Menlo Park, CA 94025

Expert 8/7
Magic7 Software Company
101 First Street, Suite 237
Los Altos, CA 95022

Hiview, Maud
Decision Analysis Unit
The London School of Economics
Houghton Street
London WC2A 2AE
England

Ismault
Department of Systems Engineering
Thornton Hall
University of Virginia
Charlottesville, VA 22903-2442

Lightyear
Thoughtware, Inc.
Suite 1000A
2699 South Bayshore Drive
Coconut Grove, FL 33133

Logical Decision
Logical Decisions
1014 Wood Lily Drive
Golden, CO 80401

MAUA
Booz, Allen & Hamilton Inc.
1953 Gallows Road
Vienna, VA 22180

Prefcalc
Euro-Decision
BP 57-78530 Buc, France

Smartedge
Haviland-Lee, Inc.
P.O. Box 5036
Pasadena, CA 91107

Treeval
Decision Analysis Group
University of Southern California
Los Angeles, CA 90089

Visa
Department of Management Studies
University of Strathclyde
Livingstone Tower
26 Richmond St.
Glasgow G1 1XH, Scotland

PAYOFF MATRIX ANALYSIS

Decision Analysis Techniques
Lionheart Press, Inc.
P.O. Box 4056
Allentown, PA 18105

WEIGHTED SCORING METHODS

Analytic Hierarchy Process
Armada Systems
18 Dana Crescent
Thornhill, Ontario LAJ 2R5, Canada

Best Choice 3
Sterling Castle Software
702 Washington Street, Suite 174
Marina del Rey, CA 90292

COMPARE!
Blue Blazer
Smoke Tree Village, Suite D
1729 E. Palm Canyon Drive
Palm Springs, CA 92264

Criterium
Sygenex
15446 Bel-Red Road, Suite 450
Redmond, WA 98052

Decision
Once Begun Computations
Searsport, ME 04974

Decision Analysis System (DAS)
Armada Systems
18 Dana Crescent
Thornhill, Ontario LAJ 2R5, Canada

Decision Analysis Techniques
Lionheart Press, Inc.
P.O. Box 4056
Allentown, PA 18105

Decision-Analyst
Business Planning Systems
Ten Pennsylvania Avenue
Rehoboth Beach, DE 19971

Decision Pad
Apian Software
P.O. Box 1224
Menlo Park, CA 94026-1224

Expert Choice
Decision Support Software Inc.
4922 Ellsworth Ave.
Pittsburgh, PA 15213

HIPRE
Santa Monica Software, Inc.
240 15th Street
Santa Monica, CA 90402

Idea Generator Plus
Experience in Software Inc.
2000 Hearst Ave.
Berkeley, CA 94704

Lightyear
Thoughtware
2699 So. Bayshore Dr., Suite 1000A
Coconut Grove, FL 33133

Logical Decision
Logical Decisions
1014 Wood Lily Drive
Golden, CO 80401

P/G%
Decision Aids Inc.
1720 Parkhaven
Champaign, IL 61820

Planning Tools
ATR Inc.
900 Artesia Blvd., Suite 110
Redondo Beach, CA 90278-2704

Priorities
Work Sciences Associates
26 Southwood Lawn Road
Highgate
London N6SSF, England

RiskPAC
Computer Security Consultants Inc.
590 Danbury Road
Ridgefield, CT 06877

REFERENCES

Buede, D. M. (1992). Software Review: Overview of the MCDA Software Market. *Journal of Multi-Criteria Decision Analysis* 1:59–61.

Buede, D. M. (1992). Superior Design Features of Decision Analytic Software. *Computers Ops. Res.* 19(1):43–57.

Buede, D. M. (1993). Aiding Insight for Decision-makers: A Survey of More Than Two Dozen Software Packages. *OR/MS Today* (April).

Buede, D. M. (1994). Aiding Insight II. *OR/MS Today* (June):62–71.

Call, H. J., and W. A. Miller (1990). A Comparison of Approaches and Implementation for Automating Decision Analysis. *Reliability Engineering and System Safety* 30:115–162.

Humphry, S. (1992). Tools Help Objectify Decision Making. *PC Week* March 9:113.

Johnson, S. M. (1992). 1992 Environmental Software Survey. *Journal of Air and Waste Management* 42(11).

Radcliff, B. (1986). Multicriteria Decision Making: A Survey of Software. *Soc. Sci. Microcomputer Rev.* 4:38–55.

Sharda, R. (1992). Linear Programming Software for Personal Computers: 1992 Survey. *OR/MS Today* (June).

APPENDIX

B

TRAINING PROGRAMS

The decision aids presented in this book vary in the level of complexity and ease of use. Some require detailed training, while others can be learned by reading a book. If the analyst has no experience using the aid, we recommend acquiring one of the books listed in the Suggestions for Further Study" section of the decision aid chapters. In addition, training may be necessary. This appendix contains a list of organizations that provide seminars or training classes for some of the decision aids. This list is not exhaustive. Included are courses of which we were aware in May 1994. The reader may also want to contact business schools or operations research or management science departments at local colleges or universities. Many professional organizations are also aware of seminars and training programs geared to their profession.

ANALYTICAL HIERARCHY PROCESS

Expert Choice: Decision Analysis for Your World
Expert Choice, Inc.
4922 Ellsworth Ave.
Pittsburgh, PA 15213

Making Complex Decisions: The Analytical Hierarchy Process (AHP) and the Expert Choice Software
University of Maryland
Instructional Television System (ITV)
2104 Engineering Classroom Bldg.
College Park, MD 20742

DECISION ANALYSIS

Basic Decision Analysis Concepts and Techniques
Advanced Topics in Decision Analysis Practice
DPL Software Training Workshop

Applied Decision Analysis, Inc.
2710 Sand Hill Road
Menlo Park, CA 94025

Decision Analysis
Spreadsheet Decision Modeling with DPL
Stanford Instructional Television
Network (SITN)
Stanford University
401 Durand Building
Stanford, CA 94305-4036

Decision Analysis: A How To Do It Practical Course
Institute of Safety and Systems
Management
University of Southern California
Los Angeles, CA 90089-0021

Decision Analysis with Supertree
Strategic Decisions Group
2440 Sand Hill Road
Menlo Park, CA 94025-6900

Decision and Risk Analysis, Inc.
P.O. Box 1777
Chadds Ford, PA 19317-1777

Management Science in Spreadsheets
The University of Chicago
Graduate School of Business
1101 East 58th Street
Chicago, IL 60637

KEPNER–TREGOE DECISION ANALYSIS

Kepner–Tregoe, Inc.
P.O. Box 704
Princeton, NJ 08542

MATHEMATICAL PROGRAMMING

Many course offerings
Stanford Instructional Television
Network (SITN)
Stanford University
401 Durand Building
Stanford, CA 94305-4036

Management Science in Spreadsheets
University of Chicago
Graduate School of Business
1101 East 58th Street
Chicago, IL 60637

MULTIATTRIBUTE UTILITY ANALYSIS

Decision Analysis: A How To Do It Practical Course
Institute of Safety and Systems
Management
University of Southern California
Los Angeles, CA 90089-0021

Multiattribute Utility Analysis
Applied Decision Analysis, Inc.
2710 Sand Hill Road
Menlo Park, CA 94025

PAYOFF MATRIX ANALYSIS

CATALYST
Electric Power Research Institute
IESD Division
3412 Hillview Avenue
P.O. Box 10412, Palo Alto, CA 94303

APPENDIX C

TOPICAL BIBLIOGRAPHY

This bibliography is organized using the following topical categories:

General Decision Making (including bibliographies)
Analytical Hierarchy Process
Applications of Decision Aids
Behavioral Aspects of Decision Making
Chemical Process Safety and Decision Making
Cost–Benefit and Cost-Effectiveness Analysis
Decision Analysis
Decision Support Systems
Economics
Expert Judgment
Expert Systems
Future Research and Development
Fuzzy Sets
Game Theory
Group Decision Making
Issues Surrounding the Use of Decision Aids
Kepner–Tregoe Decision Analysis
Mathematical Programming
Multiattribute Utility Analysis
Multiple Criteria Decision Making
Operations Research/Management Science
Other Decision Aids
Payoff Matrix Analysis
Probability and Statistics
Risk Analysis/Risk Assessment
Social Choice Theory
Software
Utility Theory
Value of Life
Voting
Weighted Scoring Methods

GENERAL DECISION MAKING (INCLUDING BIBLIOGRAPHIES)

Adelman, L. (1982). Involving Users in the Development of Decision-Analytic Aids: The Principal Factor in Successful Implementation. *Journal of the Operational Research Society* 33(4):333–342.
Algie, J., and W. Foster. (1987). Effective Decision Making. *International Journal of Manpower* 8(2):20–28.

Baram, M.S., and R. I. Field (1983). Law, Risk Management, and the Influence of the Psychological and Decision Sciences. Paper presented at the NATO–NSF Advanced Study Institute on Technology Assessment, Environmental Impact Assessment and Risk Analysis, Les Arcs, France.

Bari, R. A. (1986). Decision Making and Probabilistic Risk Assessment. *Nuclear Engineering and Design* 93(2–3):341–348.

Barnard, J. (1992). Successful CEOs Talk about Decision Making. *Business Horizons* 35(5):70–74.

Baybutt, P. (1983). Procedures for the Use of Risk Analysis in Decision Making. Pages 261–277 in S. Hartwig (ed.), *Heavy Gas and Risk Assessment II, Proceedings of the 2nd Symposium on Heavy Gases and Risk Assessment, Frankfurt-am-Main, May 25–26, 1982.* Dordrecht, Netherlands and Boston: D. Reidel.

Brown, R. V., and J. W. Ulvila (1977). *Selecting Analytic Approaches For Decision Situations (Rev. Ed.) Vol. I: An Overview of Methodology. Vol II. Case Studies. Vol. III. Appendices (Report 77-7-5).* McLean, Virginia: Decisions and Designs, Incorporated.

Clausen, M. J., D. W. Fraley, and R. S. Denning (1980). "Improved Methods for Incorporating Risk in Decision Making" (Report PNL-3523). Battelle Pacific Northwest Laboratories.

Collingridge, D. (1982). *Critical Decision Making.* New York: St. Martin's Press.

Emory, C., and P. Niland (1968). *Making Management Decisions.* Boston: Houghton Mifflin.

Gass, S. I. (1985). *Decision Making, Models and Algorithms.* New York: John Wiley and Sons.

Gehm, F. (1984). Techniques for Making Decisions Under Uncertainty. *Journal of Futures Markets* 4(1):65–73.

Gradstein, M., and S. Nitzan. (1988). Participation, Decision Aggregation and Internal Information Gathering in Organizational Decision Making. *Journal of Economic Behavior & Organization* 10(4):415–431.

Hahn, G. J. (1969). *A Categorized Bibliography on Decision and Risk Analysis.* Schenectady, NY: General Electric Research and Development Center.

Hill, P. H., H. A. Bedau, R. A. Chechile, W. J. Crochetiere, B. L. Kellerman, D. Ounjian, S. G. Pauker, S. B. Pauker, and J. Z. Rubin (1978). *Making Decisions: A Multidisciplinary Introduction.* Reading, MA: Addison-Wesley.

Janis, I. L. (1977). *Decision Making.* New York: Free Press.

Kahneman, D., and A. Tversky. (1983). Choices, Values, and Frames. *American Psychologist* 39:341–350.

Kersten, G. E., and S. Szpakowicz (1994). Decision Making and Decision Aiding: Defining the Process, Its Representations, and Support. *Group Decision and Negotiation* 3(2):237–261.

Kunruether, H. (1983). *Risk Analysis and Decision Processes.* Berlin: Springer-Verlag.

Lathrop, J. W., and J. Linnerooth (1983). The Role of Risk Assessment in a Political Decision Process. In P. Humphreys, O. Svenson, and A. Vari (eds.), *Analyzing and Aiding Decision Processes.* Amsterdam: North-Holland.

Lindley, D. V. (1985). *Making Decisions.* London: John Wiley and Sons.

Linstone, H. A. (1984). *Multiple Perspectives for Decision Making.* New York: Elsevier Science Pub. Co.

Merkhofer, M. W. (1987). *Decision Science and Social Risk Management: A Comparative Evaluation of Cost–Benefit Analysis, Decision Analysis, and Other Formal Decision-Aid-*

ing Approaches. Dordrecht, Netherlands: D. Reidel (Boston: Kluwer Academic Publishers Group).
Odiorne, G. S. (1969). *Management Decisions by Objectives*. Englewood Cliffs, NJ: Prentice-Hall.
Schlaifer, R. O. (1969). *Analysis of Decisions Under Uncertainty*. New York: McGraw-Hill.
Shafer, G. (1988). Savage Revisited. Chapter 10 in D.E. Bell et al. (eds.), *Decision Making: Descriptive, Normative, and Prescriptive Interactions*. New York: Cambridge University Press.
Shanteau, J., and T. R. Stewart. (1992). Why Study Expert Decision Making? Some Historical Perspectives and Comments. *Organizational Behavior & Human Decision Processes* 53(2):95–106.
Simon, H. A., G. B. Dantzig, R. Hogarth, C. Plott, H. Raiffa, T. C. Schelling, K. A. Shepsle, R. Thaler, A. Tversky, and S. Winter. (1987). Decision Making and Problem Solving. *Interfaces* 17(5):11–31.
Ulvila, J. W., and R. V. Brown. (1982). Decision Analysis Comes of Age. *Harvard Business Review* 60(5):130–141.
Watson, S. R., and D. M. Buede (1987). *Decision Synthesis: The Principles and Practice of Decision Analysis*. Cambridge: Cambridge University Press.

ANALYTIC HIERARCHY PROCESS

Bard, J. F. (1992). A Comparison of the Analytic Hierarchy Process with Multiattribute Utility Theory: A Case Study. *IIE Transactions* 24(5):111–121.
Basak, I. (1988). When To Combine Group Judgments and When Not To in the Analytic Hierarchy Process: a New Method. *Mathematical and Computer Modelling* 10(6):395–404.
Belton, V., and T. Gear. (1983). On a Shortcoming of Saaty's Method of Analytic Hierarchies. *Omega* 11(3):228–230.
Cambron, K. E., and G. W. Evans. (1991). Layout Design Using the Analytic Hierarchy Process. *Computers & Industrial Engineering* 20(2):211–229.
Deland, D., M. R. Holliway, and M. Ohlrich. (1982). Evaluating Plant Locations by the Analytic Hierarchy Process. Pages 372–374 in *Proceedings of the 14th Annual Meeting of the American Institute for Decision Sciences* Vol. 2. Atlanta, GA, USA: American Institute for Decision Sciences.
Dyer, J. S. (1990). Remarks on the Analytic Hierarchy Process. *Management Science* 36(3):249–258.
Dyer, R. F., and E. H. Forman. (1992). Group Decision Support with the Analytic Hierarchy Process. *Decision Support Systems* 8(2):99–124.
Emshoff, J. R., and T. L. Saaty. (1982). Applications of the Analytic Hierarchy Process to Long Range Planning Processes. *European Journal of Operational Research* 10(2):131–143.
Gass, S. I. (1985). *Decision Making, Models and Algorithms*. New York: John Wiley and Sons.
Golden, B. L., E. A. Wasil, and P. T. Harker (eds). (1989). *The Analytic Hierarchy Process: Applications and Studies*. New York: Springer.

Golden, B. L., E. A. Wasil, and D. E. Levy (1989). Applications of the Analytic Hierarchy Process, a Categorized Annotated Bibliography. In Golden, B. L., E. A. Wasil, and P. T. Harker (eds.), *The Analytic Hierarchy Process: Applications and Studies*, New York: Springer.

Hanratty, P. J., and B. Joseph. (1992). Decision-Making in Chemical Engineering and Expert Systems: Application of the Analytic Hierarchy Process to Reactor Selection. *Computers & Chemical Engineering* 16(9):849–860.

Harker, P. T. (1989). The Art and Science of Decision Making: the Analytic Hierarchy Process. Pages 3–36 in B. L. Golden, E. A. Wasil, and P. T. Harker (eds.), *The Analytic Hierarchy Process: Applications and Studies*, New York: Springer.

Harker, P. T., and L. G. Vargas. (1987). The Theory of Ratio Scale Estimation: Saaty's Analytic Hierarchy Process. *Management Science* 33(11):1383–1403.

Hosseini, J. C., and R. L. Armacost. (1992). Using the Analytic Hierarchy Process for Strategic and Policy Decisions in Engineering Management. Pg. 677 in D. F. Kocaoglu and K. Niwa (eds.) *Technology Management: The New International Language: Proceedings of the 1991 Portland International Conference on Management of Engineering and Technology -PICMET '91*. Piscataway, NJ, USA: IEEE Service Center (IEEE cat n 92CH3048-6).

Saaty, T. L. (1980). *The Analytic Hierarchy Process*. New York: McGraw-Hill International Book Co.

Saaty, T. L. (1986). Axiomatic Foundation of the Analytic Hierarchy Process. *Management Science* 32(7):841–855.

Saaty, T. L. (1987). Rank Generation, Preservation, and Reversal in the Analytic Hierarchy Decision Process. *Decision Sciences* 18(2):157–177.

Saaty, T. L. (1989). Group Decision Making and the AHP. In B. L. Golden, E. A. Wasil, and P. T. Harker (eds.), *The Analytic Hierarchy Process: Applications and Studies*,New York: Springer.

Saaty, T. L. (1990). *Decision Making for Leaders*. Belmont, CA: Lifetime Learning Publications.

Saaty, T. L., and L. G. Vargas. (1987). Uncertainty and Rank Order in the Analytic Hierarchy Process. *European Journal of Operational Research* 32(1):107–117.

Schoner, B., and W. C. Wedley. (1989). Ambiguous Criteria Weights in AHP: Consequences and Solutions. *Decision Sciences* 20(3):462–475.

Vargas, L. G. (1987). Priority Theory and Utility Theory. *Mathematical Modelling* (9):381–385.

Vargas, L. G. (1990). An Overview of the Analytic Hierarchy Process and Its Applications. *European Journal of Operational Research* 48(1):2–8 (September 5).

Winkler, R. L. (1990). Decision Modeling and Rational Choice: AHP and Utility Theory. *Management Science* 36(3):247–248.

Zahedi, F. (1986). The Analytic Hierarchy Process—A Survey of the Method and Its Applications. *Interfaces* 16(4):96–108.

Zahir, M. S. (1991). Incorporating the Uncertainty of Decision Judgements in the Analytic Hierarchy Process. *European Journal of Operational Research* 53(2):206–216 (July 25).

APPLICATIONS OF DECISION AIDS

Allen, T. R. (1986). Using Decision Analysis in Physical Protection Planning. *Nuclear Materials Management (Journal of the Institute of Nuclear Materials Management) v. 15, INMM 1986 Annual Meeting,* New Orleans, LA, USA, 1986, June 22, 251–255.

Atz, H. (1981). *Decision-Making in LNG Terminal Siting: Wilhelmshaven, F.R.G. Draft Report.* Laxenburg, Austria: International Institute For Applied Systems Analysis.

Baldwin, A. L., and E. H. Webb. (1984). Application of Decision Analysis at Gulf Research and Development Company. Pages 503–509 in *International Congress on Technology & Technology Exchange: Technology & the World Around Us.* Pittsburgh, PA, USA: International Technology Institute. Also available from IEEE Service Cent (Cat n 84CH2195-6), Piscataway, NJ, USA.

Banai-Kashani, R. (1989). New Method for Site Suitability Analysis. The Analytic Hierarchy Process. *Environmental Management* 13(6):685–693.

Benjamin, J. R. (1983). Risk and Decision Analyses Applied to Dams and Levees. *Structural Safety* 1(4):257-268.

Boyd, D. W., D. Cohan, D. S. Wilson, E. V. Niemeyer, and W. C. Trench (1985). *Proceedings of the American Power Conference: Managing PCB Risks From Transformer Fires.* Palo Alto: Electric Power Research Institute.

Brice, H., and T. Wegner. (1989). A Quantitative Approach to Corporate Social Responsibility Programme Formulation. *Managerial & Decision Economics* 10(2):163–171.

Briggs, T., P. L. Kunsch, and B. Mareschal. (1990). Nuclear Waste Management. An Application of the Multicriteria PROMETHEE Methods. *European Journal of Operational Research* 44(1):1–10.

Charnes, A., and W. W. Cooper (1961). *Management Models and Industrial Applications of Linear Programming, Vols. I and II.* New York: John Wiley and Sons.

Current, J., and M. O'Kelly. (1992). Locating Emergency Warning Sirens. *Decision Sciences* 23(1):221–234.

Fleishman, and M.S. Hogh (1989). *The Use of Cost Benefit Analysis in Evaluating the Acceptability of Industrial Risks: An Illustrative Case Study.* Moor Lane, London: BP International Limited Britannic House.

Golden, B. L., E. A. Wasil, and D. E. Levy. (1989). Applications of the Analytic Hierarchy Process, A Categorized Annotated Bibliography. In B. L. Golden, E. A. Wasil, and P. T. Harker (eds.), *The Analytic Hierarchy Process: Applications and Studies,* New York: Springer.

Hanna, R. C. (1979). Apply Decision Theory to Hazard Evaluation. *Hydrocarbon Processing* 133–150.

Hayhurst, G. (1987). *Mathematical Programming Applications.* New York: Macmillan.

Ichiishi, T., A. Neyman, and Y. Tauman (eds.) (1990). *Game Theory and Applications.* San Diego: Academic Press.

Keeney, R. L., and K. Nair. (1977). Nuclear Siting Using Decision Analysis. *Energy Policy* 5(3):223–231.

Kunreuther, H. C., J. Linnerooth, J. Lathrop, A. Hermann, S. Macgill et al. (1983). *Risk Analysis and Decision Processes: The Siting of Liquified Energy Gas Facilities in Four Countries.* New York: Springer-Verlag.

Mauney, D. A. (1993). Economic Optimization of Multiple Component Replacement/Inspection in the Power System Environment. In *Reliability and Risk in*

Pressure Vessels and Piping: Presented at the 1993 Pressure Vessels and Piping Conference, Denver CO, July 25–29, 1993. ASME, PVP Vol. 251.

Merkhofer, M. W., and H. Call. (1989). A Multiattribute Utility Analysis Model for Ranking Superfund Sites. In *Superfund 1988, Proceedings of the 9th National Conference*, Silver Spring, Maryland, November 28–30, 1988, sponsored by the Hazardous Materials Control Research Institute. Washington, DC: HMCRI. (pp. 44–65).

Mishan, E. J., and T. Page. (1979). The Methodology of Cost–Benefit Analysis: With Particular Reference to the Ozone Problem. Social Studies Working Paper 249. Pasadena, California: California Institute of Technology.

Moore, L. R. (1986). Kepner–Tregoe as a Safety Methodology. *Professional Safety* 31(6):37–41.

Noonan, F. (1986). Decision Analysis for Risk Management of Fire Safety Hazards. *Fire Safety Science, Proceedings of the First International Symposium* (pp. 621–625). Washington, DC: Hemisphere Publ. Corp.

Paté-Cornell, M. E. (1985). Fire Risks in Oil Refineries: Economic Analysis of Camera Monitoring. *Risk Analysis* 5(4):277–288.

Saaty, T. L. (1987). The Analytic Hierarchy Process. *Risk Analysis* 7(2):159–172.

Saaty, T. L. (1989). Group Decision Making and the AHP. In B. L. Golden, E. A. Wasil, and P. T. Harker (eds.), *The Analytic Hierarchy Process: Applications and Studies*, New York: Springer.

Smith, A. E., P. C. McNamee, and M. W. Merkhofer. (1982). "Development of Decision Analysis Methodology for Health Risk Assessment: An Illustrative Application to Alternative Air Quality Standards for Carbon Monoxide." Final Report 3432. Menlo Park, CA: SRI International.

Tee, A. B., M. D. Bowman, and K. C. Sinha. (1988). Fuzzy Mathematical Approach for Bridge Condition Evaluation. *Civil Engineering Systems* 5(1):17–24.

Van, A. C. (1986). Evaluation of the Risk Problem and the Selection of the Optimum Risk Management Solution. *Fire Safety Science, Proceedings of the First International Symposium* (pp. 627–636). Washington, DC, USA: Hemisphere Publishing Corp.

van Steen, J. F. J. (1987). A Methodology for Aiding Hazardous Materials Transportation Decisions. *European Journal of Operational Research* 32(2):231–244.

von Winterfeldt, D. (1987). Value Tree Analysis: An Introduction and An Application To Offshore Oil Drilling. In P. Kleindorfer and H. Kunreuther (eds.), *Insuring and Managing Hazardous Risks: From Seveso To Bhopal and Beyond* (pp. 349–377). New York: Springer-Verlag.

Wells, G. E. (1982). Use of Decision Analysis in Imperial Group. *Journal of the Operational Research Society* 33(4):313–318.

Zahedi, F. (1986). The Analytic Hierarchy Process—A Survey of the Method and Its Applications. *Interfaces* 16(4):96–108.

BEHAVIORAL ASPECTS OF DECISION MAKING

Antonides, G., and N. L. van der Sar. (1990). Individual Expectations, Risk Perception and Preferences in Relation to Investment Decision Making. *Journal of Economic Psychology* 11(2):227–245.

Arbel, A., and R. M. Tong. (1982). On the Generation of Alternatives in Decision Analysis Problems. *Journal of the Operational Research Society* 33(4):377–387.
Bell, D., H. Raiffa, and A. Tversky (eds.) (1988). *Decision Making: Descriptive, Normative, and Prescriptive Interactions*. New York: Cambridge University Press.
Brown, R. V. (1990). Assessment Uncertainty Technology for Making and Defending Risky Decisions. *Journal of Behavioral Decision Making* 3(3):213–228.
Fischhoff, B., S. Lichtenstein, P. Slovic, S.L. Derby, and R.L. Keeney (1981). *Acceptable Risk*. New York: Cambridge University Press.
Geweke, J. (1992). *Decision Making Under Risk and Uncertainty*. Dordrecht, Netherlands: Kluwer Academic Publishers.
Hasling, J. (1975). *Group Discussion and Decision Making*. New York: Crowell.
Hogarth, R. M. (1980). *Judgement and Choice*. Chichester, NY: J. Wiley.
Kahneman, D., P. Slovic, and A. Tversky (eds.) (1982). *Judgments Under Uncertainty: Heuristics and Biases*. Cambridge, England: Cambridge University Press.
Paté-Cornell, M. E., and P. S. Fischbeck. (1992). Aversion to Epistemic Uncertainties in Rational Decision Making: Effects on Engineering Risk Management. In Y. Y. Haimes, D. A. Moser, and E. Z. Stakhiv (eds.), *Risk-Based Decision Making in Water Resources V, Proceedings of the 5th [Engineering Foundation] Conference on Risk-Based Decision Making in Water Resources, Santa Barbara, CA, Nov. 3–8, 1991* (pp. 200–218). New York: American Society of Civil Engineers.
Patton, B. R. (1989). *Decision-Making Group Interaction*. 3rd Ed. New York: Harper & Row.
Russo, J. E., and P. J. H. Schoemaker. (1992). Managing Overconfidence. *Sloan Management Review* 33(2):7–17.
Simon, H. A., G. B. Dantzig, R. Hogarth, C. Plott, H. Raiffa, T. C. Schelling, K. A. Shepsle, R. Thaler, A. Tversky, and S. Winter. (1987). Decision Making and Problem Solving. *Interfaces* 17(5):11–31.
Slovic, P., B. Fischhoff, and S. Lichtenstein. (1984). Behavioral Decision Theory Perspectives On Risk and Safety. *Acta Psychologica* 56:183–203.
Starr, C., and C. Whipple. (1984). A Perspective on Health and Safety Risk Analysis. *Management Science* 30(4):452–463.
Tversky, A., and D. Kahneman (1981). The Framing of Decisions and the Psychology of Choice. *Science* 211 (January 30):453–458.
von Winterfeldt, D. (1988). Expert Systems and Behavioral Decision Research. *Decision Support Systems* 4(4):461–471.
Whyte, G. (1991). Decision Failures: Why They Occur and How to Prevent Them. *The Academy of Management Executive* [Mississippi State, Miss.] 5(3):23–31.

CHEMICAL PROCESS SAFETY AND DECISION MAKING

AIChE/CCPS (1989). *Guidelines for Chemical Process Quantitative Risk Analysis*. New York: American Institute of Chemical Engineers.
Arendt, J. S. (1990). Using Quantitative Risk Assessment in the Chemical Process Industry. *Reliability Engineering & System Safety* 29(1):133–149.

Bley, D., S. Kaplan, and D. Johnson. (1992). Strengths and Limitations of PSA. Where We Stand. *Reliability Engineering & System Safety* 38(1–2):3–26.

Burk, A. F., and W. L. Smith. (1990). Process Safety Management Within Du Pont. *Plant/Operations Progress* 9 (October).

Griffiths, R. F. (1984). Chemical Plant Risk Assessment: Uncertainties and Development Needs. *Environmental International* 10(5/6):523–530.

Heinold, D., D. Simth, and B. Schwab. (1988). Evaluating Potential Impacts From Accidental Releases of Toxic Chemicals. *Environmental Progress* 7(2):116–122.

Kuritz, S. J. (1992). A Holistic Approach to Process Safety. *Occupational Health & Safety* 61(10):28–32.

Paté-Cornell, M. E., and R. Boykin. (1987). Probabilistic Risk Analysis and Safety Regulation in the Chemical Industry. *Journal of Hazardous Materials* 15:97–122.

Schreiber, S. (1991). Guidelines for Technical Management of Chemical Process Safety. *Plant/Operations Progress* 10(2):65–68.

Sweeney, J. C. (1992). Measuring Process Safety Management. *Plant/Operations Progress* 11(2):89–98.

COST–BENEFIT ANALYSIS AND COST-EFFECTIVENESS ANALYSIS

Anandarup, R. (1984). *Cost–Benefit Analysis: Issues and Methodologies*. Baltimore: Johns Hopkins University Press.

Anderson, E. V. (1993). Chlorine Producers Fight Back Against Call for Chemical's Phaseout. *Chemical & Engineering News* 7(9):11–12.

Anderson, L. G., and R. F. Settle (1977). *Benefit–Cost Analysis: A Practical Guide*. Lexington, MA: Lexington Books.

Baram, M. S. (1980). Cost–Benefit Analysis: An Inadequate Basis for Health, Safety, and Environmental Regulatory Decision Making. *Ecology Law Quarterly* [Berkeley, CA: School of Law, Univ. of Calif.] 8:473–531.

Bentkover, J. D., V. T. Covello, and J. Mumpower. (eds.), *Benefits Assessment: The State of The Art*. Dordrecht, The Netherlands: D. Reidel.

Birkner, L. R., and L. S. Salzman. (1986). Assessing Exposure Control Strategy Cost–Effectiveness. *American Industrial Hygience Association* 47(1):50–54.

Dasgupta, A. K. (1972). *Cost–Benefit Analysis: Theory and Practice*. London: Macmillan.

Environmental Protection Agency (1987). *EPA's Use of Benefit–Cost Analysis: 1981–1986*.

Fleishman, and M.S. Hogh (1989). *The Use of Cost Benefit Analysis in Evaluating the Acceptability of Industrial Risks: An Illustrative Case Study*. Moor Lane, London: BP International Limited Britannic House.

Frost, M. J. (1975). *How to Use Cost Benefit Analysis in Project Appraisal*. 2nd Ed. New York: Wiley.

Goyal, R. K. (1986). Probabilistic Risk Analysis: Two Case Studies from the Oil Industry. *Professional Safety* 31(7):11–19.

Gramlich, E. M. (1981). *Benefit–Cost Analysis of Government Programs*. Englewood Cliffs, NJ: Prentice Hall.

Gramlich, E. M. (1990). *A Guide to Benefit–Cost Analysis. 2nd Ed.* Englewood Cliffs, NJ: Prentice-Hall.

Merkhofer, M. W. (1987). *Decision Science and Social Risk Management: A Comparative Evaluation of Cost–Benefit Analysis, Decision Analysis, and Other Formal Decision-Aiding Approaches*. Dordrecht, Netherlands: D. Reidel (Boston: Kluwer Academic Publishers Group).

Mishan, E. J. (1988). *Cost–Benefit Analysis. 4th ed.* London: Unwin Hyman.

O'Brien, B., and J. L. Viramontes (1994). Willingness to Pay: A Valid and Reliable Measure of Health State Preference? *Medical Decision Making* 14(3):289–297.

Prest, A. R., and R. Turvey. (1965). Cost–Benefit Analysis: A Survey. *Economic Journal* 75:683–735.

Riggs, J. L., and T. M. West (1986). *Essentials of Engineering Economics, 2nd Ed.* New York: McGraw-Hill.

Sarin, R. K. (1983). Social Decision Analysis of the Earthquake Safety Problem: The Case of Existing Los Angeles Buildings. *Risk Analysis* 3(1):35–50.

Sassone, P. G., and W. A. Schaffer (1978). *Cost–Benefit Analysis: A Handbook.* New York: Academic Press.

Schulze, W. D., and A. V. Kneese. (1981). Risk in Benefit–Cost Analysis. *Risk Analysis* 1(1):81–88.

Siegel, M. L. (1991). A Spreadsheet for Benefit–Cost Analysis. *Government Finance Review* 7(6):40–43.

Singer, N. M. (1976). *Public Microeconomics: An Introduction to Government Finance, 2nd Ed.* Boston: Little, Brown.

Starr, C. (1969). Social Benefit Versus Technological Risk: What Is Our Society Willing To Pay for Safety? *Science* 165:1232–1238.

Sugden, R., and A. Williams (1978). *The Principles of Practical Cost–Benefit Analysis.* Oxford: Oxford University Press.

Thompson, M. S. (1980). *Benefit–Cost Analysis for Program Evaluation.* Beverly Hills, CA: Sage Publications.

Thuesen, G. J., and W. J. Fabrycky (1989). *Engineering Economy, 7th Ed.* Englewood Cliffs, NJ: Prentice-Hall.

Ulvila, J. W., and B. B. Thompson. (1988). A Computerized Risk Analysis Aid. *Computers & Operations Research* 15(2):123–134.

U.S. House document, 93d Congress (1975). *Evaluation of Techniques for Cost–Benefit Analysis of Water Pollution Control Programs and Policies (report of Administrator of E.P.A. in compliance with Public Law 92-500 Federal Water Pollution Control Act Amendment of 1972* (December, 1974). Washington, D.C: U.S. Government Printing Office.

White, J. A., M. H. Agee, and K. E. Case (1989). *Principles of Engineering—Economic Analysis.* New York: Wiley.

DECISION ANALYSIS

Allen, T. R. (1986). Using Decision Analysis in Physical Protection Planning. *Nuclear Materials Management (Journal of the Institute of Nuclear Materials Management) v. 15, INMM 1986 Annual Meeting,* New Orleans, LA, USA, 1986, June 22 (pp. 251–255).

Baird, B. F. (1989). *Managerial Decisions Under Uncertainty: An Introduction to the Analysis of Decision Making.* New York: John Wiley and Sons.

Behn, R. D., and J. W. Vaupel (1982). *Quick Analysis for Busy Decision Makers*. New York: Basic Books, Inc.

Bell, D., H. Raiffa, and A. Tversky (eds.) (1988). *Decision Making: Descriptive, Normative, and Prescriptive Interactions*, New York: Cambridge University Press.

Boyd, D. W., D. Cohan, D. S. Wilson, E. V. Niemeyer, and W. C. Trench (1985). *Proceedings of the American Power Conference: Managing PCB Risks From Transformer Fires*, Palo Alto: Electric Power Research Institute.

Brown, R. V., and J. W. Ulvila (1988). Does a Reactor Need a Safety Backfit? Case Study on Communicating Decision and Risk Analysis Information to Managers. *Risk Analysis* 8(2):271–282.

Bunn, D. W. (1984). *Applied Decision Analysis*. New York: McGraw-Hill.

Cazalet, E. G. (1984). Forecasting for Energy and Chemical Decision Analysis. *American Institute of Chemical Engineers, 1984 Summer National Meeting (Preprints). American Institute of Chemical Engineers, National Meeting 1984 Summer*. New York: AIChE, Pap n 49d, 9p.

Clemen, R. T. (1991). *Making Hard Decisions: An Introduction to Decision Analysis*. Boston: PWS-KENT Publishing Co.

Corner, J. L., and C. W. Kirkwood (1991). Decision Analysis Applications in the Operations Research Literature, 1970–1989. *Operations Research* 39(2):206–219.

Covello, V. T. (1987). Decision Analysis and Risk Management Decision Making: Issues and Methods. *Risk Analysis* 7(2):131–140.

EPRI (1992). "CATALYST—A Group Process for Strategic Decision Making: Facilitator's Guidebook." Prepared by Applied Decision Analysis and The Brattle Group. EPRI TR-100583, Research Project 2631, Final Report. Palo Alto, CA: Electric Power Research Institute.

Genest, C., and J. V. Zidek. (1986). Combining Probability Distributions: a Critique and an Annotated Bibliography. *Statistical Science* 1(1):114–148.

Hanna, R. C. (1979). Apply Decision Theory to Hazard Evaluation. *Hydrocarbon Processing* 133–150.

Holloway, C. A. (1979). *Decision Making Under Uncertainty: Models and Choices*. Englewood Cliffs, NJ: Prentice-Hall.

Howard, R. A. (1968). The Foundations of Decision Analysis. *IEEE Transactions on Systems Science and Cybernetics* SSC-4(3):211–219.

Howard, R. A. (1980). An Assessment of Decision Analysis. *Operations Research* 28(1):4–27.

Howard, R. A., and J. E. Matheson (eds.) (1989). *Readings on the Principles and Applications of Decision Analysis*. 2 vols. Menlo Park, CA: Strategic Decisions Group.

Jensen, D. D., A. E. Tome, and W. P. Darby. (1989). Applying Decision Analysis to Determine the Effect of Smoke Detector Laws on Fire Loss in the United States. *Risk Analysis* 9(1):79–89.

Keeney, R., and H. Raiffa (1976). *Decisions with Multiple Objectives: Preferences and Value Tradeoffs*. New York: John Wiley and Sons.

Linstone, A. H., and M. Turoff (eds.) (1975). *The Delphi Method: Techniques and Applications*, Reading, MA: Addison-Wesley.

Luce, R. D., and H. Raiffa (1957). *Games and Decisions*. New York: Wiley.

Mauney, D. A. (1993). Economic Optimization of Multiple Component Replacement/Inspection in the Power System Environment. In *Reliability and Risk in*

Pressure Vessels and Piping: Presented at the 1993 Pressure Vessels and Piping Conference, Denver CO, July 25–29, 1993. ASME, PVP Vol. 251.

Merkhofer, M. W. (1987). *Decision Science and Social Risk Management: A Comparative Evaluation of Cost–Benefit Analysis, Decision Analysis, and Other Formal Decision-Aiding Approaches.* Dordrecht, Netherlands: D. Reidel (Boston: Kluwer Academic Publishers Group).

Merkhofer, M. W. (1987). Quantifying Judgmental Uncertainty: Methodology, Experiences, and Insights. *IEEE Transactions on Systems, Man, and Cybernetics* SMC-17(5):741–752.

Miller, A. C., and T. R. Rice (1984). Discrete Approximations of Probability Distributions. *Management Science* 29(3):352–362.

Newendorp, P. D. (1975). *Decision Analysis for Petroleum Exploration.* Tulsa, OK: PenWell Pub. Co.

Noonan, F. (1986). Decision Analysis for Risk Management of Fire Safety Hazards. *Fire Safety Science, Proceedings of the First International Symposium* (pp. 621–625). Washington, DC: Hemisphere Publ. Corp.

Noonan, F., and C. A. Vidich (1992). Decision Analysis for Utilizing Hazardous Waste Site Assessments in Real Estate Acquisition. *Risk Analysis* 12(2):245–251.

North, D. W. (1968). A Tutorial Introduction To Decision Theory. *IEEE Transactions on Systems, Man, and Cybernetics* SMC-4:211–219.

Oliver, R. M., and J. Q. Smith (eds.) (1990). *Influence Diagrams, Belief Nets, and Decision Analysis.* New York: John Wiley and Sons.

Owen, D. L., and Strategic Decisions Group. (1981). The Design of Hazardous Products. *IEEE Transactions on Systems, Man, and Cybernetics* SMC-11:277–288.

Paté-Cornell, M. E. (1985). Fire Risks in Oil Refineries: Economic Analysis of Camera Monitoring. *Risk Analysis* 5(4):277–288.

Pratt, J., H. Raiffa, and R. Schlaifer (1965). *Introduction to Statistical Decision Theory.* New York: McGraw-Hill.

Raiffa, H. (1968). *Decision Analysis: Introductory Lectures on Choices under Uncertainty.* New York: Random House.

Samson, D. (1988). *Managerial Decision Analysis.* Homewood, Illinois: Irwin.

Savage, L. J. (1954). *The Foundations of Statistics.* New York: John Wiley and Sons, Inc.

Schlaifer, R. O. (1969). *Analysis of Decisions Under Uncertainty.* New York: McGraw-Hill.

von Winterfeldt, D., and W. Edwards (1986). *Decision Analysis and Behavioral Research.* Cambridge: Cambridge University Press.

Watson, S. R., and D. M. Buede (1987). *Decision Synthesis: The Principles and Practice of Decision Analysis.* Cambridge: Cambridge University Press.

Winkler, R. L. (1990). Decision Modeling and Rational Choice: AHP and Utility Theory. *Management Science* 36(3):247–248.

DECISION SUPPORT SYSTEMS

Adelman, L. (1984). Real-Time Computer Support for Decision Analysis in a Group Setting: Another Class of Decision Support Systems. *Interfaces* 14(2):75–83.

Baybutt, P. (1986). Decision Support Systems and Expert Systems for Risk and Safety Analysis. *Hazards in the Process Industries: Hazards IX. Institution of Chemical Engi-*

neers Symposium Series Vol. 97 (pp. 261–267). Rugby, England: Institute of Chemical Engineers (EFCE Event n 341).

Belardo, S., and J. Harrald. (1992). A Framework for the Application of Group Decision Support Systems to the Problem of Planning for Catastrophic Events. *IEEE Transactions on Engineering Management*, 39(4):400–411.

Clark, P., and C. B. Chapman. (1987). The Development of Computer Software for Risk Analysis: A Decision Support System Development Case Study. *European Journal of Operational Research* 29(3):252–261.

Dyer, R. F., and E. H. Forman. (1992). Group Decision Support with the Analytic Hierarchy Process. *Decision Support Systems* 8(2):99–124.

Elam, J. J., and J. C. Henderson. (1983). Knowledge Engineering Concepts for Decision Support System Design and Implementation. *Information & Management* 6(2):109–114.

Eom, H. B., and S. M. Lee. (1990). Decision Support Systems Applications Research. A Bibliography (1971–1988). *European Journal of Operational Research* 46(3):333–342.

Gibson, M. L. (1986). Major Multiple-Criteria Decision Making (MCDM) Approaches in Decision Support Systems (DSS). *Proceedings of the Nineteenth Hawaii International Conference on System Sciences. Volume 1a: Architecture, Decision Support Systems and Knowledge-Based Systems, Special Topics* (pp. 434–443). Honolulu, HI, USA: Hawaii International Conference on System Sciences.

Holsapple, C. W., and A. B. Whinston (eds.) (1987). *Decision Support Systems: Theory and Application*. New York: Springer-Verlag.

Iz, P. H. (1992). Two Multiple Criteria Group Decision Support Systems Based on Mathematical Programming and Ranking Methods. *European Journal of Operational Research* 61(1,2):245–253 (August 25).

Keen, P. G. W., and S. S. Morton (1978). *Decision Support Systems: An Organizational Perspective*. Reading, Massachusetts: Addison-Wesley.

Lewandowski, A. (1989). SCDAS—Decision Support System for Group Decision Making: Decision Theoretic Framework. *Decision Support Systems* 5(4):403–423.

Lewandowski, A., and I. Stanchev (eds.) (1989). *Methodology and Software for Interactive Decision Support, Proceedings of International Workshop*. New York: Springer-Verlag.

Ligeza, A. (1988). Expert Systems Approach to Decision Support. *European Journal of Operational Research* 37(1):100–110.

Ozier, W. (1989). Risk Quantification Problems and Bayesian Decision Support System Solutions. *Information Age* 11(4):229–234.

Seils, H. L. (1982). Do 'Decision Support Systems' Really Support? *Computerworld* 16(26):SR/38 (June 28).

Vogel, D. R., and J. F. Jr. Nunamaker. (1990). Group Decision Support System Impact: Multi-Methodological Exploration. *Information & Management* 18(1):15–28.

White, C. C. (1990). Survey on the Integration of Decision Analysis and Expert Systems for Decision Support. *IEEE Transactions on Systems, Man and Cybernetics* 20(2):358–364.

ECONOMICS

Arrow, K. J. (1974). *Essays in the Theory of Risk Bearing*. New York: Elsevier Scientific Publishing Company.

Dobbs, I. M. (1985). Shadow Prices, Consistency and the Value of Life. *Journal of Public Economics* 27(2):177–193.

Grant, E. I., W. G. Ireson, and R. S. Leavenworth (1982). *Principles of Engineering Economy*. New York: John Wiley & Sons.

Longmore, D. R. (1989). The Persistence of the Payback Method: A Time-Adjusted Decision Rule Perspective. *Engineering Economist* 34(3):185–194.

Riggs, J. L., and T. M. West (1986). *Essentials of Engineering Economics, 2nd Edition*. New York: McGraw-Hill.

Singer, N. M. (1976). *Public Microeconomics: An Introduction to Government Finance, 2nd Edition*. Boston: Little, Brown and Company.

Von Neumann, J., and O. Morgenstern (1947). *Theory of Games and Economic Behavior, Ed.* 2. Princeton, New Jersey: Princeton University Press.

EXPERT JUDGMENT

Anderson, B. F., D. H. Deane, K. R. Hammond, G. H. Mcclelland, and J. C. Shanteau (1981). *Concepts in Judgment and Decision Research*. New York, New York: Praeger Publishers.

Chhibber, S., G. Apostolakis, and D. Okrent. (1992). Taxonomy of Issues Related to the Use of Expert Judgments in Probabilistic Safety Studies. *Reliability Engineering & System Safety* 38(1–2):27–45.

Clemen, R. (1986). Calibration and the Aggregation of Probabilities. *Management Science* 32(3):312–314.

Fischhoff, B. (1989). Eliciting Knowledge for Analytical Representation. *IEEE Transactions on Systems, Man and Cybernetics* 19(3):448–461.

Freeling, A. N. S. (1981). Reconciliation of Multiple Probability Assessments. *Organizational Behavior and Human Performance* 28(3):395–414.

French, S. (1986). Calibration and the Expert Problem. *Management Science* 32(3):315–321.

Glorfeld, L. W., and G. C. Fowler. (1988). A Multivariate Methodology for Simultaneously Capturing and Clustering Judgment Policies. *Decision Sciences* 19(3):504–520.

Hampton, J. M., P. G. Moore, and H. Thomas. (1973). Subjective Probability and Its Measurement. *Journal of The Royal Statistical Society, Series A* 136:21–42.

Hershey, J. C., H. C. Kunreuther, and P. J. H. Schoemaker. (1982). Sources of Bias in Assessment Procedures for Utility Functions. *Management Science* 28(8):936–954.

Huber, G. P. (1974). Methods For Quantifying Subjective Probabilities and Multiattribute Utilities. *Decision Sciences* 5:430–458.

Johnson, E. J., and D. A. Schkade. (1989). Bias in Utility Assessments: Further Evidence and Explanations. *Management Science* 35(4):406–424.

Kahneman, D., P. Slovic, and A. Tversky (eds.) (1982). *Judgments Under Uncertainty: Heuristics and Biases*. Cambridge, England: Cambridge University Press.

Kahneman, D., and A. Tversky. (1979). Intuitive Prediction: Biases and Corrective Procedures. *TIMS Studies in Management Science* 22:340–358.

Keeney, R. L., and D. von Winterfeldt. (1991). Eliciting Probabilities from Experts in Complex Technical Problems. *IEEE Transactions on Engineering Management* 38(3):191–201.

Lindley, D. V. (1983). Reconciliation of Probability Distributions. *Operations Research* 31(5):866–880.

Lindley, D. V. (1986). Another Look at an Axiomatic Approach to Expert Resolution. *Management Science* 32(3):303–306.

Merkhofer, M. W. (1987). Quantifying Judgmental Uncertainty: Methodology, Experiences, and Insights. *IEEE Transactions on Systems, Man, and Cybernetics* SMC-17(5):741–752.

Morris, P. A. (1974). Decision Analysis Expert Use. *Management Science* 20(9):1233–1241.

Morris, P. A. (1986). Observations on Expert Aggregation. *Management Science* 32(3):321–328.

Murphy, A. H., and R. L. Winkler. (1977). Reliability of Subjective Probability Forecasts of Precipitation and Temperature. *Applied Statistics (Journal of Royal Statistical Society, Section C)* 26:41–47.

Otway, H., and D. von Winterfeldt. (1992). Expert Judgment in Risk Analysis and Management: Process, Context, and Pitfalls. *Risk Analysis* 12(1):83–97.

Russo, J. E., and P. J. H. Schoemaker. (1992). Managing Overconfidence. *Sloan Management Review* 33(2):7–17.

Sanders, N. R., and L. P. Ritzman. (1991). On Knowing When to Switch from Quantitative to Judgemental Forecasts. *International Journal of Operations & Production Management* 11(6):27–37.

Schervish, M. J. (1986). Comments on Some Axioms for Combining Expert Judgments. *Management Science* 32(3).306–312.

Shanteau, J., and T. R. Stewart. (1992). Why Study Expert Decision Making? Some Historical Perspectives and Comments. *Organizational Behavior & Human Decision Processes* 53(2):95–106.

Shields, T. J., G. W. H. Silcock, H. A. Donegan, and Y. A. Bell. (1987). Methodological Problems Associated with the Use of the Delphi Technique. *Fire Technology* 23(3):175–185.

Svenson, O. (1989). On Expert Judgments in Safety Analyses in the Process Industries. *Reliability Engineering & System Safety* 25(3):219–256.

Tait, J. (1988). The Role of Values in Quantitative Decision-Making. *Journal of the Operational Research Society* 39(7):669–674.

Tversky, A., and D. Kahneman. (1974). Judgment Under Uncertainty: Heuristics and Biases. *Science* 185:1124–1131.

von Winterfeldt, D. (1988). Expert Systems and Behavioral Decision Research. *Decision Support Systems* 4(4):461–471.

Wall, K. D., and C. Correia. (1989). A Preference-Based Method for Forecast Combination. *Journal of Forecasting* 8(3):269–292.

Wallsten, T. S., and D. V. Budescu. (1983). Encoding Subjective Probabilities: A Psychological and Psychometric Review. *Management Science* 29(2):151–173.

Winkler, R. L. (1981). Combining Probability from Dependent Information Sources. *Management Science* 27(4):479–488.

Winkler, R. L. (1986). Expert Resolution. *Management Science* 32(3):298–303.

Wright, G., and P. Ayton. (1987). Eliciting and Modelling Expert Knowledge. *Decision Support Systems* 3(1):13–26.

EXPERT SYSTEMS

Adiga, S., and J. K. Cochran. (1985). Decision Analysis Approach to Modeling Conflicting and Inexact Reasoning in Rule-Based Expert Systems. *IEEE 1985 Proceedings of the International Conference on Cybernetics and Society* (pp. 983–987). New York, NY, USA: IEEE. Available from IEEE Service Cent (Cat n 85CH2253-3), Piscataway, NJ, USA.

Arendt, J. S., and M. L. Casada. (1987). Prisim—An Expert System for Process Risk Management. *Papers Presented at the American Institute of Chemical Engineers 1987 Spring National Meeting. American Institute of Chemical Engineers, National Meeting*, New York, NY, USA: AIChE.

Baybutt, P. (1986). Decision Support Systems and Expert Systems for Risk and Safety Analysis. *Hazards in the Process Industries: Hazards IX. Institution of Chemical Engineers Symposium Series* Vol. 97 (pp. 261–267). Rugby, England: Institute of Chemical Engineers (EFCE Event n 341).

Chau, P. Y. K. (1991). Better Decision Making Through Expert Systems for Management. *SAM Advanced Management Journal* 56(4):13–18.

Davidson, J. M. (1988). Application of Expert Systems Technology in Process Engineering. *Improved Plant Performance Through Instrumentation Technical Papers Composing the Proceedings of the 1988 Spring Conference and Exhibit. Instrumentation in the Chemical and Petroleum Industries* Vol. 20 (pp. 85–91). Research Triangle Park, NC, USA: ISA.

Finn, G. A., and K. F. Reinschmidt. (1990). Expert Systems in Operation and Maintenance. *Chemical Engineering* 97(2):5.

Hanratty, P. J., and B. Joseph. (1992). Decision-Making in Chemical Engineering and Expert Systems: Application of the Analytic Hierarchy Process to Reactor Selection. *Computers & Chemical Engineering* 16(9):849–860.

Heino, P., J. Suokas, and I. Karvonen. (1988). Expert System in Process Design—Analysis of Process Safety and Reliability. *Proceedings of the International Workshop on Artificial Intelligence for Industrial Applications* (pp. 225–231). New York, NY, USA: IEEE. Available from IEEE Service Cent (cat n 88CH2529-6) Piscataway, NJ, USA.

Henrion, M., J. S. Breese, and E. J. Horvitz. (1991). Decision Analysis and Expert Systems. *AI Magazine* 12(4):64–91.

Jablonowski, M. (1992). Expert Systems for Risk Management. *Risk Management* 39(9):56–60.

Kastner, J. K., and S. J. Hong. (1984). A Review of Expert Systems. *European Journal of Operational Research* 18(3):285–292.

Keeney, R. L. (1988). Value-Driven Expert Systems for Decision Support. *Decision Support Systems* 4(4):405–412.

Ligeza, A. (1988). Expert Systems Approach to Decision Support. *European Journal of Operational Research* 37(1):100–110.

Mitra, G., and UNICORN Consultants (eds.) (1986). *Computer Assisted Decision Making*. Amsterdam: North-Holland.

Mumpower, J. L., L. D. Phillips, O. Renn, and V. R. R. Uppuluri (eds.) (1987). *Expert Judgment and Expert Systems*. Heidelberg, West Germany: Springer-Verlag.

O'Leary, D. E. (1988). Methods of Validating Expert Systems. *Interfaces* 18(6):72–79.

von Winterfeldt, D. (1988). Expert Systems and Behavioral Decision Research. *Decision Support Systems* 4(4):461–471.
Wright, G., and P. Ayton. (1987). Eliciting and Modelling Expert Knowledge. *Decision Support Systems* 3(1):13–26.
Zimmerman, H. (1987). *Fuzzy Sets, Decision Making and Expert Systems*. Boston: Kluwer Academic Publishers.

FUTURE RESEARCH AND DEVELOPMENT

Bell, D. E. (1982). Potential Contributions to Decision Analysis. *Decision Sciences* 13(4) (October).
Bouyssou, D., P. Perny, M. Pirlot, A. Tsoukias, and P. Vincke (1993). A Manifesto for the New MCDA Era. *Journal of Multi-Criteria Decision Analysis* 2:125–127.
Covello, V. T., and J. Mumpower. (1985). Risk Analysis and Risk Management: An Historical Perspective. *Risk Analysis* 5(2):103–120.
Dyer, J. S., P. C. Fishburn, R. E. Steuer, J. Wallenius, and S. Zionts. (1992). Multiple Criteria Decision Making, Multiattribute Utility Theory: The Next Ten Years. *Management Science* 38(5):645–654.
Fishburn, P. C. (1982). Comments.... *Decision Sciences* 13(4) (October).
Hogarth, R. M. (1982). From Romanticism to Precision to.... *Decision Sciences* 13(4) (October).
Johnson, E. L., and G. L. Nemhauser. (1992). Recent Developments and Future Directions in Mathematical Programming. *IBM Systems Journal* 31(1):79–93.
Keeney, R. L. (1982). Potential Research Topics in Decision Analysis. *Decision Sciences* 13(4) (October).
March, J. G., and Z. Shapira. Behavioral Decision Theory and Organizational Decision Theory. In G. Ungson and D. Braustein (eds.), *New Directions In Decision Making*,New York: Kent.
Morris, P. A. (1982). Thoughts on Decision Analysis Research. *Decision Sciences* 13(4):551–553.
Simon, H. A., G. B. Dantzig, R. Hogarth, C. Plott, H. Raiffa, T. C. Schelling, K. A. Shepsle, R. Thaler, A. Tversky, and S. Winter (1987). Decision Making and Problem Solving. *Interfaces* 17(5):11–31.
Winkler, R. L.(1982). Research Directions in Decision Making under Uncertainty. *Decision Sciences* 13(4):517–533.

FUZZY SETS

Andersson, L. (1988). Theory of Possibility and Fuzzy Sets. *Document—Swedish Council for Building Research n 8 1988*. 167pp.
Barrett, C. R., and P. K. Pattanaik. (1989). Fuzzy Sets, Preference and Choice: Some Conceptual Issues. *Bulletin of Economic Research* 41(4):229–253.
Chen, S. (1992). *Fuzzy Multiple Attribute Decision Making*. Berlin: Springer-Verlag.

Chun, M., and K. Ahn. (1992). Assessment of the Potential Applicability of Fuzzy Set Theory to Accident Progression Event Trees with Phenomenological Uncertainties. *Reliability Engineering & System Safety* 37(3):237–252.

Dong, W. M., H. C. Shah, and F. S. Wong. (1985). Fuzzy Computations in Risk and Decision Analysis. *Civil Engineering Systems* 2(4):201–208.

Dubois, D., and H. Prade. (1989). Fuzzy sets, probability and measurement. *European Journal of Operational Research* 40(2):135–154.

Kacprzyk, J., and M. Fedrizzi (eds.) (1990). *Multiperson Decision Making Models Using Fuzzy Sets and Possibility Theory*. Dordrecht: Kluwer Academic Publishers.

Kickert, W. J. M. (1978). *Fuzzy Theories on Decision-Making*. Leiden: Nijoff.

Klir, G. J., and T. A. Folger (1988). *Fuzzy Sets, Uncertainty, and Information*. Englewood Cliffs, N.J.: Prentice-Hall.

Lai, Y. (1992). *Fuzzy Mathematical Programming*. Berlin: Springer-Verlag.

Lee, C. C. (1990). Fuzzy Logic in Control Systems. *IEEE Transactions on Systems, Man, and Cybernetics* 20(2):404–435.

Liang, G., and M. J. Wang. (1991). Fuzzy Multi-criteria Decision-making Method for Facility Site Selection. *International Journal of Production Research* 29(11):2313–2330.

Novak, V., et al. (eds.) (1992). *Fuzzy Approach to Reasoning and Decision-Making*. Dordrecht: Kluwer Academic Publishers.

Preyssl, C. (1986). *'Fuzzy Risk' Analysis Theory and Application*. Laxenburg, Austria: International Institute For Applied Systems Analysis.

Rao, J. R., R. N. Tiwari, and B. K. Mohanty. (1988). Preference Structure on Alternatives and Judges in a Group Decision Problem—A Fuzzy Approach. *International Journal of Systems Science* 19(9):1795–1811.

Seo, F., and M. Sakawa. (1985). Fuzzy Multiattribute Utility Analysis for Collective Choice. *IEEE Transactions on Systems, Man and Cybernetics* 15(1):45–53.

Siskos, J. (1982). A Way to Deal with Fuzzy Preferences in Multi-Criteria Decision Problems. *European Journal of Operational Research* 10(3):314–324.

Vaija, P. (1987). Application of Fuzzy Methods to Process Safety Control and to Accident Analysis. *Acta Polytechnica Scandinavica, Chemical Technology and Metallurgy Series* 180:26.

Yager, R. R. et al. (eds.) (1987). *Fuzzy Sets and Applications: Selected Papers by L. A. Zadeh*. New York: John Wiley.

Zadeh, L. A. (1965). Fuzzy Sets. *Information and Control* 8:338–365.

Zadeh, L. A. (1975). The Concepts of a Linguistic Variable and Its Application to Approximate Reasoning. *Information Science* 8(3):199–249.

Zadeh, L. A. (1978). Fuzzy Sets as a Basis for a Theory of Possibility. *Fuzzy Sets and Systems* 1:3–28.

Zadeh, L. A. (1989). Knowledge Representation in Fuzzy Logic. *IEEE Transactions on Knowledge and Data Engineering* 1(1):89–100.

Zimmerman, H. (1987). *Fuzzy Sets, Decision Making and Expert Systems*. Boston: Kluwer Academic Publishers.

Zimmerman, H. (1990). *Fuzzy Set Theory and its Applications*. : Kluwer Academic Publishers.

Zimmerman, H.-J., L. A. Zadeh, and B. Gaines (eds.) (1984). *Fuzzy Sets and Decision Analysis*. Amsterdam: North-Holland.

GAME THEORY

American Mathematical Society Short Course (1981). *Game Theory and its Applications.* Providence, RI: American Mathematical Society.

Aumann, R. J. (1989). Game Theory. In J. Eatwell, M. Milgate, and P. Newman (eds.), *The New Pelgrave Game Theory,*New York: W. W. Norton.

Bierman, H. S. (1993). *Game Theory With Economic Applications.* Reading, MA: Addison-Wesley.

Bogardi, I., and F. Szidarovsky. (1976). Application of Game Theory in Water Management. *Applied Mathematical Modelling* 1:11–20.

Chatterjee, K. (1985). Disagreement in Bargaining: Models with Incomplete Information. In *Game-Theoretic Models of Bargaining,* New York: Cambridge University Press.

Davis, M. (1983). *Game Theory: An Introduction.* New York: Basic Books.

Hati, S. K., and R. G. Lamb. (1987). Application of Game Theory in the Design of Optimal Air Pollution Control Measures. *Atmospheric Environment* 21(8):1833–1841.

Luce, R. D., and H. Raiffa *Games and Decisions, Introduction and Critical Survey.* New York: John Wiley and Sons.

May, F. B. (1970). *Introduction to Games of Strategy.* Boston: Allyn & Bacon.

Nash, J. F. Jr. (1951). Non-cooperative Games. *Annals of Mathematics* 54:289–295.

Peleg, B. (1984). *Game Theoretic Analysis of Voting in Committees.* Cambridge: Cambridge University Press.

Selman, V., and C. I. Bartfeld. (1985). Game Theory Applied to Simulation of Reliability. *Proceedings of the 1985 Summer Computer Simulation Conference* (pp. 568–572). New York, NY, USA: North-Holland.

Shubik, M. (1987). *Game Theory in the Social Sciences, Vols. I and II.* Cambridge, Mass.: MIT Press.

Zagare, F. C. (1984). *Game Theory: Concepts and Applications (Sage University Paper Series No. 41).* Newbury Park, CA: Sage Publications.

GROUP DECISION MAKING

Fisher, B. A. (1990). *Small Group Decision Making. 3rd Ed.* New York: McGraw-Hill.

Fox, W. M. (1987). *Effective Group Problem Solving. 1st Ed.* San Francisco: Jossey-Bass.

Gouran, D. S. (1982). *Making Decisions in Groups.* Glenview, Ill.: Scott Foresman.

Hipel, K. W., and K. J. Radford (1994). Introduction to the Special Issue on Decision Making Under Conditions of Conflict. *Group Decision and Negotiation* 3(2):167–168.

Hwang, C., and M. Lin (1987). *Group Decision Making Under Multiple Criteria; Methods and Applications.* New York: Springer-Verlag.

Iz, P. H., and L. R. Gardiner (1993). Analysis of Multiple Criteria Decision Support Systems for Cooperative Groups. *Group Decision and Negotiation* 2:61–79.

Kacprzyk, J., and M. Fedrizzi (eds.) (1990). *Multiperson Decision Making Models Using Fuzzy Sets and Possibility Theory.* Dordrecht: Kluwer Academic Publishers.

Keeney, R. L., and C. W. Kirkwood. (1975). Group Decision Making Using Cardinal Social Welfare Functions. *Management Science* 22(4):430–437.

Kersten, G., R. Kulikowski, and Z. Nahorski (1994). Special Issue: Negotiation Modeling and Support. *Group Decision and Negotiation* 3:7–10.

Kleindorfer, P. R. (1982). Group Decision Making Methods for Evaluating Social and Technological Risks. In H.C. Kunreuther, and E.V. Ley (eds.), *The Risk Analysis Controversy: An Institutional Perspective*, New York: Springer-Verlag.

Linstone, H. A., and M. Turoff (eds.) (1975). *The Delphi Method: Techniques and Applications*. Reading, MA: Addison-Wesley.

Lockett, G., B. Hetherington, P. Yallup, M. Stratford, and B. Cox. (1986). Modelling a Research Portfolio Using AHP: A Group Decision Process. *R & D Management* 16(2):151–160.

Nitzan, S. (1985). *Collective Decision Making*. Cambridge: Cambridge University Press.

Olsen, S. A. (ed.) (1982). *Group Planning and Problem Solving Methods in Engineering Management*. New York: John Wiley & Sons.

Pinsonneault, A., and K. L. Kraemer. (1990). The Effects of Electronic Meetings on Group Processes and Outcomes: An Assessment of the Empirical Research. *European Journal of Operational Research* 46(2):143–161 (May 25).

Saaty, T. L. (1989). Group Decision Making and the AHP. In B. L. Golden, E. A. Wasil, and P. T. Harker (eds.), *The Analytic Hierarchy Process: Applications and Studies*,New York: Springer.

Spencer, J. K. (1985). Nominal Group Technique Clarifies and Balances Issues. *Fund Raising Management* 16(2):42–47.

ISSUES SURROUNDING THE USE OF DECISION AIDS

Brown, R. V. (1982). "Towards A Prescriptive Organization Theory of Decision Aiding For Risk Management: Phase I: Conceptual Development" (Technical Report No. 82-7). Falls Church, VA: Decision Science Consortium, Incorporated.

Covello, V. T., P. M. Sandman, and P. Slovic (1988). *Risk Communication, Risk Statistics, and Risk Comparisons: A Manual For Plant Managers*. Washington, D.C.: Chemical Manufacturers Association.

Covello, V. T., D. von Winterfeldt, and P. Slovic. (1986). Communicating Scientific Information about Health and Environmental Risks: Problems and Opportunities from a Social and Behavioral Perspective. In V.T. Covello, A. Moghissi, and V.R.R. Uppuluri (eds.), *Uncertainties In Risk Assessment and Risk Management*, New York: Plenum Press.

Covello, V. T., D. von Winterfeldt, and P. Slovic. (1987). Communicating Risk Information to the Public. In J.C. Davies, V.T. Covello, and F. Allen (eds.), *Risk Communication*, Washington, D.C.: The Conservation Foundation.

Cox, L. A., Jr. and P. F. Ricci. (1989). Legal and Philosophical Aspects of Risk Analysis. In D.J. Paustenbach (ed.), *The Risk Assessment of Environmental and Human Health Hazards*,: John Wiley and Sons.

Cyert, R., and J. March (1963). *A Behavioral Theory of The Firm*. Englewood Cliffs, NJ: Prentice Hall.

Fischhoff, B. (1985). Managing Risk Perceptions. *Issues in Science and Technology* 2(1):83–96.

Guzzo, R. A. (ed.) (1982). *Improving Group Decision Making in Organizations*. New York: Academic Press.

Horowitz, I. (ed.) (1989). *Organization and Decision Theory*. Boston: Kluwer Academic Publishers.

Keeney, R. L. (1984). Ethics, Decision Analysis, and Public Risk. *Risk Analysis* 4(2):117–129.

Keeney, R. L., and R. L. Winkler. (1985). Evaluating Decision Strategies for Equity of Public Risks. *Operations Research* 33(5):955–970.

Kleindorfer, P. R. (1982). Group Decision Making Methods for Evaluating Social and Technological Risks. In H.C. Kunreuther and E.V. Ley (eds.), *Risk Analysis Controversy: An Institutional Perspective, Proceedings of a Summer Study on Decision Processes and Institutional Aspects of Risk* (pp. 181–193), Berlin, West Germany and New York: Springer-Verlag.

Krumm, F. V., and C. F. Rolle (1992). Management and Application of Decision and Risk Analysis in Du Pont. *Interfaces* 22(6):84–93.

Kunreuther, H. C., and E. Ley. (1982). Decision Processes and Institutional Aspects of Risk. In *Proceedings of a Summer Study on Decision Processes and Institutional Aspects of Risk*, June 22–26, 1981. Laxenburg, Austria: International Institute for Applied Systems Analysis.

Kunreuther, H. C., and E. V. Ley (eds.) (1982). *Risk Analysis Controversy: An Institutional Perspective, Proceedings of a Summer Study on Decision Processes and Institutional Aspects of Risk*. Berlin, West Germany and New York: Springer-Verlag.

MacLean, D. (1982). Risk and Consent: Philosophical Issues for Centralized Decisions. *Risk Analysis* 2(2):59–67.

March, J. G., and Z. Shapira. (1987). Managerial Perspectives on Risk and Risk Taking. *Management Science* 33(11):1404–1418.

Merkhofer, M. W., and W. M. Saade (1978). "Decision Flexibility in a Learning Environment." Menlo Park, CA: SRI International.

Rayner, S., and R. Cantor. (1987). How Fair is Safe Enough? The Cultural Approach to Societal Technology Choice. *Risk Analysis* 7(1):3–9.

Rodgers, W. H. J. (1987). Guerrilla Decisionmaking: Judicial Review of Risk Assessments. *Journal of Hazardous Materials* 15(1 / 2):205–217.

Starr, C. (1985). Risk Management, Assessment and Acceptability. *Risk Analysis* 5(2):57–102.

Sweeney, J. C. (1992). Measuring Process Safety Management. *Plant/Operations Progress* 11(2):89–98.

Vari, A., and J. Vecsenyi. (1984). Selecting Decision Support Methods in Organisations. *Journal of Applied Systems Analysis* 11(April):23–36.

KEPNER–TREGOE DECISION ANALYSIS

Anonymous. (1986). Decision Support: Rational Management, At Last. *Business Software Review* 5(5):9–10.

Kepner, C. H., and B. B. Tregoe (1981). *The New Rational Manager*. London: John Martin Publishing Ltd.

Moore, L. R. (1986). Kepner–Tregoe as a Safety Methodology. *Professional Safety* 31(6): 37–41.

MATHEMATICAL PROGRAMMING

Ackoff, R. L. (1956). The Development of Operations Research as a Science. *Operations Research* 4:265–295.

Bazaraa, N. S., and C. M. Shetty (1979). *Nonlinear Programming: Theory and Algorithms*. New York: Wiley.

Bellman, R. (1957). *Dynamic Programming*. Princeton, New Jersey: Princeton University Press.

Bellman, R., and S. E. Dreyfus (1962). *Applied Dynamic Programming*. Princeton, New Jersey: Princeton University Press.

Bracken, J., and G. P. McCormick (1968). *Selected Applications of Nonlinear Programming*. New York: Wiley.

Charnes, A., and W. W. Cooper (1961). *Management Models and Industrial Applications of Linear Programming, Vols. I and II*. New York: John Wiley and Sons.

Cook, T. M., and R. A. Russell (1985). *Introduction to Management Science*. 3rd ed. Englewood Cliffs, NJ: Prentice-Hall.

Daellenbach, H. G., J. A. George, and D. C. McNickle (1983). *Intoduction to Operations Research Techniques*. Boston: Allyn and Bacon, Inc.

Dantzig, G. B. (1963). *Linear Programming and Extensions*. Princeton, New Jersey: Princeton University Press.

Dreyfus, S. E., and A. M. Law (1977). *The Art and Theory of Dynamic Programming*. New York: Academic Press.

Floquet, P., L. Pibouleau, and S. Domenech. (1988). Mathematical programming tools for chemical engineering process design synthesis. *Chemical Engineering and Processing* 23(2):99–113.

Gass, S. I. (1985). *Linear Programming: Methods and Applications*. New York: McGraw-Hill, Inc.

Gould, F. J., G. D. Eppen, and C. P. Schmidt (1991). *Introductory Management Science*. Englewood Cliffs, NJ: Prentice-Hall.

Hillier, F. S., and G. J. Lieberman (1990). *Introduction to Mathematical Programming*. New York: McGraw-Hill, Inc.

Ignizio, J. P. (1976). *Goal Programming and Extensions*. Lexington, MA: Heath.

Johnson, E. L., and G. L. Nemhauser. (1992). Recent Developments and Future Directions in Mathematical Programming. *IBM Systems Journal* 31(1):79–93.

Lasdon, L. S., and A. D. Warren. (1980). Survey of Nonlinear Programming Applications. *Operations Research* 28(5):1029–1073.

Lee, S. M. (1972). *Goal Programming for Decision Analysis 1st Ed.* Philadelphia: Auerbach Publishers.

Lee, S. M. (1976). *Linear Optimization for Management*. New York: Petrocelli/Charter.

Mulvey, J. M. (ed.) (1982). *Evaluating Mathematical Programming Techniques, Proceedings of a Conference held at the National Bureau of Standards, January 5–6, 1981 (1982)*..New York: Springer-Verlag.

Ozan, T. (1986). *Applied Mathematical Programming for Production and Engineering Management*. Englewood Cliffs, NJ: Prentice-Hall.

Rhinehart, R. R., and J. D. Beasley. (1987). Dynamic Programming for Chemical Engineering Applications. *Chemical Engineering* 94(18):113–119.

Roberts, S. M. (1964). *Dynamic Programming in Chemical Engineering and Process Control.* New York: Academic Press.

Salkin, H. M. (1975). *Integer Programming.* Reading, MA: Addison-Wesley.

Ulungu, E. L., and J. Teghem (1994). Multi-objective Combinatorial Optimization Problems: A Survey. *Journal of Multi-Criteria Decision Analysis* 3(2):83–104.

Wagner, H. M. (1975). *Principles of Operations Research (2nd ed.).* Englewood Cliffs, NJ: Prentice-Hall.

Weingartner, H. M. (1965). *Mathematical Programming and the Analysis of Capital Budgeting Problems.* Englewood Cliffs, NJ: Prentice-Hall.

Williams, H. P. (1990). *Model Building in Mathematical Programming.* Chichester, England: Wiley.

MULTIATTRIBUTE UTILITY ANALYSIS

Anandalingam, G. (1987). A Multiple Criteria Decision Analytic Approach for Evaluating Acid Rain Policy Choices. *European Journal of Operational Research* 29(3):336–352.

Bard, J. F. (1992). A Comparison of the Analytic Hierarchy Process with Multiattribute Utility Theory: A Case Study. *IIE Transactions* 24(5):111–121.

Bell, D., R. Keeney, and H. Raiffa (eds.) (1977). *Conflicting Objectives in Decisions.* New York: John Wiley and Sons.

Bohn, M. P. (1992). "Decision Making Under Uncertainty: An Investigation into the Application of Formal Decision-Making Methods to Safety Decisions" (NUREG/CR-5906). Division of Safety Issues Resolution, Office of Nuclear Regulatory Research, U.S. Nuclear Regulatory Commission.

Dyer, J. S., P. C. Fishburn, R. E. Steuer, J. Wallenius, and S. Zionts. (1992). Multiple Criteria Decision Making, Multiattribute Utility Theory: The Next Ten Years. *Management Science* 38(5):645–654.

Edwards, W. (1977). How To Use Multi-Attribute Utility Measurement for Social Decision Making. *IEEE Transactions On Systems, Man, and Cybernetics* SMC-7:326–339.

Edwards, W., and J. R. Newman (1982). *Multiattribute Evaluation (Sage University Paper Series No. 26).* Newbury Park, CA: Sage Publications.

Fishburn, P. C. (1984). Multiattribute Nonlinear Utility Theory. *Management Science* 30(11):1301–1310.

French, S. (1986). *Decision Theory: An Introduction to the Mathematics of Rationality.* New York: Halsted Press.

Gardiner, P. J., and W. Edwards. (1975). Public Values: Multiattribute-Utility Measurement for Social Decision Making. In M.F. Kaplan, and S. Schwartz (eds.), *Human Judgement and Decision Processes,* New York: Academic Press.

Howard, R. A. (1975). Social Decision Analysis. *Proceedings of the IEEE* 63(3):359–371.

Keeney, R. L. (1975). Examining Corporate Policy Using Multiattribute Utility Analysis. *Sloan Management Review* 17(1):63–76.

Keeney, R., and H. Raiffa (1976). *Decisions with Multiple Objectives: Preferences and Value Tradeoffs.* New York: John Wiley and Sons.

Keeney, R. L., and R. L. Winkler. (1985). Evaluating Decision Strategies for Equity of Public Risks. *Operations Research* 33(5):955–970.

Keeney, R. L., R. K. Sarin, and R. L. Winkler. (1984). Analysis of Alternative National Ambient Carbon Monoxide Standards. *Management Science* 30(4):518–528.

Kirkwood, C. W. (1992). Estimating the Impact of Uncertainty on a Deterministic Multiattribute Evaluation. *Management Science* 38(6):819–826.

Merkhofer, M. W. (1987). *Decision Science and Social Risk Management: A Comparative Evaluation of Cost–Benefit Analysis, Decision Analysis, and Other Formal Decision-Aiding Approaches*. Dordrecht, Netherlands: D. Reidel (Boston: Kluwer Academic Publishers Group).

Merkhofer, M. W., and R. Keeney (1987). A Multiattribute Utility Analysis of Alternative Sites for the Disposal of Nuclear Waste. *Risk Analysis* 7(2):173–194.

Merkhofer, M. W., T. A. Cotton, and T. P. Longo (1988). A Program Optimization System For Aiding Decisions to Fund the Cleanup of Hazardous Waste Sites at Department of Energy Defense Facilities. In *Superfund 1988, Proceedings of the 9th National Conference*, Silver Spring, Maryland, November 28–30, 1988, sponsored by the Hazardous Materials Control Research Institute (HMCRI).

Torrance, G. W., M. H. Boyle, and S. P. Horwood. (1982). Application of Multiattribute Utility Theory To Measure Social Preferences For Health States. *Operations Research* 30(6):1043–1069.

U.S. Department of Energy (1986). "A Multiattribute Utility Analysis of Sites Nominated For Characterization For The First Radioactive-Waste Repository—A Decision-Aiding Methodology" (DOE/RW-OO74). Office of Civilian Radioactive Waste Management. Washington, D.C.

von Winterfeldt, D. (1987). Value Tree Analysis: An Introduction and an Application to Offshore Oil Drilling. In P. Kleindorfer, and H. Kunreuther (eds.), *Insuring and Managing Hazardous Risks: From Seveso To Bhopal and Beyond* (pp. 349–377). New York: Springer-Verlag.

von Winterfeldt, D., and W. Edwards (1986). *Decision Analysis and Behavioral Research*. Cambridge: Cambridge University Press.

Wenstop, F. E., and A. J. Carlsen. (1988). Ranking Hydroelectric Power Projects with Multicriteria Decision Analysis. *Interfaces* 18(4):36–48.

Zeleny, M. (1982). *Multiple Criteria Decision Making*. New York: McGraw-Hill.

Zionts, S. (1992). Some Thoughts on Research in Multiple Criteria Decision Making. *Computers & Operations Research* 19(7):567–570.

MULTIPLE CRITERIA DECISION MAKING

Arrow, K. J. (1986). *Social Choice and Multicriterion Decision-Making*. Cambridge, MA: MIT Press.

Bogetoft, P., and P. Pruzan (1991). *Planning with Multiple Criteria: Investigation, Communication, Choice*. New York: North-Holland.

Dyer, J. S., P. C. Fishburn, R. E. Steuer, J. Wallenius, and S. Zionts. (1992). Multiple Criteria Decision Making, Multiattribute Utility Theory: The Next Ten Years. *Management Science* 38(5):645–654.

Fandel, G., J. Spronk, and B. Matarazzo (eds.) (1985). *Multiple Criteria Decision Methods and Applications*. New York: Springer-Verlag.

Hwang, C. L. (1979). *Multiple Objective Decision Making Methods and Applications*. Berlin: Springer-Verlag.
Lee, J. K., and E. G. Jr. Hurst. (1988). Multiple-Criteria Decision Making Including Qualitative Factors: The Post-Model Analysis Approach. *Decision Sciences* 19(2): 334–352.
Ozernoy, V. M. (1992). Choosing the "Best" Multiple Criteria Decision-Making Method. *INFOR* 30(2):159–171 (May).
Roy, B. (1990). Decision-Aid and Decision-Making. *European Journal of Operational Research* 45(2/3):324–331.
Seo, F. (1988). *Multiple Criteria Decision Analysis in Regional Planning*. Dordrecht, Netherlands: D. Reidel.
Siskos, J., J. Lombard, and A. Oudiz. (1986). The Use of Multicriteria Outranking Methods in the Comparison of Control Options Against a Chemical Pollutant. *Journal of the Operational Research Society* 37(4):357–371.
Tabucanon, M. T. (1988). *Multiple Criteria Decision Making in Industry*. New York: Elsevier.
Vincke, P. (1986). Analysis of Multicriteria Decision Aid in Europe. *European Journal of Operational Research* 25(2):160–168.
Vincke, P. (1992). *Multicriteria Decision-Aid*. New York: John Wiley & Sons.
Wenstop, F. E., and A. J. Carlsen. (1988). Ranking Hydroelectric Power Projects with Multicriteria Decision Analysis. *Interfaces* 18(4):36–48.
Zeleny, M. (1982). *Multiple Criteria Decision Making*. New York: McGraw-Hill.

OPERATIONS RESEARCH/MANAGEMENT SCIENCE

Ackoff, R. L., and P. Rivett (1963). *A Manager's Guide to Operations Research*. New York: Wiley.
Chacko, G. K. (1993). *Operations Research and Management Science*. New York: McGraw-Hill.
Churchman, C. W., R. L. Ackoff, and E. L. Arnoff (1957). *Introduction to Operations Research*. New York: John Wiley & Sons.
Cook, T. M., and R. A. Russell (1985). *Introduction to Management Science*. Englewood Cliffs, NJ: Prentice-Hall.
Davis, K. R., P. McKeown, and T. R. Rakes (1986). *Management Science: An Introduction*. Boston: Kent.
Fabrycky, W. J., P. M. Ghare, and P. E. Torgensen (1984). *Applied Operations Research and Management Science*. Englewood Cliffs, NJ: Prentice-Hall.
Ford, L. R. Jr., and D. R. Fulkerson (1962). *Flows in Networks*. Princeton, NJ: Princeton University Press.
Gould, F. J., G. D. Eppen, and C. P. Schmidt (1991). *Introductory Management Science*. Englewood Cliffs, NJ: Prentice-Hall.
Grayson, C. J. Jr. (1973). Management Science and Business Practice. *Harvard Business Review* 51:41–48.
Gross, D., and C. M. Harris (1974). *Fundamentals of Queuing Theory*. New York: John Wiley & Sons, Inc.

Hillier, F. S., and G. J. Lieberman (1986). *Introduction to Operations Research.* Oakland: Holden-Day, Inc.

Howard, R. A. (1971). *Dynamic Probabilistic Systems—Volume I: Markov Models.* New York: John Wiley and Sons, Incorporated.

Vatter, P. A. et al. (1978). *Quantitative Methods in Management: Text and Cases.* Homewood, IL: Richard D. Irwin Inc.

Von Lanzenauer, C. H. (1975). *Cases in Operations Research.* London, Ontario: The University of Western Ontario.

Wagner, H. M. (1975). *Principles of Operations Research (2nd ed.).* Englewood Cliffs, NJ: Prentice-Hall.

Wagner, H. M. (1988). Operations Research: a Global Language for Business Strategy. *Operations Research* 36(5):797–803.

OTHER DECISION AIDS

Azani, H., and R. Khorramshahgol. (1990). Analytic Delphi Method (ADM): A Strategic Decision Making Model Applied to Location Planning. *Engineering Costs & Production Economics* 20(1):23–28.

Bogetoft, P., and P. Pruzan (1991). *Planning with Multiple Criteria: Investigation, Communication, Choice.* New York: North-Holland.

Buckley, J. J. (1986). Stochastic Dominance: An Approach to Decision Making Under Risk. *Risk Analysis* 6(1):35–41.

Chicken, J. C., and M. R. Hayns (1989). *The Risk Ranking Technique in Decision Making.* Oxford: Pergammon Press.

Fast, J. C., and L. T. Looper. (1988). "Multiattribute Decision Modeling Techniques: A Comparative Analysis" (AFHRL-TR-88-3). Brooks Air Force Base, Texas: Air Force Human Resources Laboratory (August).

Hwang, C., and M. Lin (1987). *Group Decision Making Under Multiple Criteria; Methods and Applications.* New York: Springer-Verlag.

Hwang, C., and K. Yoon (1981). *Multi-Attribute Decision Making Methods and Applications.* New York: Springer-Verlag.

Keeney, R. L. (1994). Creativity in Decision Making with Value-Focused Thinking. *Sloan Management Review* (Summer):33–41.

Markowitz, H. M. (1959). *Portfolio Selection.* New York: John Wiley & Sons.

Massam, B. H. (1988). Multi-Criteria Decision Making Techniques in Planning. D. Diamond and J. B. McLaughlin (eds.) *Progress In Planning* Vol. 30, Part 1. New York: Pergamon Press.

Olsen, S. A. (ed.) (1982). *Group Planning and Problem Solving Methods in Engineering Management.* New York: John Wiley & Sons.

Roy, B., M. Present, and D. Silhol (1986). A Programming Method for Determining Which Paris Metro Stations Should Be Renovated. *European Journal of Operations Research* 24(2):318–334.

Roy, G. G. (1982). The Use of Multi-Dimensional Scaling in Policy Selection. *Journal of The Operational Research Society* 33(3):239–245.

Stork, P., and J. M. Viaene. (1992). Policy Optimization by Lexicographic Preference Ordering. *Journal of Policy Modeling* 14(5):655–673.

Tabucanon, M. T. (1988). *Multiple Criteria Decision Making in Industry*. New York: Elsevier.
Van, L. R. R. G. (1989). Option Analysis: Making Better Decisions Faster. *European Journal of Operational Research* 38(3):318–328.
Vincke, P. (1992). *Multicriteria Decision-Aid*. New York: John Wiley & Sons.
Walker, L. R., and M. D. Rosko. (1988). Evaluation of Health Care Service Diversification Options in Health Care Institutions and Programs by Portfolio Analysis: A Marketing Approach. *Journal of Health Care Marketing* 8(1):48–59.
Walker, W. E. (1986). Use of Screening in Policy Analysis. *Management Science* 32(4):389–402.
Wang, J., and B. Malakooti. (1992). A Feedforward Neural Network for Multiple Criteria Decision Making. *Computers & Operations Research* 19(2):151–167.
Wright, D. J. (1983). Catastrophe Theory in Management Forecasting and Decision Making. *Journal of the Operational Research Society* 34(10):935–942.

PAYOFF MATRIX ANALYSIS

Behn, R. D., and J. W. Vaupel (1982). *Quick Analysis for Busy Decision Makers*. New York: Basic Books, Inc.
Belyaev, L. S. (1990). Pay-off Matrix Technique. *Energy (Oxford)* 15(7–8):631–643.
Bunn, D. W. (1984). *Applied Decision Analysis*. New York: McGraw-Hill.
Chapel, S. (1989). CATALYST: Designing Flexible Business Strategies. *EPRI Journal* 14(7):46–49.
Doyle, M., and D. Straus (1976). *How to Make Meetings Work*. New York: Jove Books.
EPRI (1992). "CATALYST—A Group Process for Strategic Decision Making: Facilitator's Guidebook." Prepared by Applied Decision Analysis and The Brattle Group. EPRI TR-100583, Research Project 2631, Final Report. Palo Alto, CA: Electric Power Research Institute.
Hanna, R. C. (1979). Apply Decision Theory to Hazard Evaluation. *Hydrocarbon Processing* (December):133–150.
Kane, L. (1990). Selecting a distributed control system. *Hydrocarbon Processing* 69 (February).
Leemhuis, J. P. (1985). Using Scenarios to Develop Strategies. *Long Range Planning* 18(2):30–37.
Lindley, D. V. (1985). *Making Decisions*. London: John Wiley and Sons.
Martin, M. P. (1979). Decision Making: The Payoff Matrix. *Journal of Systems Management* 30(1):14–18.
Merkhofer, M. W. (1987). Quantifying Judgmental Uncertainty: Methodology, Experiences, and Insights. *IEEE Transactions on Systems, Man, and Cybernetics* SMC-17(5): 741–752.
Savage, L. J. (1954). *The Foundations of Statistics*. New York: John Wiley and Sons, Inc.
von Winterfeldt, D., and W. Edwards (1986). *Decision Analysis and Behavioral Research*. Cambridge: Cambridge University Press.

PROBABILITY AND STATISTICS

Berger, J. O. (1985). *Statistical Decision Theory and Bayesian Analysis*, 2nd ed. New York: Springer-Verlag.

Buckley, J. J. (1983). Decision Making Under Risk: A Comparison of Bayesian and Fuzzy Set Methods. *Risk Analysis* 3(3):157–168.

Hogarth, R. M. (1975). Cognitive Processes and The Assessment of Subjective Probability Distributions. *Journal of The American Statistical Association* 70:271–289.

Miller, A. C., and T. R. Rice. (1984). Discrete Approximations of Probability Distributions. *Management Science* 29(3):352–362.

Raiffa, H., and R. Schlaifer (1961). *Applied Statistical Decision Theory*. Cambridge: Harvard Business School.

Savage, L. J. (1954). *The Foundations of Statistics*. New York: John Wiley and Sons.

Schlaifer, R. (1961). *Introduction to Statistics for Business Decisions*. New York: McGraw-Hill.

Staël von Holstein, C.-A. S. (1970). Assessment and Evaluation of Subjective Probability. Stockholm: Stockholm School of Economics, Economic Research Institute.

RISK ANALYSIS AND RISK ASSESSMENT

AIChE/CCPS (1989). *Guidelines for Chemical Process Quantitative Risk Analysis*. New York: American Institute of Chemical Engineers.

AIChE/CCPS (1985). *Guidelines for Hazard Evaluation Procedures*. New York: Center for Chemical Process Safety, American Institute of Chemical Engineers.

AIChE/CCPS (1992). *Guidelines for Hazard Evaluation Procedures, 2nd Edition with Worked Examples*. New York: American Institute of Chemical Engineers.

Apostolakis, G. (1981). Bayesian Methods in Risk Assessment. In J. Lewis, and M. Becker (eds.), *Advances In Nuclear Science and Technology* 13:415–465. New York: Plenum Press.

Arendt, J. S. (1990). Using Quantitative Risk Assessment in the Chemical Process Industry. *Reliability Engineering & System Safety* 29(1):133–149.

Arendt, J. S. (1992). Managing Risk. *Chemtech* (August):490–497.

Arendt, J. S., D. K. Lorenzo, and A. F. Lusby. (1989). "A Manager's Guide to Quantitative Risk Assessment of Chemical Process Facilities" (JBFA-119-88). Knoxville, TN: JBF Associates, Inc.

Baybutt, P. (1983). Procedures for the Use of Risk Analysis in Decision Making. Pages 261–277 in S. Hartwig (ed.), *Heavy Gas and Risk Assessment II, Proceedings of the 2nd Symposium on Heavy Gases and Risk Assessment, Frankfurt-am-Main, May 25–26, 1982*. Dordrecht, Netherlands and Boston: D. Reidel.

Billington, R., and R. N. Allan (1983). *Reliability Evaluation of Engineering Systems: Concepts and Techniques*. New York: Plenum Press.

Boykin, R. F., R. A. Freeman, and R. R. Levary. (1984). Risk Assessment in a Chemical Storage Facility. *Management Science* 30(4):512–517.

Covello, V. T. (ed.) (1983). *The Analysis of Actual Versus Perceived Risk*. Society for Risk Analysis, International Workshop on the Analysis of ..., 1981. New York: Plenum Press.

Covello, V. T., and J. Mumpower. (1985). Risk Analysis and Risk Management: An Historical Perspective. *Risk Analysis* 5(2):103–120.

Covello, V. T., P. M. Sandman, and P. Slovic (1988). *Risk Communication, Risk Statistics, and Risk Comparisons: A Manual For Plant Managers*. Washington, D.C.: Chemical Manufacturers Association.

Cox, L. A. Jr., and P. F. Ricci (1989). Risk, Uncertainty, and Causation: Quantifying Human Health Risks. In D.J. Paustenbach (ed.), *The Risk Assessment of Environmental and Human Health Hazards*, John Wiley and Sons.

Cullingford, M. C., S. M. Shah, and J. H. Gittus (1988). *Implications of Probabilistic Risk Assessment*. Barking, England: Elsevier Applied Science Publisher.

Dow Chemical Company (1994). *Fire and Explosion Index—Hazard Classification Guide*, 7th edition (CEP Technical Manual series). New York: American Institute of Chemical Engineers.

European Economic Community. (1982). Major-Accident Hazards of Industrial Activities ("Sevesco Directive"). *European Economic Community, Council Directive 82-501-EEC Official Journal, L230*(5.8.1982),

European Federation of Chemical Engineers (1985). Risk Analysis in the Process Industries. *European Federation of Chemical Engineers* Vol. 45 (pp. 92). Rugby, UK: Institution of Chemical Engineers.

Fischhoff, B. (1984). Setting Standards: A Systematic Approach To Managing Public Health and Safety Risks. *Management Science* 30(7):823–843.

Fischhoff, B., S. Lichtenstein, P. Slovic, S.L. Derby, and R.L. Keeney (1981). *Acceptable Risk*. New York: Cambridge University Press.

Freeman, R. A. (1990). Centre for Chemical Process Safety Guidelines for Chemical Process Quantitative Risk Analysis. *Plant/Operations Progress* 9(4):231–235.

French, R. W., R. E. Olsen, and G. L. Peloquin. (1990). Quantified Risk as a Decision Aid. *International Conference on Safety and Loss Prevention in the Chemical and Oil Processing Industries. Institution of Chemical Engineers Symposium Series n 120* (pp. 461–469). Rugby, England: Inst of Chemical Engineers, Davis Building.

Gibson, S. B. (1980). Hazard Analysis and Risk Criteria. *Chemical Engineering Progress* 76:46–50.

Griffiths, R. F. (1984). Chemical Plant Risk Assessment: Uncertainties and Development Needs. *Environmental International* 10(5/6):523-530.

Helmers, E. N., and L. C. Schaller. (1982). Calculated Process Risks and Hazards Management. *Plant/Operations Progress* 1(3):190.

Henley, E. J., and H. Kumamoto (1981). *Reliability Engineering and Risk Assessment*. Englewood Cliffs, NJ: Prentice Hall.

Hoerger, F. (1984). The Need for Risk Assessment of Chemicals in Corporate Decision Making. In J.V. Rodricks and R.G. Tardiff (eds.), *Assessment and Management of Chemical Risks*,Washington, D.C.: American Chemical Society.

Joschek, K. T. (1983). Risk Assessment in the Chemical Industry. *Plant/Operations Progress* 2(January 1):1–5.

Keller, D. B. (1992). Putting Risk Assessment in Perspective. *Professional Safety* 37(2):35–58.

Kletz, T. A. (1976). The Application of Hazard Analysis to Risks to the Public at Large. *World Congress of Chemical Engineering*, July.

Kletz, T. A. (1978). Practical Applications of Hazard Analysis. *Chemical Engineering Progress* 74(10):47–53.

Kunruether, H. (1983). *Risk Analysis and Decision Processes.* Berlin: Springer-Verlag.

Lees, F. P. (1980). *Loss Prevention in the Process Industries.* 2 vols. London and Boston: Butterworths.

McKone, J. E. (1986). The Implicit Valuation of Environmental Cancer by the United States Regulatory Agencies. *Toxics Law Reporter* 1(16) (September 24).

Ormsby, R. W. (1982). Process Hazards Control at Air Products. *Plant/Operations Progress* 1:141–144.

Paté-Cornell, M. E., and R. Boykin. (1987). Probabilistic Risk Analysis and Safety Regulation in the Chemical Industry. *Journal of Hazardous Materials* 15:97–122.

Raiffa, H. (1982). Science and Policy: Their Separation and Integration in Risk Analysis. *Risk Analysis Controversy: An Institutional Perspective, Proceedings of a Summer Study on Decision Processes and Institutional Aspects of Risk* (pp. 27–37). Berlin, Germany and New York, NY, USA: Springer-Verlag.

Raman, R. (1990). Risk Perceptions and Problems in Interpreting Risk Results. *Eighteenth Australasian Chemical Engineering Conference Part 2 (of 2). CHEMECA '90* (pp. 848–855). Auckland, New Zealand: University of Auckland School of Engineering.

Rowe, W. D. (1977). *An Anatomy of Risk.* New York: John Wiley and Sons.

Schreiber, S. (1991). Guidelines for Technical Management of Chemical Process Safety. *Plant/Operations Progress* 10(2):65–68.

Sheridan, P. J. (1990). Risk Assessment: The Path of Uncertainty. *Occupational Hazards* 52(10):68–72.

Sherif, Y. S. (1991). On Risk and Risk Analysis. *Reliability Engineering & System Safety* 31(2):155–178.

Shortreed, J. (1990). Recent Advances in Research and Future Requirements. *Plant/Operations Progress* 9(3):198–203.

Slovic, P., B. Fischhoff, and S. Lichtenstein. Perceived Risk: Psychological Factors and Social Implications. In F. Warner, and D.H. Slater (eds.), *The Assessment and Perception of Risk.* London: The Royal Society.

Starr, C., and C. Whipple. (1984). A Perspective on Health and Safety Risk Analysis. *Management Science* 30(4):452–463.

Sweeney, J. C. (1992). Measuring Process Safety Management. *Plant/Operations Progress* 11(2):89–98.

U.K. Health and Safety Commission (1991). *Major Hazards Aspects of the Transport of Dangerous Substances.* London, England: UK Health and Safety Executive.

U.K. Health and Safety Executive. *Quantified Risk Assessment: Its Input To Decision Making.* London: HMSO.

U.S. Nuclear Regulatory Commission (1985). "Probabilistic Safety Analysis Procedures Guide" (NUREG/DR-2815). Nuclear Regulatory Commission, Washington, DC (August).

U.S. Nuclear Regulatory Commission (1987). "Reactor Risk Reference Document" (NUREG-1150, Vol. 1). Office of Nuclear Regulatory Research, Washington, DC (February).

Van Kuijen, C. J. (1987). Risk Management in the Netherlands: A Quantitative Approach. *UNIDO Workshop on Hazardous Waste Management and Industrial Safety,* Vienna, 1987, June 22.

Waterstact Groningen Provincial (1979). *Pollution Control and the Use of Norms in Groningen—Criteria for Risks Related to Dangerous Goods.* Groningen, Netherlands (April).

SOCIAL CHOICE THEORY

Arrow, K. J. (1963). *Social Choice and Individual Values* (2nd. Ed). New York: Wiley.
Arrow, K. J. (1986). *Social Choice and Multicriterion Decision-Making*. Cambridge, MA: MIT Press.
Bonner, J. (1986). *Introduction to the Theory of Social Choice*. Baltimore, MD: John Hopkins University Press.
Elster, J., and A. Hylland (eds.) (1986). *Foundations of Social Choice Theory*. New York: Cambridge University Press.
Epstein, L. G., and U. Segal. (1992). Quadratic Social Welfare Functions. *Journal of Political Economy* 100(4):691–712.
Fishburn, P. C. (1973). *The Theory of Social Choice*. Princeton: Princeton University Press.
Keeney, R. L. (1988). Structuring Objectives for Problems of Public Interest. *Operations Research* 36(3):396–405.
Kelly, J. S. (1988). *Social Choice Theory*. Berlin: Springer-Verlag.
Merkhofer, M. W. (1987). *Decision Science and Social Risk Management: A Comparative Evaluation of Cost–Benefit Analysis, Decision Analysis, and Other Formal Decision-Aiding Approaches*. Dordrecht, Netherlands: D. Reidel (Boston: Kluwer Academic Publishers Group).
Murakami, Y. (1968). *Logic and Social Choice*. London: Routledge & K. Paul.
Otway, H. J. (1975). "Risk Assessment and Social Choices" (IIASA Research Memorandum RM-75-2). Laxenburg, Austria: International Institute For Applied Systems Analysis.
Pattanaik, P. K. (1971). *Voting and Collective Choice*. Cambridge, England: University Press.
Rayner, S., and R. Cantor. (1987). How Fair is Safe Enough? The Cultural Approach to Societal Technology Choice. *Risk Analysis* 7(1):3–9.
Sen, A. K. (1970). *Collective Choice and Social Welfare*. San Francisco, CA: Holden-Day.
Starr, C. (1969). Social Benefit versus Technological Risk: What is Our Society Willing to Pay for Safety? *Science* 165:1232–1238.

SOFTWARE

Anonymous (1986). Decision Support: Rational Management, At Last. *Business Software Review* 5(5):9–10.
Belanger, R. (1984). Can't Decide? *Microcomputing* 8(5):94–100.
Bonini, C. P. (1980). *Computer Models for Decision Analysis*. Palo Alto, CA: Scientific Press.
Buede, D. M. (1992). Software Review: Overview of the MCDA Software Market. *Journal of Multi-Criteria Decision Analysis* 1:59–61.
Buede, D. M. (1992). Superior Design Features of Decision Analytic Software. *Computers & Operations Research* 19(1):43–58.
Buede, D. M. (1993). Aiding Insight for Decision-Makers: A Survey of More Than Two Dozen Software Packages. *OR/MS Today* 20(2):52–60.
Call, H. J., and W. A. Miller. (1990). A Comparison of Approaches and Implementation for Automating Decision Analysis. *Reliability Engineering and System Safety* 30:115–162 (later reprinted in G. E. Apostolakis and P. Kafka (eds.), *The Role and Use of*

Personal Computers in Probabilistic Safety Assessment and Decision Making, London and New York: Elsevier.)

Canada, J. R., and G. L. Hodge. (1988). Microcomputer Software Costing Less than $1000 for Economic and Multi-Attribute Decision Analysis. *Engineering Economist* 33(2):130–144.

Dobias, A. P. (1990). Designing a Mouse Trap Using the Analytic Hierarchy Process and Expert Choice. *European Journal of Operational Research* 48(1):57–65 (September 5).

Friend, J. (1992). New Directions in Software for Strategic Choice. *European Journal of Operational Research* 61(1/2):154–164 (August 25).

Grauer, M., and A. P. Wierzbicki (eds.) (1984). *Interactive Decision Analysis. Proceedings of an International Workshop on Interactive Decision Analysis and Interpretive Computer Intelligence*, Laxenburg, Austria, 1983. New York: Springer.

Harbeson, P. J. (1988). Making Model Decisions with Decision Modeling. *Managers Magazine* 63(6):21–23.

Humphress, G. B. (1984). VA—Value Assessment Software for Multiple Criteria Decision Making by Technical Groups. In *Engineering Databases: Software for On-Line Applications. Presented at the 1984 ASME Pressure Vessels and Piping Conference and Exhibition. American Society of Mechanical Engineers, Pressure Vessels and Piping Division* (pp. 99–105). New York: ASME (Publication) PVP Vol. 96.

Humphry, S. (1992). Tools Help Objectify Decision Making. *PC Week* (March 9):113.

Johnson, S. M. (1992). 1992 Environmental Software Survey. *Journal of Air and Waste Management* 42(11):1516–1535.

Jones, J. M. (1986). Decision Analysis Using Spreadsheets. *European Journal of Operational Research* 26(3):385–400.

Lewandowski, A., and I. Stanchev (eds.) (1989). *Methodology and Software for Interactive Decision Support, Proceedings of International Workshop*. New York: Springer-Verlag.

McDaniel, H. (1970). *Decision Table Software*. Princeton, NJ: Brandon Systems Press.

Mitra, G., and UNICORN Consultants (eds.) (1986). *Computer Assisted Decision Making*. Amsterdam: North-Holland.

Modak, P. M., and J. F. Dhoondia. (1988). A Computer-Assisted Package For Minimum Risk Layout For Process Industries. *Ind. Environ.* 11(3):34–38.

Nagel, Stuart S. (1992). *Applications of Decision-Aiding Software*. Basinstoke, Hampshire: Macmillan.

Radcliff, B. (1986). Multicriteria Decision Making: Survey of Software. *Soc. Sci. Microcomputer Rev.* 4:38–55.

Rawlins, M. G. (1982). Graphics Software for Decision-Making. *Mini/Micro 82, Conference Record*,El Segundo, Calif, USA: Publ by Electronic Conventions Inc. Distributed by Western Periodicals Co., North Hollywood, Calif, USA.

Rousseau, W. R., and J. E. Matheson (1967). *Computer Programs For Decision Analysis*. Menlo Park, California: Stanford Research Institute.

Schlaifer, R. O. (1971). *Computer Programs For Elementary Decision Analysis*. Boston, Massachusetts: Harvard Business School Division of Research.

Sharda, R. (1992). Linear Programming Software for Personal Computers: 1992 Survey. *OR/MS Today* 19(3):44–60.

Stanley, L. (1992). Risk Analysis: Using Lotus 1-2-3 with @RISK Add-In. *Journal of Cash Management* 12(1):45–48.

Strassmann, P. A. (1989). Executive Software: Decisions, Decisions. *Inc.* 11(11):175–176.

Varis, O., and J. Kettunen. (1988). DAVID Influence Diagram Processing System in Environmental Management. *Environmental Software* 3(2):81–84.

Williams, A. J. (1983). "Decisions, Decisions.....". *Journal of the Operational Research Society* 34(4):319–330.

UTILITY THEORY

Ayres, R. U., and M. S. Sandilya. (1987). Utility Maximization and Catastrophe Aversion: A Simulation Test. *Journal of Environmental Economics & Management* 14(4):337–370.

Bell, D. E. (1988). One-Switch Utility Functions and a Measure of Risk. *Management Science* 34(12):1416–1424.

Bell, D. E., and P. H. Farquhar (1986). Perspectives on Utility Theory. *Operations Research* 34(1):179–183.

Brick, I. E., and H. S. Jagpal. (1984). Utility Theory, Value Maximization and the Quality Decision Under Uncertainty. *International Economic Review* 25(2):369–377.

Brock, H. W. (1980). The Problem of Utility Weights in Group Preference Aggregation. *Operations Research* 28(1):176–187.

Chang, S. H., and K. Y. Ra. (1984). Improved Cost–Benefit Criterion Using Utility Theory. *1984 Annual Meeting—American Nuclear Society. Transactions of the American Nuclear Society* Vol. 48 (pp. 523–525). La Grange Park, IL, USA: ANS.

Daboni, L. et al. (eds.) (1986). *Recent Developments in the Foundations of Utility and Risk Theory*,Dordrecht: D. Reidel Pub. Co.

Daniels, R. L., and L. R. Keller. (1992). Choice-Based Assessment of Utility Functions. *Organizational Behavior & Human Decision Processes* 52(3):524–543.

Dillinger, A. M., W. E. Stein, and P. J. Mizzi. (1992). Risk Averse Decisions in Business Planning. *Decision Sciences* 23(4):1003–1008.

Edwards, W. (ed.) (1992). *Utility Theories: Measurements and Applications*. Hingham, MA: Kluwer Academic Publishers.

Epstein, L. G., and U. Segal. (1992). Quadratic Social Welfare Functions. *Journal of Political Economy* 100(4):691–712.

Farmer, T. A. (1987). Testing the Robustness of Multiattribute Utility Theory in an Applied Setting. *Decision Sciences* 18(2):178–193.

Farquhar, P. H. (1984). Utility Assessment Methods. *Management Science* 30(11):1283–1300.

Fischer, G. W. (1976). Multidimensional Utility Models For Risk and Riskless Choice. *Organizational Behavior and Human Performance* 17(1):127–146.

Fischer, G. W. (1979). Utility Models For Multiple Objective Decisions: Do They Accurately Represent Human Preferences? *Decision Sciences* 10(3):451–479.

Fishburn, P. C. (1982). *Multiattribute Nonlinear Utility Theory*. Murray Hill, NJ: Bell Telephone Laboratories, Inc.

Fishburn, P. C. (1988). *Nonlinear Preference and Utility Theory*. Baltimore: John Hopkins University Press.

Friedman, M., and L. J. Savage. (1952). The Expected-Utility Hypothesis and The Measureability of Utility. *Journal of Political Economics* 60:463–474.

Gardenfors, P., and N. Sahlin (1988). *Decision, Probability, and Utility*. New York: Cambridge University Press.

Hershey, J. C., H. C. Kunreuther, and P. J. H. Schoemaker. (1982). Sources of Bias in Assessment Procedures for Utility Functions. *Management Science* 28(8):936–954.

Keller, L. R. (1989). Role of Generalized Utility Theories in Descriptive, Prescriptive, and Normative Decision Analysis. *Information and Decision Technologies* 15(4):259–271.

Quiggin, J. (1993). *Generalized Expected Utility Theory*. Boston: Kluwer Academic Publishers.

Scott, D., and R. Sathianathan. (1991). How Much Risk Can Your Company Stand? *Risk Management* 38(6):85–88.

Stigum, B. P., and F. Wenstop (1983). *Foundations of Utility and Risk Theory with Applications*. Dordrecht: D. Reidel.

Vargas, L. G. (1987). Priority Theory and Utility Theory. *Mathematical Modelling* (9):381–385.

Watson, S. R., and D. M. Buede (1987). *Decision Synthesis: The Principles and Practice of Decision Analysis*. Cambridge: Cambridge University Press.

Winkler, R. L. (1990). Decision Modeling and Rational Choice: AHP and Utility Theory. *Management Science* 36(3):247–248.

VALUE OF LIFE

Acton, J. P. (1976). The Value of Life: An Overview and Critique of Alternative Measures and Measurement Techniques. *Law and Contemporary Problems* 40:46–72.

Arthur, W. P. (1981). The Economics of Risks to Life. *American Economic Review* 71(1):54–64.

Bailey, M. J. (1980). *Reducing Risks to Life: Measurements of the Benefits*. Washington, D.C.: American Enterprise Institute for Public Policy Research.

Barrager, S. M., B. R. Judd, and D. W. North (1976). "The Economic and Social Costs of Coal and Nuclear Electric Generation." Menlo Park, CA: Decision Analysis Department, SRI International.

Conley, B. C. (1976). The Value of Human Life in the Demand for Human Safety. *American Economic Review* 66(1):45–55.

Fried, C. (1969). The Value of Life. *Harvard Law Review* 82(7):1415–1437.

Gillette, C. P., and T. D. Hopkins. (1988). Federal Agency Valuations of Human Life: A Report to the Administrative Conference of The United States (July).

Graham, J. D., and J. W. Vaupel. (1981). The Value of Life: What Difference Does It Make? *Risk Analysis* 1(1):89–95.

Hapgood, F. (1979). Risk–Benefit Analysis: Putting a Price on Life. *Atlantic* 243:33–38.

Howard, R. A. (1979). "Life and Death DecisIon Analysis" (Research Report EES-DA-79-2). Stanford, CA: Dept. of Engineering–Economic Systems, Stanford University (December).

Howard, R. A. (1984). On Fates Comparable to Death. *Management Science* 30(4):407–422.

Howard, R. A., J. E. Matheson, and D. Owen. (1978). The Value of Life and Nuclear Design. Proceedings of the Topical Meeting on Probabilistic Analysis of Nuclear Reactor Safety, American Nuclear Society.

Jones-Lee, M. W. (1976). *The Value of Life: An Economic Analysis*. Chicago: University of Chicago Press.

Jones-Lee, M. W. (ed.) (1982). *The Value of Life and Safety*, Amsterdam: North-Holland.

Jones-Lee, M. W., M. Hammerton, and P. R. Philips. (1985). The Value of Safety: Results of A National Sample Survey. *Economic Journal* 95(377):49–72.

Kahn, S. (1986). Economic Estimates of the Value of Life. *IEEE Technology and Society Magazine* 5(2):24–31.

Keeler, E. B., and S. Cretin. (1983). Discounting of Life-Saving and Other Nonmonetary Effects. *Management Science* 29(3):300–306.

Keller, B. (1984). What is the Audited Value of Life? New York *Times*, Friday 8(N) A24(L) Column 4 Vol. 134 (October 26).

Leigh, J. P. (1987). Gender, Firm Size, Industry, and Estimates of the Value-of-Life. *Journal of Health Economics* 6(3):255–273.

Linnerooth, J. Reevaluating the Value of Life: Practical Considerations. Presented at ORSA/TIMS Conference, Los Angeles.

Linnerooth, J. (1979). The Value of Human Life: A Review of The Models. *Economic Inquiry* 17:52–74.

Rappaport, E. (1974). "Economic Analysis of Life-And-Death Decision Making" (School of Engineering and Applied Sciences. Report No. ENG 7478). Berkeley: University of California.

Rhoads, S. E. (1978). How Much Should We Spend To Save A Life? *Public Interest* 51(Spring):74–92.

Rowe, W. D. (1977). *An Anatomy of Risk*. New York: John Wiley & Sons.

Starr, C. (1969). Social Benefit Versus Technological Risk: What Is Our Society Willing To Pay for Safety? *Science* 165:1232–1238.

Thaler, R., and S. Rosen. (1976). The Value of Saving of Life: Evidence From The Labor Market. In N. Terlecky (ed.), *Household Production and Consumption*. New York: Columbia University Press.

U.S. Office of Management and Budget. (1992). Budget of the United States Government, Fiscal Year 1992 (pp. Section IX.C. Part Two-370). : U.S. Government Printing Office.

Viscusi, W. K. (1978). Labor Market Valuations of Life and Limb: Empirical Evidence and Policy Implications. *Public Policy* 26:339–386.

Viscusi, W. K.(1983). Alternative Approaches to Valuing the Health Impacts of Accidents: Liability Law and Prospective Evaluations. *Law and Contemporary Problems* 47(4):49–68.

Viscusi, W. K. (1986). The Valuation of Risks to Life and Health: Guidelines for Policy Analysis. In J. D. Bentkover, V. T. Covello, and J. Mumpower (eds.), *Benefits Assessment: The State of the Art* (pp. 193–210). Dordrecht, The Netherlands: D. Reidel.

Weinstein, M., D. Shepard, and J. Pliskin. (1980). The Economic Value of Changing Mortality Probabilities: A Decision-Theoretic Approach. *Quarterly Journal of Economics* 94.

Zeckhauser, R. (1975). Procedures For Evaluating Lives. *Public Policy* 23:419–464.

VOTING

Arrow, K. J. (1963). *Social Choice and Individual Values (2nd ed)*. New York: Wiley.

Black, D. (1958). *Theory of Committees and Elections*. Cambridge: Cambridge University Press.

Brams, S., and P. Fishburn (1983). *Approval Voting*. Boston: Birkhauser.
Cook, W. D., and M. Kress. (1985). Ordinal Ranking with Intensity of Preference. *Management Science* 31(1):26–32.
Dummet, M. (1984). *Voting Procedures*. Oxford: Clarendon Press.
Enelow, J. M. (1984). *The Spatial Theory of Voting*. Cambridge: Cambridge University Press.
Fain, J. R., and J. B. Dworkin. (1991). Voting Rules and Election Outcomes: When Does the True Majority Win? *Journal of Labor Research* 12(3):247–260.
Farquharson, R. (1969). *Theory of Voting*. New Haven: Yale University Press.
Fishburn, P. C. (1973). *The Theory of Social Choice*. Princeton: Princeton University Press.
Gibbard, A. (1973). Manipulation of Voting Schemes: a General Result. *Econometrica* 41:587–601.
Hwang, C., and M. Lin (1987). *Group Decision Making Under Multiple Criteria; Methods and Applications*. New York: Springer-Verlag.
Merrill, S. (1988). *Making Multicandidate Elections More Democratic*. Princeton, NJ: Princeton University Press.
Mueller, D. C. (1989). *Public Choice II*. New York: Cambridge University Press.
Niemi, R. G., and W. H. Riker. (1976). The Choice of Voting Systems. *Scientific American* 234(6):21–27.
Nurmi, H. (1986). Problems of Finding Optimal Voting and Representation Systems. *European Journal of Operational Research* 24(1):91–98.
Nurmi, H. (1987). *Comparing Voting Systems*. Boston: D. Reidel Publishing Company.
Pattanaik, P. K. (1971). *Voting and Collective Choice*. Cambridge: Cambridge University Press.
Riker, W. H. (1982). *Liberalism Against Populism*. San Francisco: W. H. Freeman.
Satterthwaite, M. (1975). Strategy-Proofness and Arrow's Conditions. *Journal of Economic Theory* 10:187–217.
Sen, A. K. (1970). *Collective Choice and Social Welfare*. San Francisco, CA: Holden-Day.
Steen, L. A. (1980). Essay in Martin Gardner's column. *Scientific American* 238:16–26B.
Straffin, P. D. (1980). *Topics in the Theory of Voting*. Boston: Birkhauser.

WEIGHTED SCORING METHODS

Acton, W. H., and J. F. Courtright. (1990). Adapting SMART Methodology for Use with Subjective Rating Data. *Proceedings of the Human Factors Society 34th Annual Meeting—Orlando '90* (pp. 645–648). Santa Monica, CA, USA: Human Factors Society, Inc.
Barron, H., and C. P. Schmidt. (1988). Sensitivity Analysis of Additive Multiattribute Value Models. *Operations Research* 36(1):122–127.
Bell, D., H. Raiffa, and A. Tversky (eds.) (1988). *Decision Making: Descriptive, Normative, and Prescriptive Interactions*, New York: Cambridge University Press.
Belton, V. (1986). A Comparison of the Analytic Hierarchy Process and a Simple Multi-Attribute Value Function. *European Journal of Operational Research* 26(1):17–21.
Belton, V., and T. Gear. (1983). On a Shortcoming of Saaty's Method of Analytic Hierarchies. *Omega* 11(3):228–230.

Carr, K. R. (1990). Expert Systems for Implementing Group Consensus Opinion in Complex, Weighted Decisions. *Proceedings of the ISA '90 International Conference and Exhibition Part 4 (of 4). Advances in Instrumentation* Vol. 45 (pp. 1951–1966). Research Triangle Park, NC, USA: ISA Services Inc.

Dyer, J. S. (1990). Remarks on the Analytic Hierarchy Process. *Management Science* 36(3):249–258.

Fast, J. C., and L. T. Looper. (1988). "Multiattribute Decision Modeling Techniques: A Comparative Analysis" (AFHRL-TR-88-3). Brooks Air Force Base, Texas: Air Force Human Resources Laboratory (August).

Gass, S. I. (1985). *Decision Making, Models and Algorithms*. New York: John Wiley and Sons.

Gershon, M. (1984). The Role of Weights and Scales in the Application of Multiobjective Decision Making. *European Journal of Operational Research* 15(2):244–250.

Golden, B. L., E. A. Wasil, and P. T. Harker (eds). (1989). *The Analytic Hierarchy Process: Applications and Studies*. New York: Springer.

Golden, B. L., E. A. Wasil, and D. E. Levy. (1989). Applications of the Analytic Hierarchy Process, A Categorized Annotated Bibliography. In B. L. Golden, E. A. Wasil, and P. T. Harker (eds.), *The Analytic Hierarchy Process: Applications and Studies*, New York: Springer.

Kepner, C. H., and B. B. Tregoe (1981). *The New Rational Manager*. London: John Martin Publishing Ltd.

Khorramshahgol, R., and V. S. Moustakis. (1988). Delphic Hierarchy Process (DHP): A Methodology for Priority Setting Derived from the Delphi Method and Analytical Hierarchy Process. *European Journal of Operational Research* 37(3):347–354.

Krawiec, F. (1984). Evaluating and Selecting Research Projects by Scoring. *Research Management* 27(2):21–25.

Moody, P. E. (1983). *Decision Making: Proven Methods for Better Decisions*. New York: McGraw-Hill.

Moore, L. R. (1986). Kepner–Tregoe as a Safety Methodology. *Professional Safety* 31(6):37–41.

Saaty, T. L. (1980). *The Analytic Hierarchy Process*. New York: McGraw-Hill.

Saaty, T. L. (1986). Axiomatic Foundation of the Analytic Hierarchy Process. *Management Science* 32(7):841–855.

Saaty, T. L. (1989). Group Decision Making and the AHP. In B. L. Golden, E. A. Wasil, and P. T. Harker (eds.), *The Analytic Hierarchy Process: Applications and Studies*, New York: Springer.

Saaty, T. L. (1990). *Decision Making for Leaders*. Belmont, CA: Lifetime Learning Publications.

Stillwell, W. G., D. von Winterfeldt, and R. S. John (1987). Comparing Hierarchical and Nonhierarchical Weighting Methods for Eliciting Multiattribute Value Models. *Management Science* 33(4):442–450.

Vansnick, J. (1986). On the Problem of Weights in Multiple Criteria Decision Making (the Noncompensatory Approach). *European Journal of Operational Research* 24(2): 288–294.

Watson, S. R., and D. M. Buede (1987). *Decision Synthesis: The Principles and Practice of Decision Analysis*. Cambridge: Cambridge University Press.

Winkler, R. L. (1990). Decision Modeling and Rational Choice: AHP and Utility Theory. *Management Science* 36(3):247–248.

INDEX

Printed and bound in the UK by
CPI Antony Rowe, Eastbourne

Printed and bound by CPI Group (UK) Ltd, Croydon, CR0 4YY

16/09/2026

14963446-0001